EVOLUTION, ANIMAL 'RIGHTS',

& THE ENVIRONMENT

ἐν πᾶσι γὰρ τοῖς φυσιχοῖς ἔνεστί τι θαυμαστόν

[In all of nature there is cause for wonder.

—ARISTOTLE, *Parts of Animals,* I, v, 645a]

EVOLUTION, ANIMAL 'RIGHTS',
THE ENVIRONMENT

James B. Reichmann, S.J.

The Catholic University of America Press • Washington, D.C.

The paper used in this publication meets the minimum requirements of
American National Standards for Information Science—Permanence of
Paper for Printed Library materials, ANSI Z39.48-1984.
∞

LIBRARY OF CONGRESS CATALOGING-IN-PUBLICATION DATA
Reichmann, James B., 1923–
        Evolution, animal 'rights', and the environment / James B.
Reichmann.
        Includes bibliographical references (p.) and index.
        1. Animal rights—Philosophy. 2. Human-animal relationships.
        3. Evolution (Biology) 4. Ecology. I. Title.
        HV4708.R45 1999
        179'.3—dc21
        98-47146
        ISBN 0-8132-0931-5 (cl.: alk. paper)
        ISBN 0-8132-0954-4 (pbk.: alk. paper)

To the memory of my Mother and Father

Contents

Acknowledgments

There are a number of people whom I should like to acknowledge for the assistance they have rendered in the preparation of this book. First of all I am deeply grateful to Jeanette Ertel, formerly of Loyola University Press, who read through the first draft of the manuscript and provided many detailed, critically constructive comments which proved of great help in improving the manuscript. I am likewise grateful for the helpful and encouraging comments of Professor Theodore R. Vitali of St. Louis University and of Professor Vincent M. Colapietro of Pennsylvania State University, who read a later draft of most of the definitive form of the manuscript. I am also grateful for the comments and suggestions of a third anonymous reader. I wish also to acknowledge my indebtedness to my colleague Professor J. Patrick Burke for his comments on the penultimate draft of the manuscript and his continuing support during its period of gestation. I wish likewise to thank my former colleague, the current President of Gonzaga University, Robert J. Spitzer, S.J., for his initial and continuing encouragement for the undertaking and pursuit of this project. I am similarly indebted to my compatriot Joseph O. McGowan, S.J., for his warm interest in and support of this undertaking even during its darker days. I also wish to express my gratitude to Dr. David J. McGonagle, Director of The Catholic University of America Press, who has been most supportive and instrumental in successfully guiding the manuscript past Scylla and Charybdis on its way to publication. My greatest indebtedness, however, is owing to Kate Reynolds, our departmental secretary, and to Susan

Needham of The Catholic University of America Press, who edited the manuscript. The former skillfully and with consummate patience brought the initial manuscript safely through its many computer drafts, while the latter, through her deep interest in and commitment to the project and her unmatched pursuit of clarity and logical precision, offered innumerable suggestions and raised many questions aimed at eliminating ambiguity—a Herculean task no doubt when dealing with a book which would ambition *philosophical* status! Most of the above, I believe, were 'cheerfully' heeded. For what remains, of course, I alone stand responsible.

Finally, I am grateful to Jiggs and to Faust I and II, who, though not human, taught me a lot about what it means to be human.

My thanks, too, to the following publishers and authors who have kindly granted permission to quote materials from their publications.

Animal Minds and Human Morals: The Origins of the Western Debate by Richard Sorabji. Cornell University Press, 1993. Copyright 1993 by Cornell University Press. Reprinted by permission.

Beast and Man: The Roots of Human Nature by Mary Midgley, 1978; used by permission of the American publisher, Cornell University Press.

The Blind Watchmaker: Why the Evidence of Evolution Reveals a Universe Without Design by Richard Dawkins. Copyright 1996, 1987, 1986 by Richard Dawkins. Reprinted by permission of W. W. Norton & Company, Inc.

The Case for Animal Rights by Tom Regan, published by the University of California Press, 1983.

Darwinism Defended: A Guide to the Evolution Controversies. London: Addison-Wesley, 1982. Copyright 1982 by Michael Ruse, and reprinted with his permission.

Dinosaur in a Haystack: Reflections in Natural History, by Stephen Jay Gould, copyright 1995 by Stephen Jay Gould. New York: Harmony Books, 1995. Reprinted by permission of the publisher.

Earth's Insights: A Survey of Ecological Ethics from the Mediterranian Basin to the Australian Outback by J. Baird Callicott, published by the Unversity of California Press, 1994.

An Environmental Proposal for Ethics: The Principle of Integrity by Laura Westra, published by Rowman and Littlefield, 1994.

EVOLUTION, ANIMAL 'RIGHTS', & THE ENVIRONMENT

Introduction

Among the remarkable developments of the twentieth century, the widespread attention that rights issues have commanded at all levels of society must surely rank near the top of anyone's list. Never before has the issue of rights attracted such widespread interest or stirred so much controversy. Indisputably, there are justifiable grounds for characterizing the twentieth century as the century of rights.

In one form or another rights talk has become a commonplace not only among legislators and leaders of state, but among private citizens of all walks of life, and at every level of society. A casual perusal of the daily newspaper or a glimpse at the evening television news provides more than enough evidence to validate this claim. Few today are the areas of social intercourse where a lively interest is not shown in a rights question of one form or another.

Media news is all but saturated with rights issues. We continually hear and read about the rights of the working man and woman, the rights of minorities, property rights, rights of parents, of children, of the elderly, rights of the state or government, the right to free speech and free assembly, the right to vote, to a fair trial, the rights of victims of crimes, of criminals, of the imprisoned, of the poor, the physically disadvantaged, rights of students, of teachers, of institutions, the right to a just wage, to an education, health care, to travel freely, to transact business, the right to privacy, to have an abortion, the right to life, liberty and the pursuit of happiness. The list is endless, and many of the above alleged rights fall safely within the parameters of 'inalienable rights' guaranteed by the Constitution and the Bill of Rights. Especially noteworthy is the common admission that inalienable rights apply universally and equally to all humans, irrespective of age, gender, race, occupation, religion, or nationality. The foregoing are rights accredited to all individuals *precisely* because they are human, and not because of any particular skill or characteristic, whether native or learned, an individual might possess.

Until recently, in the Western world, rights were traditionally and all but universally recognized as attributable to humans alone. Today, new winds are blowing, and we increasingly are hearing a call to extend rights to the nonhuman animal and even, on occasion, to the environment as well. Though still perhaps a minority position, increasingly contemporary life scientists and philosophers argue that the planet upon which we live belongs no more properly to us humans than it does to the animals. As a consequence, it is alleged, we humans ought to extend to them also at least some of the rights we extend to ourselves; others there are who are very vocal in contending that the environment itself is to be viewed as a subject of rights. There is, in short, a rising antipathy to a view, often characterized as elitist, which accords a privileged status to humans, often at great cost both to the nonhuman animal and to the environment. This view is often stigmatized as *anthropocentrist*. As a corrective measure, many environmentalist ethicists opt for a position they denominate as *biocentrist*. According to this view, humans are to accept their place as merely one among many other life forms inhabiting planet earth, all of which compete for their share in this planet's coveted but increasingly limited resources.

The present study strives to address this line of reasoning, and to examine both the proposed extension of rights to the nonhuman animal and to the environment as well as the several assumptions underlying such an enterprise. Such scrutiny seems in order, for these issues are indeed complex and imply presuppositions that frequently are subtly hidden and, more often than not, singularly controversial. Admittedly, the animal 'rights' issue can at first glance appear only marginally philosophical. Yet serious reflection on the implications of extending 'rights' to the nonhuman animal, and *a fortiori* to nonliving segments of the environment, reveals that the rights question is a watershed issue bringing in its train weighty implications for the very nature and status not only of nonhuman creatures but of humans themselves.

Further, upon closer examination, the rights issue reveals itself to be as elusive as it is complex. As one seeks methodically to trace the ethical roots of rights to their source, one is soon obliged to plunge further into the opaque thicket of experience than any merely sensory awareness is capable of facilitating, let alone achieving. Little wonder, then, that the rights question is and has been a question much discussed in recent years as well as singularly contested. That is precisely why any serious inquiry into the phenomenon of rights cannot avoid being philosophical in nature, and why the present study is concerned with raising and addressing three seminal questions: What are rights? Who has them? and Why?

The present study, then, undertakes as preparatory to an investigation of rights themselves, the uncovering of the metaethical grounds of rights theory, with special emphasis on the controverted issue of whether creatures other than humans can and should be considered authentic subjects of rights. The book will argue forcefully, therefore, that before assigning rights to this or that individual or group, whether human or not, we need to be clear about *what* it is we are assigning.

The openly discordant nature of the ongoing debate among environmental ethicists regarding the rights issue—particularly the grounds upon which one can effectively argue for rights for living forms other than the human—dramatically underscores both its complexity and the consequent need to attend meticulously to the presuppositions of any rights theory. The Book will argue that a major source of this discordancy is that animal rights proponents as well as environmental ethicists practically unanimously, albeit diversely, appear to strive to base their rights theories upon a foundation of classical Darwinian evolutionary theory. The present work, contrarily, contends that the only anvil upon which an inalienable rights theory can be forged is natural law theory.

Because of the inescapably philosophical nature of the questions relating to the existence of animal and environmental rights, and because of their deep-rootedness in subsurface soil well beyond the superficial gaze of merely statistical, quantitative analysis, such questions can be satisfactorily addressed only through a balanced comparative study of the human and the nonhuman animal, a study wherein one carefully examines and evaluates the various kinds and levels of activity of each of them. Any position taken regarding the relationship between the human and the nonhuman animal of necessity rests, it is claimed, on basic assumptions effectively shaping the response one gives to the questions: *What does it mean to be human, and why have humans traditionally been said to possess rights?*

Accordingly, the present study first investigates—as they refer both to the human and to the nonhuman animal—the meaning of *species,* and the basis upon which rests the affirmation that things differ in kind one from another; the nature of knowing, the phenomenon of language, behavioral patterns, and finally, the issue of behavioral freedom. Only then will a formal consideration be undertaken regarding the question of *rights* and *who* has them. This study, then, first focuses on the core issues of any authentic philosophical assessment of the human person. What one holds the human person to be will ultimately prove, it will be argued, the deciding factor in resolving the complex question of animal rights, for it would be naïve to trade here exclusively on conventional wisdom by simply assuming humans to possess rights without first having examined more closely what a right is and why humans are said to possess them.

Animal rights protagonists for the most part seek to minimize the differences between the human and the nonhuman animal, in an effort to render coherent their claim that if humans have rights, animals ought to have them, too. In other words, the argument most often put forward is that the characteristics found in the human and upon which a rights theory is based, are also found to exist in at least some of the nonhuman animals. To claim otherwise, it is argued, is to indulge in anthropocentrism. Customarily this anthropocentrist viewpoint, which trumpets the human as superior to other life-forms, is assailed as the root cause of the denial of rights to nonhuman animals. In a recent study discussing this very point, Richard Sorabji has traced the genesis of the anthropocentrist view to none other than Aristotle. He feels that the latter was too much driven "by scientific interest in reaching his decision that animals lack reason" (*Animal Minds and Human Morals: The Origins of the Western Debate,* 1993, p. 2). Sorabji goes on to concede that the issue of animal intelligence is one of far-reaching import and remains a matter of continuing contemporary debate. "It was," he says, "a crisis both for the philosophy of mind and for theories of morality, and the issues raised then are still being debated today" (p. 7). Thus, though the debate is very much a contemporary one, it ought not be concluded that it is of recent origin. Over the years variant strains of the animal rights position have appeared. One might cite the teachings of Pythagoras, and perhaps even those of Plato, who, as Sorabji remarks "supplied some material for either side of the debate" (p. 178). The most explicit early advocates in the West, however, were Plutarch, Plotinus, and Porphry. In the East, especially in India and other areas deeply influenced by Buddhism and its attendant views regarding the nonhuman animal, vegetarianism was often held to be a moral imperative, although even in the West Plotinus and Plutarch were staunch defenders of this view as well.

Today perhaps more than at any time previous the attention of humans has been riveted on the complexity of the alignment between themselves and their environing world. After unknown millenia during which humans had inhabited planet earth, this past century with its breathtaking technological advances has raised anew, with a marked sense of urgency, the perennial questions touching on the nature of the human and his place in the world. High on the list of these questions is that of rights, their nature and their origin. This seems owing to the increased awareness modern technology has given us that this small planet we call earth has a limited capacity for satisfying the human appetite for the consumption of its primary goods. Closely related to questions pertaining to humans and the environment are those which focus more narrowly on the relation the human has to the nonhuman animal.

Several centuries ago, beginning especially with the Enlightenment, the unlimited perfectibility of the human was commonly accepted in intellectual circles. Yet, ironically, it has been the accelerated pace of human progress during the intervening decades which has brought us humans face to face with the startling realization that the world in which we live places significant limitations on our ability to exploit to the fullest our plans for the achievement of a utopia on earth. Less than four decades ago, when a human first ventured beyond the field of gravity of planet earth and, after a space voyage of some 239,00 miles, stepped on the surface of the moon, this spectacular achievement spelled the beginning of a new era. At this juncture the naive assumptions underlying the theorem of unlimited human perfectibility were brought fully into the sunlight.

Profoundly moving as was the first astronaut's view from the moon's surface of the blue planet earth, without visible support, floating and twisting lazily in empty space, it was also a highly sobering vision, bringing in its train the instant demise of the myth of the eternal perfectibility of man. Our twentieth-century space ventures, both manned and unmanned, had afforded humankind a fateful glimpse of the fragility of their planetary world. Since then, becoming progressively aware of the radical contingency of his status as an earth dweller, the contemporary human, incontestably conscious of the need to reassess his situation, is now more benignly disposed to accept a less sanguine account both of himself and of his planetary world. Paradoxically, while the knowledge of the universe has dramatically expanded, the human's own lived world has noticeably shrunk in size and scope. The fallout from this new self-awareness as, awestruck, we viewed our planet home as an alien world waiting to be explored, was 'earthshaking'. The future of our world and of the human race itself became highly problematical. A new tone of seriousness began to underlie human speculation, often mixed with doubt and perplexity.

There thus no longer appears on the broad horizon of human expectations, even obscurely, the visage of a confident, self-complacent humanity. There is, indeed, it seems, an ever-increasing awareness that the future can not deliver on its earlier promise of someday providing an earthly utopia. In fact, the mirrored vision of an earthly utopia has become of late noticeably astigmatic. This latter-day recognition, reluctant and belated as it may have been, of the radical finitude of human existence and indeed of the human's earthly habitation has been for contemporary generations a quite unexpected development, and, in many repects, for many of us, a rude awakening.

The human had been viewed as the indisputable world's master and the controller of his own destiny. There were no predictable limits to what the human might eventually accomplish. His vision of the future was forever brightened by a firm belief in the revered doctrine of the eternal perfectibility of humankind. But exerting perhaps an even more lasting impact on the already wounded human psyche was the failure, both unexpected and exasperating, of any of the numerous space probes to uncover the slightest trace of life anywhere beyond the pale of planet earth. Even our nearest neighbor, the moon, proved to be utterly barren of life, and, more recently, numerous space probes of Venus and Mars as well as the far away planets, Jupiter and Saturn and their satellite moons, have uncovered nothing more than lifeless planets; heavenly bodies utterly barren of life. Yet scientists still hold out the hope that eventually some evidence will give sign that at some time previous at least primitive life in some form did exist on one or more of these planets. And now, as the truth of these sobering developments seeps into our collective consciousness, we humans are all but reduced to silence as we wonder at the uniqueness of our tiny planet-world, which, however small comparatively, is the only place where at present one can confidently affirm that life does or did exist. Perhaps the supreme irony here is that it is precisely through the wondrous achievement of modern science that the sanguine expectations of the Enlightenment thinkers, creators of the myth of the unlimited perfectibility of man, and their progeny, have met with devastating disappointment.

From all of which it seems reasonable to surmise that the likelihood that planet earth is indeed the life center, not only of our solar system but of the entire universe, has contributed in a significant way to the raising of some very basic questions concerning the stability of life on our planet. With this the astronomer Neil F. Comins would seem to concur. In his highly interesting and informative book, *What If the Moon Didn't Exist* (1993), he comments: "As of this writing, we astronomers have not yet identified a single Earth-like planet around any other star. Even when such planets are found, we must determine which, if any of them, have suitable atmospheres and surfaces for human habitation" (p. 183). Still more recently, two respected scientists, Peter Ward, a paleontologist, and Donald Brownlee, noted astronomer, have authored a book, *Rare Earth* (2000), in which they conclude that there is little likelihood that life in any form can exist outside of planet earth.

Our own century, most particularly the past half-century, has witnessed a spectacular amount of attention being given to the human person, his status on planet earth and particularly his relationship to the other biota or living beings that share this planet with us. The incomparably intricate networking we have come to recognize, especially during these latter five or so decades, among the vast number of life forms existing on and within our planet seems to have brought into bold relief the reality of the interdependence of all of these life-forms. Indeed, this interdependence is seen to extend to the inorganic world as well. Our oceans, our atmosphere, our rivers, our forests all play crucial roles in the maintenance of the delicate balance essential to the existence of all life forms on our planet, including not only the human and nonhuman animals but the environment as well. Fortunately, the contemporary younger generation is unquestionably much better informed on these issues than were any of its preceding generations. The recently acquired awareness of the compressed status of earth as life's sole habitat for countless species—humans, plants, animals—brings new questions urgently to the fore. This point, too, is raised by professor Comins. As geneticists and scientists struggle the better to understand the ecological structure of the world of nature, critical ethical questions arise. "At a time," he writes, "when the human race and indeed the rest of nature is struggling for survival, this issue of culling the most threatening species would take on a new meaning. In that case it could be argued that it is 'unnatural' to let these breeds survive. But where would we draw the line? . . . Do we have the right to do that?" (p. 203).

The questions spawned by recent scientific findings as well as the continued and accelerated development of modern technology all seem to lead us back to the rights issue. "What are rights, who has them, and why?" are clearly questions of supreme concern to contemporary man. As the humanly habitable portion of our universe seems more and more identifiable with planet earth alone, and as one comes to realize more poignantly with the passing of each day the inability of our planet to sustain indefinitely our increasing demands upon its limited resources, whole clusters of pressing moral questions arise. Though many of these questions are not new ones, others are, and the resolution of those that are not new takes on a newfound urgency. A half-conscious general awareness that this is indeed the case has brought the rights issue to center stage. It appears that innumerable groups as well as individuals of all ages and occupations are now obsessed with asserting and vindicating, through legal means if necessary, their perceived 'rights'. The term is on everyone's lips and is used in so many different contexts—many patently conflictual—that one is led to conclude that this has become in our society a 'rogue' term, often overladen with a variety of ambivalent meanings. Rights, in short, have become in modern society a new, ethical Tower of Babel.

On this premise the present study rests. It is an attempt by means of philosophic inquiry to address the question of rights in general; it seeks to explore what they are, who may properly be said to have them, and why. We acknowledge from the outset that this is a very intricate issue, and that if one is to avoid the impass presented by a kalaidoscopic understanding of rights—an understanding that permits viewing them as whatever seems to be advantageous to oneself—one must be prepared to confront very difficult, perhaps insoluble, complex problems.

As is true of all ethical issues, rights theory entails in its affirmation a goodly number of underlying assumptions. These hidden presuppositions need to be examined, if one is to come to a coherent, unambiguous understanding of what a right is. Mary Midgley, in her book *Evolution as a Religion: Strange Hopes and Stranger Fears* (1985), has pointed out, correctly I believe, that the impersonal objectivity of science cannot "simply extend to total absence of meaning. It cannot demand—as is sometimes suggested—that all facts should be treated as equally important. Facts have to be connected up somehow, and in every system of connection some are more important than others" (p. 134). Midgley then concludes, "The kind of importance they [scientific findings] have, the kind of coloring they take on, will be determined by the general world-picture which the enquirer accepts." The present work proceeds on the premise that rights theory must be examined against the broad backdrop of "a general world picture," which the present author takes as designating a philosophy of nature together with a philosophy of the human person.

Consequently, before addressing directly the question of 'rights' and who has them, we will undertake in the first chapter of this study a preliminary consideration of nature as it is understood by nominalists and by those who support evolutionary theory as explanatory of the origin of all living things. In this chapter the question of what constitutes a species and how precisely one species may be said to be distinguished from others, will be examined, as will the coherence of an evolutionary theory of the origin and development of living things that excludes all teleological considerations.

The second chapter undertakes the formulation of a definition of knowing, and that type of knowing that is termed *reasoning,* and asks in what sense knowing is applicable to the nonhuman as well as to the human animal. The third chapter examines the issue of freedom: What are the conditions for its possibility? Does the behavior of the nonhuman animal exhibit the characteristics endemic to freedom as found in the human? The fourth chapter examines the phenomenon of language, seeking to clarify its nature: In what sense may the nonhuman animal be said to be capable of communicating through language?

The considerations contained in these early chapters provide a detailed propaedeutic to the seminal consideration of rights, which is taken up in chapter five. Here the nature of rights is closely examined together with its correlative, 'obligation'. This latter could well be the thorniest element in the 'question of rights' issue, whether for the human or for the nonhuman animal, but nonetheless it is critically relevant to the formulation of a natural, or inherent, rights theory. A further question is here raised: Is it coherent to speak of the nonhuman animal as having rights? And yet more specifically, is it consistent to speak of such 'rights' as natural, if they are viewed as deriving from classic evolutionary theory? Finally, the chapter concludes with reflections on the viability of advocating philosophical vegetarianism to be morally obligatory.

The sixth chapter is devoted to an examination of rights as they relate to the environment. Some animal 'rights' theorists seek to extend the application of an inherent rights theory to include the world of nonliving things. Four contemporary ethical environmentalists, whose positions regarding environmental rights pretty much cover the full range of viable options, are outlined and critically examined. In a certain sense one can say that the animal rights cause has been often closely allied to environmentalism. Not infrequently one finds that advocates of animal rights are in the forefront of those actively supporting the protection of the environment—rivers, forests, mountains, lakes, and streams—as a full-fledged ethical imperative.

The seventh chapter is devoted to examining the place of the animal in the world of the human. Here the focus is on the positive "social" contributions that animals make. These considerations include the keeping of animals as pets and the nature of the friendships many humans, children and the aged in particular, often form with animals of a variety of species.

The book concludes with an epilogue, which sketches a tentative theology of the nonhuman animal; that is, it inquires into what can be gleaned from Scripture regarding the nature of animals and their place in the schema of the created universe. The question of 'philosophical' vegetarianism is also reexamined here, this time from a theological perspective.

It is, finally, our contention that the rights question in general, and the animal-environmental rights question in particular, inherently center and rest on the definition of human personhood. To ignore or deny this is irrevocably to commit, not only the issue of animal rights, but the environmental and human rights issues as well, to an inexorable limbo of ambiguity. Thus it can be acknowledged that the abiding conviction underlying this study is that the various issues that have been examined, including especially those of species, nature, intellection, language, and freedom, are inseparably related to rights issues in all their forms and applications, since these uniquely coalesce to form one splendid seamless ethical garment. Inseparably related to and dependent upon the foregoing, the issues relating to animal rights and environmental ethics cannot viably be treated in isolation from the broader context of the human person, which alone can give them meaning.

1 The Phenomenon of Things Living

The human is the most complex and behaviorally sophisticated of all living things. Insofar as his inborn psychic powers and the complexity of his physical constitution are concerned, the human has no peer. The human's powers of knowing and the scope of his knowledge exhibit greater variety and breadth than the behavior of any other animal reveals. The cellular structure of the human is markedly more intricate than that of any other living thing. No other animal plans for or consciously seeks out an emotional experience. In the number of tasks it performs, the human is clearly unexcelled even by those animals exhibiting complex forms of social behavior. The everyday network of social relationships of the average human is incomparably more intricate than that of other animal species.

Yet, all this notwithstanding, the question hovering over our contemporary world with greater insistence than ever before is simply: How and to what extent does the human differ from other animals? This searching question lies at the center of contemporary philosophic inquiry and for this reason is worthy of being referred to as a watershed issue. Any response given it is certain to influence the flow and direction of other significant questions about the human. This question forces confrontation with the totality of human experience, for no areas of human experience or activity remain untouched by the extensive fallout resulting from one's response.

Since humans are capable of multiple and enormously varied experiences, which open them up to and put them in contact with their environing world, it is bizarre and utterly unrealistic to assume that humans can find their identity by disregarding the outer world by being narcissistically preoccupied with 'the self'. If humans expect to know themselves they must first turn their attention to things in the world and abandon the dream-like quality of introversion inspired by Descartes, who began his philosophic inquiry with the allegedly pre-suppositionless, introverted "I think." To know oneself, one must turn beyond one's self, for without interaction with his world, man remains but an empty shell without direction, without meaning. When the Delphic Oracle bade Socrates "Know thyself," she was not anticipating the Cartesian introspective method, but rather introducing him to the paradox of human achievement, which is, simply stated, that left to ourselves we can accomplish nothing; to learn who we are, we must gain knowledge of our world. We are not unilaterally its master, for in some sense it is ours.

THE DEFINITION OF MAN

The response of the Greek philosopher Aristotle to the question "What is man?" was "Man is a 'rational animal'." Did Aristotle's definition come to him, as it were, out of the 'blue' as something akin to a Kantian a priori? The truth of the matter is that this classic, succinct definition, so familiar not only to philosophers but to every educated person, emerged only after a painstaking, thorough sifting of countless data of experience both everyday and scientific. Aristotle's brilliant synthesis at once envelops and transforms those experiences of his own lifetime into a vision composed of thought greater than the sum of its parts. Aristotle's definition of the human can be compared to the celebrated Ravenna mosaics of the Basilica of San Vitale: while composed of miniature tiles of various colors they are, as an artistic work, infinitely more than the myriad tile pieces. And indeed, such a piece of work is man.

Yet, it avails one little to be acquainted with Aristotle's definition of the human without simultaneously seeking to relive the very experiences that led the latter to his creative synthesis. If we are to avoid using 'rational animal' as a mere verbal slogan, we must distill from the words that which binds them together as an indissoluble unity, thereby uncovering their full meaning.

THE NATURE OF DEFINING

As the foregoing suggests, a definition is a creative forging together of several or many experiential acts. The unity of definition thus results from an immaterialization of experience and is, accordingly, the product of a trans-temporal and trans-spacial activity. It is not an activity performed only by the highly educated, but something everyone does as a matter of course, for the act of defining is really nothing more than the commonplace act of understanding. Commonplace, that is, for the human, for it is an achievement uniquely human.

As a synthesis of a limited field of experience, the definition amounts to a condensed generalization obtained by isolating those particularities of experience which, though present within the overall experience, do not, as particularities, effectively control the actual shape given to the experience itself. We are constrained to proceed in this fashion if we are to succeed in rendering our unimaginably large number and variety of experiences manageable, and hence productive. Intellectual knowing is basically a matter of condensing and unifying our experience. That is to say, if we are to succeed in deriving meaning from our experience and thus exploit and utilize it in a manner beneficial to both ourselves and others, we must come to grasp to some degree its harmony and interrelatedness. It is only thus that classifications of experience can occur, that hypotheses can be formulated and scientific advances made. This is of course the same as interpreting our experience, and is the underlying meaning given to the term *hermeneia* as employed by Aristotle. It is in this sense that the human is the interpreter of nature, standing at the crossroads between the material world and the world of pure intelligence.

The human, then, though he does not give meaning to the world, does mediate or interpret its meaning by discovering and revealing it. While this is to some an unwelcome generalization, it is nevertheless an inescapable law of human knowing which, like Aristotle's first principles of reason, though it be denied with the lips, is implicitly affirmed whenever one attempts its denial.

It is therefore both comforting and reassuring to hear one such as Charles Darwin—whose reverence for the value of empirical facts is indisputable—acknowledge with complete candor that the whole aim of collecting facts is to draw from them generalizations that are themselves 'non-factual', in the sense that they transcend the limitations of the very facts that spawn them. "How odd it is," Darwin remarks, "that anyone should not see that all observation must be for or against some view if it is to be of any service" (*Letters of Charles Darwin,* i, p. 176, as quoted by Jonathan Howard in *Darwin,* 1982, p. 91). The view to which he refers is of course nothing more than a general or universalized judgment that is meant to coalesce a multitude of facts experienced separately into a community of meaning. Facts that remain in isolation are literally meaningless. There would be no point at all to amassing facts were it not one's intent to find in them either verification or refinement of a particular point of view.

FACTS AND GENERALIZATION

In commenting on Darwin's remark Jonathan Howard observes with equal acuity:

> To the theoretical scientist observations are subservient to explanations. It is the ability of a general statement or argument to subsume innumerable facts which appeals: a fact is only of interest in so far as it is included within or excluded from an argument. (*Darwin,* p. 91)

Howard is emphasizing how in some manner the universal must precede the singular judgment if there are indeed to be any such things as facts, for the latter cannot be experienced save as instances of a larger whole in which they somehow share. Howard concludes his comment by further pointing out, correctly I believe, that "The theoretical scientist may look under stones, but he does so not merely because he might find something, but because he already has a clear expectation of finding a particular thing" (pp. 91–92). Precisely.

A generalization, then, is a transformed constellation of facts, and it is with a view to the formulating of a fresh and more unrestricted generalization that the singular facts are gathered in the first place. But this is possible only on the condition that prior generalizations have set one upon the search for further 'facts' with the view to generalizing yet further still.

It should not be surprising, therefore, that according to this underlying schema of human knowing, different individuals, while having the same or similar experiences, may 'discover' or claim to have found quite divergent sets of facts. This radically "pluralistic" phenomenon accounts for the different philosophies that individual philosophers have formulated, which make up what we commonly refer to as the history of philosophy. Thus, depending upon how one reads the 'facts', which includes determining what indeed the facts are, different explanations and world pictures have emerged. And this has proven to be above all true with regard to those 'facts' which surround the human, and which enable one to give answer to the questions, "who is man?" and "what is his origin?"

The reputation of the philosopher seeking to respond to questions involving the human is, as it were, exposed to a double jeopardy, since all such questions and responses reflectively affect the one responding. The philosopher is commenting not just about something wholly extraneous to self but about his or her own self as well. There is inevitably an autobiographical dimension to every response one makes about the human, for one cannot help but speak about oneself in speaking about the human. But facts alone do not and cannot speak for themselves. They must be interpreted, for events become facts only upon being assumed into the manifold which set the inquirer seeking after them in the first place.

One comes to grasp what something such as an animal is by studying it in its natural habitat, noting what activities it performs and how it relates to and interacts with its environment. The behavioral patterns its activities trace unerringly and inevitably expose the underlying nature that is their source. Hence when it comes to the question of who man is, one's response is not only conditioned, but is literally controlled and shaped, by one's observation of the behavior of the human. There is simply no other way of ascertaining what man is capable of doing than by observing him doing it. The power to act becomes manifest and palpable to us only when it finds expression in action, and the manifested power in turn reveals the nature of the doer.

CAPACITIES AND ACTUALIZATION

By employing this simple but fundamental criterion, then, we are able to differentiate between various kinds of organisms. Thus, for example, we readily conclude that this particular animal, which we have not observed before, can fly—not because of its appearance, for example that it has feathers or what we later come to identify as wings—but for the plain fact that we observe it in flight. As long as the animal remains on the ground we have no assurance that it is capable of flight. That it can fly we come to recognize simply by observing it flying. A similar experience can be had when observing a certain species of insects with whose behavior we are unfamiliar. We are unable to affirm with certainty, as we watch it crawl along a window ledge, that it can fly, and if it can, we learn this only if it takes to flight—for example, as we are about to grasp it. If the insect flies off we learn, only then, that it is capable of flight.

Powers are known in their acts; without observing the latter we can only conjecture the kinds of behavior of which the creature is capable. Only on the basis of experience, whereby we observe the activities of things, are we in a position to learn something about their natures, i.e., the kinds of creatures they are. Their activities are strictly proportionate to their natures, so that the latter are mediated to us only through their activities, which are to us an overt expression of their underlying nature. In this sense it can be justly affirmed that the activities of things provide us with the hermeneutic for determining the nature of things. Hence the distinctions we make between different kinds of organisms are wholly dependent upon the singular dynamic thrust of each one as exhibited through its activities, or behavior.

It follows that mere description, which relies solely on the superficial and hence observable structure of things, is quite incapable of unfolding for us their latent inner structure. Only the activity of things, in which is manifested the hidden nature from which the activity flows, can provide this. That is to say, we should never have come to understand the 'purpose' of birds' wings had we never observed a bird in flight, nor would we comprehend the utility of a fish's gills had we not

[handwritten marginalia: observation is crucial for knowledge]

first remarked the fish's ability to 'breathe' while in wa
be able to extract oxygen from it.

Now it is precisely this ability to penetrate th
things—much as an x-ray penetrates the chest c
skeletal structure within—which we identify as an intellective ac..
in the next two chapters we shall conduct a fairly detailed investigation
of the nature of intellective knowing, so critically fundamental to the
purposes of this study, we will not dwell further on it here. For the
present, suffice it to say that the position one takes regarding the nature
of human understanding and human reasoning inevitably must play a
crucially important role in analyzing the question of 'rights' and identifying precisely who has them. Further, our remarks pertaining to sense
observation and the intellective act have been set forth here to make
manifest to the reader a hidden presupposition of the Darwinian claims
regarding the origin of species: the quiet, all too often 'unremarked' assumption that to sense or perceive and to understand are activities fundamentally equivalent one to the other. As the preceding reflections
suggest, we maintain that the intellective acts of understanding and reasoning transcend the world of sensory experience in that they deal with
the general and universal, even though the contents of these acts derive
from sensory experience.

DARWIN AND REASONING

Now as long as what is claimed to be true—such as the size or
shape of this tree—can be directly monitored by the senses, there is little or no disagreement among realist philosophers and scientists, or indeed anyone prepared to accept the reality of sensory perception as a
viable source of knowledge. It is when one leaves the secure arena of
sense observation, however, and ventures forth into the (for many) less
familiar terrain of perception and pure intellective knowing, that controversy has flourished. This has certainly proved to be the case with
regard to inquiry into the nature of the human and the rich behavioral
repertoire of humans. Should one hazard to affirm that the human is
'an animal that reasons', it immediately becomes problematical for
many as to what will serve as evidence that indeed the human does or

can reason. Indeed, what precisely one means by reasoning is discoverable only through the simultaneous employment of sensation, as the above example of the bird and its ability to fly sought to make evident. Reasoning itself is not observable in the way that sensible things are. To state that humans are reasoning beings is to say very little, if we conceive 'reasoning' to be something equally shared between humans and other animals. What do we mean, then, by intellective reasoning, and what is there about it that sets a reasoning being apart from other beings that lack reason? What are the criteria one might employ enabling one to conclude with assurance that some animals are capable of reasoning while others are not?

Not surprisingly, it is at this point that viewpoints vary. It is alleged by some that no clear lines of demarcation can be drawn between the human and at least some species of the nonhuman animal. Of these, some will claim that certain of the 'higher' animals, such as the gorilla and chimpanzee, perform activities that exhibit 'reason'. In addition, such would appear to be an implicit requirement, many will claim, of evolutionary theory itself. If one admits to an unbroken biological continuity between the human and the nonhuman animal, it manifestly involves a serious inconsistency to conclude that the ability to 'reason', whatever one might take that to mean, is restricted to the human alone. There would then be, it seems, no grounds upon which to support such a claim, for the admission of discontinuity between the human and the nonhuman animal species would render inexplicable an ontogenic dependency of the one species on the other—which, of course, the Darwinian evolutionary theory requires.

The importance of maintaining just such a continuity has been expressly acknowledged by Sarel Eimerl and Irven DeVore in their 1965 book *The Primates*. In stressing the presuppositions upon which evolutionary theory rests, these anthropologists support the view taken by Darwin that the progressive development of species can be explained only if the changes that took place occurred very gradually. "Actually all evolutionary progress," they write, "not only that of man, has taken place through a series of very tiny changes, each one at once producing some small advantage and paving the way for the next change" (p.

180). But among evolutionists it was Darwin himself who first strenuously objected to allowing any difference in kind between the nonhuman animal's 'reasoning' and the thinking processes of the human.
Writing in his notebooks he remarks cryptically: "Having proved men's
and brutes bodies are one type: almost superfluous to consider minds"
(iv, 163, quoted by Howard, p. 66). In his short but informative work,
Jonathan Howard sums up the thought of Darwin in the most candid
terms. "Darwin viewed metaphysical objections to extending human
mental qualities to the animals as 'arrogance'" (*Darwin*, p. 66). He
goes on to comment on the underlying operative principle in Darwin's
mind that rendered such an attitude possible, and he accounts for Darwin's insistence on the existence of a strict continuity between all animal life-forms, including the human:

> There was one continuous 'thinking principle' throughout the animals
> which Darwin viewed as being contingent on the presence of an organ-
> ized nervous system, and consequently 'The difference between intellect of
> man and animals is not so great as between living things without thought
> (plants) and living thing with thought (animals)'. (P. 67; the quotation
> from Darwin taken from *Notebooks on the Transmutation of Species*, i, p.
> 66)

Howard's comments and the quotation from Darwin indicate clearly enough that Darwin remained comfortable with a distinctly blurred
concept monitoring the relationship between reason and instinct. He
was quite intolerant of any metaphysical attempt to provide a further
nuancing of either of these terms, and himself made no effort to refine
his own thinking on the subject. Darwin states: "I will not attempt any
definition of instinct. It would be easy to show that several distinct
mental actions are commonly embraced by this term. . . . A little dose,
as Pierre Huber expresses it, of judgment or reason, often comes into
play, even in animals very low in the scale of nature" (*Origin of
Species*, pp. 207–8, quoted by Howard, p. 67). Returning to this subject in his last major work, Darwin reiterates his prior conviction, flatly
reaffirming his unwillingness to admit any fundamental difference between man and the animal. He also indicates that what leads him to
this conclusion is the 'truth' of evolution of the species. Darwin will

not deny that man and the animal differ, but does not grant that they differ in *kind*. He states: "*Everyone who admits the principle of evolution, must see that the mental powers of the higher animals which are the same in kind with those of man, though so different in degree, are capable of advancement*" (*The Descent of Man*, Pt. 3, chap. 21, p. 591 B; italics added). It is hard to see how Darwin could have missed the circularity of his overall argument.

EVOLUTION AND THE CONTINUITY OF SPECIES

In demanding a biological continuity between all animal species, Darwin does at the same time readily grant that, of all life-forms, the human is quite evidently the highest. Man's reason, while not unlike that of the higher animal life-forms such as the ape, monkey, wolf, or elephant, is nonetheless, in Darwin's view, appreciably more highly developed and apparently capable of yet more development. In harmony with this view, it is instructive to note, Darwin sees no inconsistency in allowing that the human has not, during his lengthy period of development, retrogressed in any major way. In his view the human's ascent from an earlier more primitive status to his present (for Darwin, mid to late nineteenth century of course) level of development has been a remarkable success story, as the human has all but unerringly ascended the ladder of evolutionary complexity and development. Further, Darwin also seems to have accepted the human of his own time as representing the zenith of human development thus far.

In the concluding fifth chapter of his *Descent of Man*, entitled "The Development of Intellectual and Moral Faculties," Darwin summarizes his view with regard to the progression and retrogression of the human:

> To believe that man was aboriginally civilized and then suffered utter degradation in so many regions, is to take a pitiably low view of human nature. It is apparently a truer and more cheerful view that progress has been much more gentle than retrogression; *that man has risen, though by slow and interrupted steps, from a lowly condition to the highest standard as yet attained by him in knowledge, morals and religion.* (Italics added)

MIDGLEY, DARWIN, AND CONTINUITY

Despite the centrality of the evolutionary development of man in the thinking of Darwin, a contemporary English philosopher, who has written extensively on this topic and is strongly supportive of the latter's general outline of evolutionary theory, has nonetheless opted for a quite different view of the evolution of species as regards the question of human progression. In her widely discussed book *Beast and Man: The Roots of Human Nature,* Mary Midgley firmly opposes the characterization of one species of animal life as higher than or superior to another. She finds this view alien to the authentic conception of Darwinist evolution, raising as it does unsettling questions about the very direction and purpose of evolution itself. Obviously her objection touches on a point highly sensitive to the entire question of evolutionary theory.

"In what sense, if any," Midgley asks, "can evolution be said to be a direction or a purpose? It is hard to see whom we could credit with such a purpose unless we call upon the Lord . . ." (p. 145). She raises the further issue, again a seminal one, as to whether evolution 'had' to develop the way it did. Would it have "failed, fallen short, or become deflected if it had followed a different path? Would there be something wrong if, for instance, birds or ants, snakes or people had never developed?" (ibid.). Midgley finds this to be an "alarming question," but feels nonetheless that "we commit ourselves to some kind of answer to it whenever we mention *higher* or *lower* life-forms" (p. 146). Further, creatures do not, she argues "'maximize' survival, nor in any clear sense aim to do so. They survive, or they don't. If that is all, perhaps we must renounce the habit of dealing in evolutionary hierarchies, must stop talking of 'higher' and 'lower' life-forms" (p. 150).

For this reason Midgley suggests that one employ the bush as a symbol of evolutionary development rather than the stately tree which stretches ever skyward. A bush grows out in all directions more or less uniformly and thus better represents the community of species that inhabit the earth (p. 160). In her view, consequently, man is not the highest of the species forms, save perhaps in his own eyes alone, but is rather one of many different living things making up the one family of all living forms, or at least of those forms we identify as 'animals'. It then becomes our responsibility (though the "why" of this remains unclear) as humans to accept our unique though limited niche within the earthly community as one among many life-forms. Our singular dignity consists in just such an acknowledgment. "We have somehow to operate as a whole," she states, "to preserve the continuity of our being. This means acknowledging our kinship with the rest of the biosphere. Our dignity arises *within* nature, not against it" (p. 196). Midgley concludes her polemic in support of the theme of continuity between the human and other animals by chiding humans for their reluctance to allow other animals a place within our world and for the human's propensity to fear that such recognition would constitute a *bona fide* threat to their dignity. "I have suggested," she remarks, "that we ought not to feel that dignity threatened by our continuity with the animal world" (ibid.).

It should be noted, however, that the expression "continuity with the animal world" is open to diverse interpretations and emphases, and is, indeed, an altogether controlling notion in any serious attempt to determine the underlying meaning of 'human person'. The continuity referred to can be understood—and it seems evident that Midgley does so understand it—as a continuity of biological descent, whereby all animal species are interrelated because all derive ultimately from a common ancestor.

It should be evident that such a view of continuity limits the difference between the human and the nonhuman animal to variations in degree and not in kind. Another possible view of continuity is that the human and nonhuman animals have a common nature, in that they both possess sensory awareness. But if this be the case, whence derives the dignity to which Midgley refers? Does it make good sense to state, as she does, that "people, as Kant rightly said, are ends in themselves" (*Evolution as a Religion*, 1985, p. 59) when 'people' connotes no more than a superficially descriptive difference between the human and nonhuman animal? How does one establish that humans alone are ends in themselves, as Kant allegedly does, if one is already committed to the

view that humans are no further up the ladder of evolutionary development than are any of the other species of animals? They are no further
up the ladder because this figure of ladder is purely fictive, a teleological assumption that gratuitously points in the direction of a greater
complexification of life-forms and a built-in tendency of hierarchical
development. That Midgley unqualifiedly rejects even the most attenu-
ated form of teleological explanation of the diversity of life-forms, she
leaves no doubt whatever; she is equally clear in claiming that her position is indistinguishable from that originally taken by Darwin himself.

> No ladders are needed for classification, linear development, or orthogene-
> sis as an ideal fifth wheel to the coach. Darwin saw no reason to posit any
> law guaranteeing the continuation of any of the changes he noted, or to
> pick out any one of them, such as increase in intelligence, as the core of the
> whole proceeding. (Pp. 34–35)

Further on, Midgley returns to her polemic against the introduction
of any form of purposefulness within the fabric of evolutionary development. She affirms that "it is really not possible to make much sense
of the notion of evolution's steady, careful progress towards this goal
without language so deeply teleological that it implies an agent" (p.
61). Doubtless many who have embraced Darwin's evolutionary theory
with enthusiasm are surprised, even shocked, by Midgley's insistence
that Darwin did not see his theory as supportive of or even compatible
with a progressivist view of nature. Much less did he view the origin of
species and their differentiation as implicative of any kind of teleological explanation or resulting hierarchical structure. At bottom, whatever
has occurred during the long period of the natural evolution of species
is to be explained as the product of mere chance. To assign purpose to
nature at any level is to indulge in fanciful creation.

Yet, that this should come as a surprise is itself surprising. Despite
the popular view to the contrary among a goodly number of committed
theists—for whom Darwin's theory provided but an intelligent, scientific explanation of the scriptural account of creation—many if not most
of the scientific leaders of our era have interpreted Darwin in exactly
the manner in which Midgley has portrayed him, and they make no effort to conceal their allegiance to his stand. Among these we might sin-

gle out Stephen J. Gould, who as a geologist and paleontologist has written and continues assiduously to write on the topic of evolution. In two recent works he gives considerable attention to the issue of evolution as progressivist and strives to dispel the—as he sees it—common misconception that there is purposefulness guiding the direction of the evolution of species. In *Full House: The Spread of Excellence from Plato to Darwin* (1996), Gould vigorously contends that 'progress' is anathema to authentic evolutionary theory. "Claims for progress," he writes, "represent a quintessential example of conventional thinking about trends as entities on the move." He adds: *"We label this trend to increase as 'progress'—and we are locked into the view that such progress must be the defining thrust of the entire evolutionary process"* (p. 146; italics added). Gould will go on to inform his reader that the "crux" of his book *(Full House)* aims at "debunking this conventional argument for progress in the history of life" (p. 168). It is Gould's further claim that the progressivists have myopically focussed exclusively on complex organisms extrapolating from the history of these, attributing their developmental characteristics to all species of organisms. He concludes: "This argument is illogical and has always disturbed the most critical consumers" (ibid.).

That Darwin's own words seem to betray him, in that he did on occasion speak of evolution as progressivist, Gould does not deny. This he attributes to the social pressures naturally brought on by the euphoric atmosphere of Victorian England to which Darwin could hardly be insensitive. Allegedly Darwin was fearful that, should the full implications of his evolutionary theory be grasped by the general populace, that would adversely affect the theory's acceptance by a society intoxicated with the wealth and prestige concomitant with England's then unchallenged status as mistress of the seas and industrial leader of the world.

In a letter dated December 4, 1872, however, and addressed to a certain Alpheus Hyatt, who had earlier supported a progressivist evolutionary theory, Darwin writes: "After long reflection, I cannot avoid the conviction that no innate tendency to progressive development exists" (quoted by Gould in *Full House,* p. 137). Gould's final assessment of Darwinism and progressive or teleologically oriented evolution is that "Darwin's denial of progress arises for a special and technical reason within his theory, and not merely from a general philosophical preference" (ibid.).

In querying Gould as to how he might account for the emergence of complex life-forms not only of our contemporary era but also those now extinct, we find his response forthright and unequivocal. It is all simply a matter of fortuitous circumstance. "Humans are here by the luck of the draw, not the inevitability of life's direction or evolution's mechanism" (p. 175). Later Gould states that even a miniscule change in the sequence of events that led to modern humans would have led to an entirely different outcome, "making history cascade down another pathway that could never have led to *Homo Sapiens,* or to any self-conscious creature" (p. 215). Though earlier he had attributed to arrogance the general reluctance of humans to accept a "luck of the draw" scenario to account for the emergence of humans from matter (pp. 136–37), he now inexplicably characterizes humans as *"glorious accidents* of an unpredictable process with no drive to complexity, not the expected results of evolutionary principles that yearn to produce a creature capable of understanding the mode of its own necessary construction" (pp. 215–16). If humans are indeed "glorious accidents," this would seem to imply that they are, irrespective of their alleged humble origin, rather superior to the other less glorious accidents that have also "occurred" along the way. Has a directional teleology surreptitiously slipped in through the back door? And in the evolutionary context, whatever meaning can the word "glorious" possess?

Can any evolutionary view that *excludes* all forms of orthogenesis, or intrinsic directionality, reasonably speak of the human as possessing a special dignity not also enjoyed by at least some other animal species (especially those capable of sharing to some degree in emotional experience)? And if this special dignity is denied to any animal life-forms at all, then it is clear that the original problem about continuity remains. What explanation can justify any such exclusionary distinction? But if no distinction is made, the concept of 'dignity' is emptied of meaning; no basis remains for according to the human any special form of respect over that accorded to other forms of animal life. Such a view cannot but have exceedingly profound repercussions for questions regarding inborn, natural human rights and obligations. We will return to this point in chapter 5, when we deal directly with the question of human and animal rights.

But to return to the issue of the continuity between the human and the nonhuman animal, there is yet another way in which the notion of continuity can be viewed. To maintain that there is a certain continuity of 'nature' between the human and the nonhuman animal does not of itself require that one subscribe to a continuity of generative dependence, i.e., that one species evolved from another. There is ample evidence that both the human and the nonhuman possess a sensory nature. It does not follow that the human and the nonhuman cannot differ in a significant way, namely, in that the nature of the human is *intellective* as well as *sentient.* The continuity between the two natures—human and nonhuman animal—is then seen to be on the level of generic, not specific, identity. That is, the sentient nature belongs to the genus—both human and nonhuman animals—but the intellective nature belongs only to the species—human. This of course is the view taken by Aristotle. It is obviously not a view congenial to Midgley, for a number of reasons, the most important being that it renders the emergence of new life-forms wholly problematical. As Midgley herself admits, there seems no way that one can account for the emergence of intelligent beings from merely sentient ones through a natural generative process, without the introduction of some superior agent, an agent responsible for the emergence of living things originally as well as for the subsequent differentiation of new life-forms.

MIDGLEY, EVOLUTION, AND RELIGION

This brings up the main theme of Midgley's more recent work, *Evolution as a Religion,* in which she develops her own view that what once fell under the umbrella term 'religion' is today in great part assumed under the spacious tent of 'science'. She contends, moreover, that the majority of contemporary scientists are confirmed evolutionists-and in this she is unquestionably right. Her focused position is, however, that even a weak teleological Darwinism—such as that promoted originally by Lamarck, Herbert Spencer, and, in our own time, William Day—while firmly denying to religion a role in the explanation of life-forms including man, ascribes to purely scientific explanations the self-same characteristics commonly attributed to religion. In effect, Midgley argues pointedly, this form of evolutionism has, in spite of protestations to the contrary, transformed itself into a crypto-religion; at the same time, by its denial of religion it has gained nothing. "As we are seeing," she writes, "extravagance is not eliminated merely by becoming anti-religious, and thoughts which are designed to be sternly reductive often compensate by strange, illicit expansions elsewhere. In fact when we encounter an especially harsh reduction, officially launched in the name of parsimony, our first question should probably be 'and what are these savings being used to pay for?'" (p. 82).

Midgley concludes by asserting that "today, a surprising number of the elements which used to belong to traditional religion have regrouped themselves under the heading of science, mainly around the concept of evolution" (p. 31). It is these religious remnants harbored by science which in her view provide the élan for the move toward an ever more perfect life-world. This concept of evolution she describes as a "vast escalator, proceeding steadily upwards from lifeless matter through plants and animals to man, and inevitably on, to higher things." The term evolution, she adds, "was coined by Lamarck and given currency by Herbert Spencer" (p. 34). Midgley quickly distances Darwin from such an understanding of evolution: "Darwin utterly distrusted the idea, which seemed to him a baseless piece of theorizing, and avoided the name. 'As far as he [Spencer] could see,' he said, 'An innate tendency to progressive developments exists.... It is curious how seldom writers define what they mean by progressive development'" (p. 34; Darwin quotation from Francis Darwin and A. C. Seward, eds., *More Letters of Charles Darwin*, pp. 338–39).

EVOLUTION AND TELEOLOGY

Indeed, one *might* reasonably wonder whether 'progressive development' has any useful meaning, if there is utterly lacking in nature both a sense of direction and a persistent upward thrust toward more perfect life-forms. Further, if there is within the nature or constitution of things no mechanism whatever whereby organisms are called upon to transform themselves or be transformed into more complex, higher life-forms, there seems no explanation available as to how such transformations might occur. In the 'natural selection' explanation of Darwin, different life-forms must somehow lie hidden in the gene pool, ready to emerge when the necessity imposed by environmental change brings about a new genetic mixture from which can result a 'modified' organism, that is, an organism hopefully better adapted to its environment and in some way or ways different internally from its immediate progenitors. This view, then, implicitly affirms that the DNA molecule itself—which provides the code directing both the organism's behavioral patterns and the irreducibly complex developmental process by which it unerringly moves from a single-celled organism to one of unimaginable complexity—must itself be present from the very start. This is most dramatically evidenced in nature's most biologically complex organism, namely, the human. A single cell, a minute organism, replicates itself repeatedly but always according to an already fully developed control pattern present within it. This control pattern is found within one of the molecules found within the cell, which is popularly referred to as the DNA molecule. To appreciate the enormous complexity of organic life and of the life of the human organism in particular, it is important to recognize that the DNA molecule is only one of approximately two hundred trillion molecules within each human cell, and that the mature, adult human comprises somewhere in the neighborhood of 100 trillion cells.

Thus, when from the zygote the human organism has at length developed to full maturity, each of the 100 trillion cells has within it a DNA molecule that is an exact replica of the DNA molecule in the original single-celled organism. What seems to be the most incredible feature of Darwinian evolutionary theory is its gratuitously contrafactual claim that such complex activity has no teleology whatever governing or guiding it, but is in the end merely the result of "the luck of the draw." Yet the biological facts themselves reveal that already in the

first cell of the organism the "entire" mature organism is already potentially present. What the organism is going to become is already
'written out' boldly in its DNA code; natural selection has nothing to
do with the creation of the code, although it is inexplicably wholly dependent on there being a code there in the first place to account for the
very possibility of a 'random mutation' occurring. Surely it makes no
sense to speak of something mutating unless it is already 'something' in
the first place. Without purposefulness latent within organic structure,
natural selection has nothing to feed on and is thus fated to experience
death by starvation.

Noting slight variations among individuals of specific populations
of living things owing mainly to sexual selection (fortuitous mating resulting in offspring with an original genetic strain) and environmental
changes, Darwin concluded that these variations gave an advantage to
some individuals while disadvantaging others in their common struggle
for existence. The organisms so favored would tend to thrive, while the
number of the others would gradually diminish. Over time, the newly
acquired favorable characteristics of those more fortunate members,
would, if inherited, result in these advantages being passed on to their
progeny. The result of this gradual process, by which the strong would
supplant the weak, would eventually but irrevocably bring about a
transformation of the species itself. Darwin identified the process by
which the evolution of new species took place as one of 'natural selection'. As remarked earlier, Darwin does not see any purposefulness behind this phenomenon of transforming species; it is a mere chance
event, wholly unpredictable in its outcome, a patchwork matrix resulting from a blind calculus of circumstance.

Yet Darwin's own assessment of the applicability of his theory of
natural selection to the world of singular events is not without its reservations. Writing to Jeremy Bentham in 1863 he grants that "in fact the
belief in Natural Selection must at present be grounded entirely on general considerations. . . . When we descend to details, we cannot prove
that any one species has changed; nor can we prove that the supposed
changes are beneficial, which is the groundwork of the theory. Nor can
we explain why some species have changed and others have not" (May

22, 1863, *Life and Letters*, III, 25, as quoted by Gertrude Himmelfarb in *Darwin and the Darwinian Revolution*, 1959, p. 366).

The 'natural selection' explanation of Darwin thus entails one significant flaw. As already pointed out, it ultimately depends on a hidden teleology underpinning all organisms in order to render coherent the emergence of 'new life-forms'. There must be lurking beneath the surface of living things a capacity for self-transformation. Different life-forms must somehow lie hidden in the capacity of the DNA gene pool, ready, so to speak, to emerge. What seems to be missing entirely from the Darwinian new-life equation is the recognition that, in order for there to be change, there must be something to be changed. As W. H. Thorpe has pointed out, no mutations can take place unless there is first something to be mutated (*Animal Nature and Human Nature*, 1974, p. 57). Furthermore, the very fact that some organisms allegedly undergo change while others do not implicitly recognizes that there is in each individual a 'nature' which grounds its stability, obstinately resisting change. Indeed, the Darwinian theory rests on the assumption that, if the environment remains relatively stable and favors the organism's continued existence, there is no reason to suspect, let alone assume, that any variation in its present 'life-style' will occur. Yet without exception our experience is that all organisms exhibit a determined 'desire' to preserve their status quo. They are all specialists in survival. Unfortunately, Darwinists in general have shown little interest in seeking to account for this commonly acknowledged characteristic of all organisms, of whatever complexity or kind.

A noteworthy exception to this phenomenon, however, is Stephen Jay Gould, who, though a committed evolutionist, has admonished many of his own colleagues for having focused almost exclusively on *progressive process* while generally overlooking the obvious fact that many species of organisms have been on the scene for a very long time, some of them since as long ago as the Cambrian period, which dates back 600 or so million years. In a very recent work, *Dinosaur in a Haystack: Reflections in Natural History* (1995), Gould makes the following remarkable comment concerning the phenomenon of permanency (stasis) as found in living things:

The stasis or nonchange of most fossil species during their lengthy geological life spans had been tacitly acknowledged by all paleontologists, but almost never studied explicitly, because prevailing theory treated stasis as uninteresting nonevidence for nonevolution. Evolution had been defined as gradual transformation in extended fossil sequences, *and the overwhelming prevalence of stasis became an embarrassing feature of the fossil record,* best ignored as a manifestation of nothing (that is, nonevolution). (Pp. 127–28; italics added)

What is astonishing, however, is that despite his strong emphasis on the persistence of stability among life-forms, Gould himself still persists in ascribing the emergence of the very organisms, so practiced in resisting change, to the mere "luck of the draw."

The issue we have been pursuing regarding the stubborn resistance that all life-forms put up in preventing their being slowly immolated to the rapacious god of evolutionary change is strikingly underscored by Michael J. Behe, recent author of *Darwin's Black Box: The Biochemical Challenge to Evolution* (1996), a work that has occasioned considerable stirrings in the scientific community. A professor of biochemistry at Lehigh University, Behe dramatically describes the biochemical complexities of, among other things, blood clotting, the immune system, the cell's manufacture of proteins, and its intricate vesicular transportation system. With regard to vesicular transport Behe indicates how critically important it is for the health of the individual human that this highly complex system function flawlessly. "A single flaw in the cell's labyrinthine protein-transport pathways is fatal. . . . Attempts at a gradual evolution of the protein transport system are a recipe for extinction" (p. 114). He goes on to indicate that to date the professional biochemical literature shows "that no one has ever proposed a detailed route by which such a [vesicular] system could have come to be. In the face of the enormous complexity of vesicular transport, *Darwinian theory is mute*" (pp. 115–16; italics added).

Similarly, Behe states: *"The scientific literature has no answers to the question of the origin of the immune system"* (p. 138; italics added). Again referring to the immune system he concludes: "As scientists we yearn to understand how this magnificent mechanism came to be, *but the complexity of the system dooms all Darwinian explanations to frustration*" (p. 139; italics added). Behe views all of the systems referred to above as "irreducibly complex" and concludes that they constitute "nasty roadblocks for Darwinian evolution; the need for minimal function greatly exacerbates the dilemma" (p. 46).

Still, what would seem to constitute the supreme irony of Darwin's 'natural selection' theory is that, according to evolutionary theory, genetic changes are attributable ultimately to an instinct for self-preservation, since the 'reason' for the adaptation to new environmental conditions is presumed to be on behalf of the individual's survival. Oddly enough, however, on this view the survival of the species is achieved only by its gradually phasing itself out, as it transforms itself or is transformed into something quite different from what it formerly had been. This would appear to be a poor man's version of a *secularized cosmic asceticism*. One must ask oneself in all candor whether such an accounting for 'evolutionary development' is logically coherent. Expressed in the plainest terms, survival—if that be the proper word—is achieved only through self-destruction. Is this not an excessive price the organism must pay for such a highly dubious reward, particularly if the newly emerging species is seen as really in no way superior to its victim host?

Yet, when the threads of evolutionary theory are untangled, this does appear to represent a fair assessment of Midgley's view of organic evolution. In seeking to account for the fact that significant adaptive changes have taken place, she is able to provide no explanation as to why living things apparently wish to survive in the first place, although an inborn resistance to destructive forces and a determined will for survival clearly constitute the basis upon which the entire evolutionary theory rests; and this would seem to apply not only to Midgley's version of evolution but to Darwin's as well, if we may assume that these do in fact differ. For how else could one construe the theory of organic development through sexual selection than by viewing the singular organism to be resolutely 'intent', as manifested through its own behavior, on promoting its individual well-being as well as working sedulously for the survival of its own species?

Indeed, were there no habitual resistance to change on the part of

the organisms allegedly undergoing transformation through sexual selection, there would hardly be a compelling reason for attempting to
explain, as Darwin and others certainly did attempt to explain, why it
was that the 'lower' life-forms have evolved into 'higher ones'. If all
life-forms actually sought 'by nature' to divest themselves of their own
specific identity and thus to transform themselves into different kinds
of things than they formerly were, they should be quite capable, it
seems reasonable to suppose, of effecting this metamorphic change very
much on their own, with much greater rapidity than allowed for by the
classical Darwinian theory of the evolution of the species by natural selection. It is further ironic that proponents of evolutionary theory, in
building their case for sexual selection as the source of species-related
change, have been able to support it only by subverting the permanency of the organic structures for whose transformation the theory itself
seeks to account.

If there is no purpose or direction to living things, then why should
they continue to struggle to live? Surely, within the ambit of the evolutionary hypothesis, they would have no 'obligation' or incentive to do
so. In fact, an inquiry into the meaning of living things then becomes
quite pointless. And it is for the salvation of such pointlessness that
Midgley argues. What seems to distress her above all is any classification of living things that arranges them in hierarchical order, particularly with the human placed at its apex. On this very score she is also
highly critical of Kant.

> [According to Kant] we can ask, "Why do animals exist?" but to ask,
> "Why does man exist?" is [for Kant] a meaningless question. What about
> that? Kant is taking nature as a pyramid, converging towards a single end.
> This will not do, not only because we are not sure who forms the purpose.
> It makes no sense to consider the enormous range of animal species as ex-
> isting as a *means* to anything, let alone us. (*Beast and Man*, p. 358)

This general leveling of all life-forms which Midgley advocates will
raise vexing questions regarding the interrelationships that do or ought
to exist between the animals and the human. At bottom, it of course
also raises sensitive questions about animal rights, but this is a matter
we will address at some length in a later chapter.

Midgley simply wants to eliminate altogether any talk of means and ends, substituting for this manner of speaking 'part' and 'whole' (p. 359). All living things together form a contiguous whole, and the human is to be ranked, in her view, alongside of but not above any other organism. Her view is shared by Eugene Linden, who points an accusing finger at the human for having allowed pride to turn his head. "Man is different," Linden affirms, "because that is the way we look at him. . . . The price of our pride has been the loss of our sense of place in nature. Now we have discovered that they [animals] have been here [on planet earth] all along" ("Talk to the Animals," *Omni*, Vol. II, no. 4 [Jan. 1980], p. 109).

For reasons to which we will attend in chapter 2, Midgley, Linden, and others who share their philosophical perspective exhibit no very clear notion of human intellection and its distinction from sensory knowing. (As a detailed investigation of this very issue will be taken up in the chapter following, there will be no need to say more about this distinction here, other than that the intellective act transcends the spacial-temporal realm, whereas the sensory act never succeeds in detaching itself from space-time.) Now the blurring of the distinction between intellection and sensing renders the grounds for adequately distinguishing the human from the animal problematical in the extreme. When the distinction becomes blurred, reasoning can more easily be assumed to be an activity common to both the animals and the human. This often will lead to the conclusion that a basic equality must exist between the nonhuman animal and the human. Thus we note for example that Midgley affirms:

> In the philosophic tradition, Reason, though not always equated with mere intellect, has usually been sharply opposed to Feeling or Desire. That has determined the attitude of those *respectable* philosophers to the related subjects of animals and human feelings. They have usually just dismissed animal activities from all comparison with human ones on the general ground that, in man, decision is a formal, rational process, while animals have only feeling. (*Beast and Man*, pp. 256–57; italics added)

Just which "respectable philosophers" Midgley is referring to is not clear, but it is obvious that she cannot be including any philosopher within the Aristotelian tradition and certainly not Thomas Aquinas, who has presented, after Aristotle, arguably the most detailed philosophic account of the similarities and differences between animal and human knowing to date. The view reflected in the above comment of Midgley is not uncommonly shared by contemporary philosophers (whom we willingly acknowledge as respectable), and it is predictably the view also of those adhering to the Darwinian explanation of the appearance of man. Since these same individuals often make appeal to the science of paleontology in support of the claim that a strict continuity exists between all organic life-forms, it is fitting that we now turn our attention to the question of man's origin.

EVOLUTION AND THE ORIGIN OF MAN

According to evolutionary theory as formulated by Darwin, all changes of life-forms came about gradually over a long period of time through miniscule but continual modification induced by the mechanics of random mutation through natural selection. It was essential to the coherency of such a theory that there be an observable continuity reflected in the behavioral patterns of humans and other primates. Otherwise, if the gap to be bridged was excessive, the emergence of new life-forms was rendered altogether improbable, and this for Darwin had become unthinkable, for, as Cassirer notes, evolution is to be taken as "a scientific hypothesis, a *regulative maxim* for our observation and classification," and hence not merely as a loose generalization of historical facts (*Essay on Man: An Introduction to a Philosophy of Human Culture,* 1944, p. 209). This helps explain why there is a notable systematic ambiguity latent in the Darwinist's efforts to identify the precise nature of intelligence in the human and why the higher vertebrates, especially the primates, are gratuitously referred to as 'being intelligent'. Anthropology and paleontology, when practiced under the nominalist banner, are unable of themselves to provide a clear distinction between intellective and sensate activities, since these disciplines willfully restrict their field of inquiry to what is observable to the senses. Accordingly, there is no room within their classificatory vocabulary to distinguish an act of sensing from an act of understanding. No viable criteria are available to them by which to draw such a distinction. Since these sciences collapse what transcends the sensible into the merely sensible, they can then regard the behavior of primates and other 'higher' animals as exhibiting basically the same or similar 'intellective' qualities as those found in the human animal.

In this manner, the distance separating the various species, and particularly the human and nonhuman primates, is significantly narrowed, and the pathway that could conceivably link them is cleared of thorny obstacles that could prevent a smooth, reasoned transition from the nonhuman animal to *homo sapiens*. Thus, such a transition now looms as not only possible but, from a superficial perspective, highly creditable as well. How, after all, if one is examining the matter through the eyes of the anthropologist, could the acts of understanding and sensing be 'seen' to differ? What palpable evidence could one appeal to that would support the claim that an intellectual act differs from a sensory one when only that is admissible as evidence which is itself sensible? This explains why the anthropologist puts so much stock in the comparative physical characteristics of the human and the brute animal. Why, that is, cranial capacity and skeletal structure is of altogether prime importance in measuring the comparative intelligence of the human and nonhuman animal.

The obvious physical similarities between the body structure of the nonhuman primates and the human lead the naturalist to conclude to a marked continuity between them and to view them as sharing a common ancestry. Because of physiological similarities, the gap between the human and the nonhuman animal (higher primates) is narrowed to the point where the emergence of *homo sapiens* from lower life-forms is considered now not only possible but apodictically certain. In short, the difference between the human and the nonhuman animal becomes one of degree only. Indeed, this is the view taken by Darwin himself and by those accepting his theory of natural selection as the primal source of variation between all animal life-forms. The one notable exception among the early advocates of evolution is that of Darwin's contemporary, the respected naturalist Alfred Russel Wallace, who, quite independently of Darwin, developed a theory of the natural selection of species in terms all but identical with those of the views advanced by

Darwin. Although Wallace warmly supported the view that new forms emerged through natural selection, he felt that the facts them-selves did not permit the 'law' of evolution to be applied to the human animal. Seeing in the natural selection explanation no satisfactory ac-count for the emergence of intellective activity in the human, Wallace felt constrained to acknowledge quite another source for the species man. With that directness and clarity of language for which he had come to be noted, Wallace wrote in 1869—much to both Darwin's and T. H. Huxley's dismay and, especially for the latter, chagrin—that natu-ral selection alone could not satisfactorily account for the biological origin of the human. Particularly disconcerting to Darwin was the fact that, prior to the 1869 statement, Wallace's views regarding evolution appeared to differ not one whit from those of Darwin himself. It is un-derstandable, therefore, that Darwin was quite shaken when he learned of Wallace's unconditioned restrictions regarding the origin of the hu-man species. There is no question, however, but that Wallace's view of the origin of the human was critically opposed to that of Darwin. The following excerpt from Wallace's essay *Contributions to the Theory of Natural Selection* succinctly and unambiguously states his case:

> But there is another class of human faculties that do not regard our fellow men, and which cannot, therefore, be thus accounted for. Such are the ca-pacity to form ideal conceptions of space and time, of eternity and infini-ty—the capacity for intense artistic feelings of pleasure, in form, color, and composition—and for those abstract notions of form and number which render geometry and arithmetic possible. *How were all or any of these fac-ulties first developed, when they could have been of no possible use to man in his early stages of barbarism?* How could 'natural selection', or survival of the fittest in the struggle for existence, at all favor the development of mental powers *so entirely removed from the material necessities of savage men,* and which even now, with our comparatively high civilization, are, in their farthest developments, in advance of the age, and appear to have rela-tions rather to the future of the race than to its actual status. (P. 352, as quoted by Wilma George in *Biologist Philosopher: Study of the Life and Writings of Alfred Russel Wallace,* pp. 72–73; italics added)

In describing Darwin's reaction to Wallace's 'modification' of the natural selection theory which he and Wallace had both made public through their writings, Wilma George supplies this terse comment:

"This was a great blow to Darwin and Huxley, who felt they had lost their most valuable ally in the campaign for evolution, natural selection, and man as an animal" (ibid., p. 72). George further adds: "He [Wallace] had ceased to believe that natural selection could account for *the whole of man*" (italics added). Showing that he fully understood the implications of these views, as Darwin certainly did also, Wallace says further on in the same 1869 essay: "I must confess, that this theory *has the disadvantage of requiring intervention of some distinct individual intelligence,* to aid in the production of what we can hardly avoid considering as the ultimate aim and outcome of all organized existence—intellectual, ever-advancing, *spiritual man*" (quoted by George, p. 73; italics added).

The fact that Wallace denied the applicability of the theory of natural selection to the genesis of the human did not go down easily with the author and the disciples of Darwinian evolution. Neither Darwin nor his followers saw any reason for making an exception in the case of man, for they did not regard the intellective powers of man as so extraordinary as to be capable of breaking the chain of continuity between the beast and the human. There was in their view no inherent reason why these 'special capacities', as Wallace referred to them, could not be the outcome of gradual changes brought about through sexual selection, which in turn were occasioned by abrupt environmental changes, which latter induced the physiological changes admittedly essential for the development of superior reasoning powers.

DARWINISM AND BIPEDALISM

Bipedalism is generally regarded by Darwinians as the genetic key setting off this domino theory of chain reaction. The arboreal progenitors of man, they speculate, for one reason or another abandoned their forest abode, exchanging it for the open savannah, either out of preference or out of necessity. In either case, they were now more dependent (so the argument goes) on their forward hands for food gathering, which required them to stand for greater lengths of time on their hind legs. Out of the practical necessity resulting from these and similar changes in living and work habits, and in accordance with the principle of the survival of the fittest, adaptations took place within their skeletal structure (evolutionists claim) permitting these primates to walk as a matter of course on their hind legs and to thus in time become bipedal.

The ability to walk on two legs further freed their forward limbs (hands) both to gather food more efficiently and to fashion and make use of tools. Reciprocally, these activities stimulated the brain cells to increase in number, thus causing a larger brain and increased intellectual capacity. In his work *Darwinism Defended* Michael Ruse describes how Darwinists understand this reciprocal process.

> In Darwinian evolution, when two things occur together, as often as not, one does not have simple cause and effect. Rather, one has a reciprocal feed-back process, with improvement in one thing leading to improvement in the other and vice versa. This was probably the case for the increase and development in human tool making and use, and for the growth of the brain *with the corresponding rise in intelligence*. The creature that was more intelligent made more efficient tools, which led to more favorable prospects of survival and reproduction. There was then a strong selective pressure back to increased abilities at tool making, *and to yet higher intelligence*. (*Darwinism Defended: A Guide to the Evolution Controversies*, 1982, p. 246; italics added)

In order to provide a rationale for the ascent of man from lower life-forms, the evolutionist must resort to such logical legerdemain. There is no other term more fitting to explain how the greater can be lifted from the hat in which only the lesser exists. In short, to justify changes that are seen as systematic changes for the better, at least in the sense that the new life-forms are biologically more complex in structure, evolutionary theory must postulate within organisms an 'inborn desire' to improve their lot. Whatever improvements do occur must always involve a genetic mutation, since only in this way could they be transmitted to future generations, with a guarantee of permanency assured. At the same time, in order to explain how such a change is possible, things themselves must not only be open to the possibility of change, but they must actually be internally oriented toward total, irrevocable transformation. This is simply a *sine qua non* condition for 'natural selection', even for it to be conceivable, since changes in environment and other such occurrences are not actually *causes* of the emergence of the transformed organism but merely *occasions* for it. They can only serve to trigger a mutation toward which the organism is already predisposed. And yet the openness and proclivity to change cannot itself be the product of previous natural selections, for to affirm this implies the denial of the necessary precondition for change, i.e., the possibility of change occurring in the first place.

Darwinian evolutionist theory demands a world of nature that is utterly and irretrievably unstable: each organism covertly seeks constantly to divest itself of itself and become something it is not. The theory in effect postulates a radical instability within each living thing—including even a pronounced suicidal tendency. But this flies in the face of the universal law of self-preservation governing all animate beings, in virtue of which each organism strenuously struggles to preserve its own identity. Each being strives to retain itself in its own mode of being, resisting death to the very last. Now the slight modifications that might occur are reasonably explainable in terms of the individual's survival; that is, through minor adaptation to environmental change, the species may improve its chances for survival. This much, sexual selection can conceivably achieve, the fossil record shows that this is exactly what has been achieved: through miniscule variations of living things, those very characteristics that most favor survival are enhanced. But evolutionist claims far exceed what the fossil evidence reveals. By his mode of reasoning the evolutionist elevates these all but imperceptible modifications to the high level of macro-developments, developments resulting in the emergence not only of more robust individuals but of entirely new species—which is to say, very different kinds of organisms.

THE DEBATE AMONG EVOLUTIONISTS

The fossil record shows that in the Precambrian period, 600 million to one and one-half billion years ago, there existed no truly complex life-forms. (It should be recalled that the consistent view of geologists is that the earth formed some four and one-half billion years ago.) Then suddenly, geologically speaking, about 600 million years ago there appears on the scene an enormous proliferation of life-forms. How does one explain the origin of this almost countless list of multicellular animals? At this point one does not, if one speaks from an authentically scientific point of view. That is, whatever position one takes in this matter must be a bald extrapolation from other more general principles. There is no sufficient evidence forthcoming from the fossil record itself.

When Darwin learned of this gap in the fossil record, he feared that it would provide those hostile to his natural selection argument a potent rejoinder. The sudden and unexplained emergence of new species remains to this day a formidable problematic for the evolutionist position. Stephen Jay Gould candidly grants as much. In his widely read book *Ever Since Darwin* (1973) Gould writes: "We may deny the Cambrian problem by casting it back upon an earlier event, but the nature and cause of this earlier episode remain as *the enigma of paleontological enigmas*" (p. 130; italics added).

Although Gould remains a committed evolutionist, he grants that all present multi-cellular life-forms came into being during the period of the Cambrian explosion, and that since that time the changes occurring among life-forms have been comparatively limited. He writes:

> Increasing diversity and multiple transitions seem to reflect a determined and inexorable progression toward higher things. But the paleontological record supports no such interpretation. There has been *no steady progress in the higher development of organic design*. We have had instead, vast stretches of little or no change and *one evolutionary burst that created the entire system*. For the first two-thirds to five-sixths of life's history, monerans (one-celled creatures) alone inhabited the earth, and we detect no steady progress from "lower" to "higher" prokaryotes. Likewise, there has been *no addition of basic designs since the Cambrian explosion filled our biosphere* (although we can argue for limited improvement *within* a few designs—vertebrates and vascular plants, for example). (*Ever Since Darwin: Reflections in Natural History*, p. 118; italics added)

In a more recent study Gould states that it has in the last few years been established "that the onychophora [air-breathing arthropods] have a fossil record extending right back to the Cambrian explosion" (*Dinosaur in a Haystack*, p. 112). He also informs us that the fossils of pentastomes (parasitic veriform arachnidians) found recently in Sweden are "entirely comparable to the moderns" and that their history "invalidates the hypothesis of their evolution from terrestrial arthropods" (pp. 117–18). He concludes that the explosion of life-forms during the Cambrian period is the key to the history of multicellular life (p. 120). These recent findings, wholly unexpected by the scientific community of paleontologists, pose what appears to be an insoluble problem for the Darwinists who have adamantly held to the orthodox 'natural selection' view that all changes in life-forms occurred slowly over periods of perhaps even hundreds of millions of years. The record seems to show that the stability of life-forms is a phenomenon that must at last be addressed anew with all seriousness.

Though neither Gould nor any other biologist or paleontologist has thus far provided an 'explanation' for the sudden surge both in the variety and complexity of animal life-forms during the Cambrian period, the vast majority continue to express confidence that an explanation will eventually be found. They are generally acceptive of the idea, along with most everyone else save strict Humeans, that every event has a cause. Yet when it comes to applying the principle to empirical phenomena, they often betray the very principle they claim to respect. It is here that the scientist and the philosopher often part ways, for the former is altogether intolerant of any application of the causal principle to singular events. Here scientists often revert to the Humean definition of cause, which interprets the cause-effect correlation as merely expressing a juxtaposition of events, without these events being necessarily connected. The scientist is understandably unable to visualize a "necessary connection" since the connection is not a physico-temporal reality. Hence the biologist, for example, takes it as a given that the 'causes' of all biological events are themselves biological in nature. That is, they insist that the cause of every event is directly proportioned to the resulting effect, and hence that, since the event is biological in nature, the cause of that event must also be biological. But as just suggested, this is in effect to deny the causal principle, for this principle is not itself an object amenable to biological scrutiny. Rather, it transcends the space-time continuum. It is here that the realist philosopher and the scientist often part company. In nontechnical terms, for the philosopher, as we here understand the term, what the eye sees is not, without qualification, what the intellect understands. As Aquinas will put it: the intellect does know material things, but immaterially (cf. *Summa Theologiae* [hereafter abbreviated *ST*] I, q. 85, a. 1, resp.).

With regard, then, to the question of the origin of living things, for the paleontologist this means that unless the record of events has been destroyed, it will always be possible, at least in theory, for the cause of every biological event to be traced and detected by biological methods. In short, there are in principle no questions stemming from biological occurrences to which the biologist or life scientist, acting as a biologist, cannot in time uncover the answers. To waver in his faith that such is indeed the case is, for the biologist, tantamount to assuming an unscientific stance, thereby denying to science its rightful place in the respected academy of knowledge.

THE SCIENTIST AND THE PHILOSOPHER

In effect, then, the biologist (or any scientist arguing in similar fashion) indirectly claims that biology (or science) is the highest tribunal of empirical knowledge to which the human can appeal. One can detect the hand of Immanuel Kant here, for he claims that no authentic philosophical approach to the empirical world is possible. Experience alone can provide us with no universal laws of nature, but only with material universals true for the most part (cf. *Critique of Pure Reason*, Introduction B3, trans. Norman Kemp Smith, 1958). Ernst Cassirer reflects this view and seems supportive of it when he writes: "Physics is concerned no longer with the actual itself, but with its structure and formal principles. . . . Order and relation have, then, become the basal concepts of physics" (*The Philosophy of Symbolic Forms*, III, 1955, p. 545). This remark and others in a similar vein have led Wilbur M. Urban to surmise that Cassirer's philosophy of symbolic forms is "apparently, merely a phenomenology, not a metaphysics" (Cassirer's "Philosophy of Language," in *Philosophy of Ernst Cassirer,* ed. Paul Arthur Schilpp, 1949, p. 428.) Many scientists of the contemporary era as well as those of the last century have been nourished on Kantian knowledge theory, which limits 'scientific' inquiry to the 'natural sciences' and relegates metaphysics, a well-intentioned but hopeless quest for transcendental knowledge, to the dustbin of transcendental illusions (B19ff.).

Yet if one gives at least equal weight to what scientists *do* as to what they *say,* we find that few if any are conspicuously faithful to their Kantian heritage. They do not hesitate to extrapolate conclusions from their observed findings and present these as part of the fabric of newly won 'scientific' knowledge. Yet extrapolations are themselves generalizations which are not directly observable. Now this procedure seems perfectly justifiable in principle, although difficult to reconcile with the concomitant refusal of the scientist to deny an analogous privilege to the philosopher. The mistrustful attitude toward the philosopher seems particularly noteworthy when the discussion involves such an issue as evolutionist theory. The evolutionist of today seems noticeably unwilling to recognize that one does not 'gather facts' without having some 'idea' as to why one is gathering them. (And recall that Darwin himself freely acknowledged this pattern of scientific proceeding.) The plain aim of studied observation is to extract from the observed data some general knowledge. This, in effect, is precisely what the scientist himself does. And the philosopher seeks to do the same, the only difference being that the scientist restricts his generalizations to classifications of things that in some way still carry the tag of temporality and spatiality, and which are thus, at least in some sense, imaginable. The philosopher's generalizations, on the other hand, are not so limited, since they deal with the intelligible aspect of things as such, and therefore cannot be expressed in terms which themselves bear the telltale mark of a sensory origin. The nature and object of the intellective act and how it both relates to and is distinct from sensory acts are points which will be taken up in more detailed fashion in the following chapter.

Thus biologists, seeking out, as biologists, the origin of life, restrict their inquiry to living things possessing some form of cellular structure or at least the capability of self-replication. Accordingly, since in the Darwinian view all contemporary organisms have themselves derived from simpler, less complex living forms, all living things are seen as sharing in, and hence are traceable back to, a common biological heredity. As Darwin observes at the end of his book *The Variation of Animals and Plants under Domestication,* "The most distinct genera and orders within the same great class—for instance mammals, birds, reptiles, and fishers—are all the descendants of one common progeni-

tor, and we must admit that the whole vast amount of difference between these forms has primarily arisen from simple variability" (Book
II, chap. 28, p. 425, as quoted by Ernst Cassirer in *An Essay on Man*,
1970, pp. 21–22). As to the origin of the pioneer one-celled creatures,
since there can obviously be no justification for the claim that they proceeded from simpler life-forms, it is simply admitted that they in turn
emerged from prior nonliving chemical structures. Such a position,
however, runs the very real risk of so diluting the distinction between
living and nonliving as to render it more logical than real. For the biologist it is, therefore, inadmissible to entertain the possibility, let alone
the likelihood, that living things came from other 'life-forms' or agents
not themselves biological in nature. The scientist would see such a step
as moving beyond the legitimate boundaries of scientific inquiry,
boundaries, withal, that were themselves self-imposed. The methodological canons of science restrict scientific inquiry to the field lying
within the parameters of space-time.

EVOLUTIONARY THEORY AND SCIENCE

Though at the present time there is no clear 'evidence' that the cascade of life-forms occurred in the manner described above, the naturalist assumes that it must have come about this way, because it is the only
'explanation' that appears to harmonize with evolutionary theory. The
theory of the evolution of living things through natural selection thus
assumes the role of an imperious measuring rod of our encompassing
experience of them. All explanatory claims regarding their nature and
origin must rigidly conform to the procrustean bed of evolutionary
theory.

The following comment by a prominent philosopher-biologist and
defender of the evolutionary realm serves as an apposite example:

> Darwinism is a theory about causes in the biological world. It tries to give
> answers to questions about the way in which organic types develop and
> change over time, showing also why organisms today are as they are. But,
> although Darwinism is a theory of biological causation, it invites questions
> beyond its own strict domain. Naturally—almost inevitably—one is led
> back in time to ask questions about ultimate origins: where did life come
> from in the first place? Thus at a very minimum, complementing Darwin-

ism, as it were, we could use a theory of the first production of life: a theory of chemical evolution perhaps? (Michael Ruse, *Darwinism Defended*, p. 156)

Perhaps not. Curiously, although Ruse describes Darwinism as "a theory of biological causation," it nonetheless, with apparent legitimacy, "invites questions beyond its own domain." What Ruse seems oblivious to is that every reflective thinker—philosopher, scientist, artist, or humanist—has at least an unarticulated philosophy of life underlying all his or her ruminations and speculations. It is a basic human characteristic to seek the big picture, and if the philosopher and others offend in stepping into arenas where they can claim no real expertise, scientists, too, are not total strangers to this same weakness. As a case in point, evolutionary theory itself wanders well beyond the field of strictly scientific knowledge. Ultimately this theory involves a philosophic claim, since it is actually seeking a unifying view of all living things, but such a view is not subject to "directly observable evidence."

Yet it appears that for Ruse there can be no serious raising of ultimate questions that do not square with the methodology of science. These are, in his view, strictly religious questions. Unfortunately, Ruse displays little interest in seeking to clarify his own understanding of the term 'religion'. The following excerpt supports this conclusion:

> Please note that I am not here denying the validity of the argument for the existence of God from the supposedly purposeful nature of the world, nor even am I denying the validity of the earlier-given causal argument for God's existence and nature. . . . All I claim, and no further argument is necessary, is that they take one out of science and into religion. (Pp. 323–34)

Doubtless this may follow if one's theory of knowing follows the Kantian model, but there are other epistemologies, which allow the raising of ultimate questions regarding the origin of the world and of the human as altogether legitimate areas of philosophic concern. It is understandable, however, how one who accepts biology as the highest of the human sciences would be closed to such a possibility. This does appear to be the shared view of the majority of contemporary biologists, who also often turn out to be confirmed evolutionists.

Thus, when it comes to reading the pre-Cambrian fossil record,

Ruse is confident that it is supportive of the classical evolutionist view, for in his estimate it pretty well satisfies the evolutionist's predictions. "In addition to indisputable pre-Cambrian organisms, the order of pre-Cambrian life seems to be what the evolutionist expects. One starts with the most primitive forms, and then works up to full-bodied multicellular organisms" (p. 311). Ruse skirts the entire question of the relation between evolution and progress. If "one starts with the most primitive forms and works up to full-bodied multicellular organisms," then how does one explain the fact that there are primitive life-forms extant today identical to organisms living during the Cambrian era some 600 million years ago?

DISCORD AMONG EVOLUTIONISTS

Ruse recognizes only too well, however, that present-day Darwinism is not a seamless garment. There are disaffected paleontologists who find difficulty in accepting what may be termed 'classical Darwinism', or what Ruse refers to as 'neo-Darwinism'. One of the principle proponents of the new, or modified evolutionary theory is the well-known geologist-paleontologist already referred to, Stephen Jay Gould of Harvard University. Gould and those who share his views read the fossil record quite differently than do the 'orthodox' (to use Ruse's own term) evolutionists. The former find the minute changes occurring through natural sexual selection alone inadequate to explain the sudden (in geological terms) proliferation of life-forms at the end of the pre-Cambrian period.

In fact, Gould's rather startling claim is that there has been no major development of living things since the Cambrian period; whatever changes have occurred since then amount to little more than modifications and slight adaptive changes to the variety of species already in place at that time. Gould writes, "The entire system of life arose during about 10 percent of its history surrounding the Cambrian explosion some 600 million years ago. . . . The world of life was quiet before and *it has been relatively quiet ever since*" (*Ever Since Darwin*, p. 118; italics added). Gould and others have described their modified Darwinian theory of evolution as the 'punctuated equilibria' theory. This theory is also sometimes referred to as 'saltationism'. This term derives from the Latin word for 'jump', *saltare*. Ruse himself is sensitive to the discordant note this new evolutionary view sounds within the previously close-knit choir of classical Darwinians, among whom he of course numbers himself. His characterization of the saltationist position is forthright and his opposition to it unyielding, even though he is able to express his opposition in a moderately humorous way. "In the past decade," he writes, "paleontology has again grown fractious, and again we find voices being raised against the Darwinian synthesis. A number of articulate and informed paleontologists, Miles Eldredge, Stephen Jay Gould, and Steven M. Stanley, to name but three, have thrown a large rock—a large saltationary rock—into the still waters of unanimity" (*The Challenge to Darwinism*, p. 210). Ruse sees that in abandoning the emphasis on 'gradualism' or minute adaptive changes prolonged over an extended period of time and continuing to the present, the saltationist opts for a rapid (in Darwinian terms) change from one species to another. This view would render otiose the slow adaptive changes through random natural selection defended by the 'orthodox' Darwinist. "Insofar," Ruse states, "as it has endorsed the notion of species selection, the theory of punctuated equilibria has always represented more than just a change of emphasis within neo-Darwinism. . . . *The position of the punctuated equilibria theorist seems to be that, with respect to any particular feature, speciation could take an organism in virtually any direction. The orthodox Darwinian could not accept this*" (p. 214; italics added). Much less of course could the orthodox Darwinist accept Steven M. Stanley's comment that he for his part tends "to agree with those who have viewed natural selection as a tautology rather than a true theory" (S. M. Stanley, *Macroevolution: Pattern and Process*, 1979, pp. 192–93; quoted by Ruse in *Darwinism Defended*, p. 214).

Perhaps even more directly confrontational is a statement Ruse quotes from Gould, which leaves little to the orthodox imagination. Gould first quotes biologist Ernst Mayr, who sums up the evolutionist theory quite unambiguously:

> "The proponents of the synthetic theory maintain that all evolution is due to the accumulation of small genetic changes, guided by natural selection,

and that trans-specific evolution is nothing but an extrapolation and magnification of the events that take place within populations and species." (Gould, "Is a New and General Theory of Evolution Emerging?" *Paleobiology* 6 [1980]: 120; quoting Mayr, *The Growth of Biological Thought* [1963], p. 586.)

Gould comments that this view, if accurate, spells the demise of the very theory it defines:

> I well remember how the synthetic theory beguiled me with its unifying power when I was a graduate student in the mid-1960's. Since then *I have been watching it slowly unravel as a universal description of evolution.* The molecular assault came first, followed quickly by renewed attention to unorthodox theories of speciation and by challenges at the level of macroevolution itself. I have been reluctant to admit it—since beguiling is often forever—but if Mayr's characterization of the synthetic theory is accurate, *then that theory, as a general proposition, is effectively dead, despite its persistence as text-book orthodoxy.* ("Emerging?" p. 120; italics added)

Gould, then, no longer allying himself with the classic, Darwinist theory of evolution, still stoutly maintains the origin of life-forms through an evolutionary, developmental process. The precise nature of this process remains, however, shrouded in obscurity. Gould himself has described it as a theory of 'punctuated equilibria' although he does not appear to have been particularly interested in providing tangible arguments in support of his claim.

Richard Dawkins, a contemporary stalwart defender of the 'classic' Darwinian position, takes note in a recent work of the views Gould and others have of Darwin's theory. He dismisses them as a mere *lis de verbis,* suggesting that Darwin himself "might well have approved [of them] if the issue had been discussed in his time" (*The Blind Watchmaker,* 1986, p. 250). Dawkins views the punctuated equilibria position as "a minor gloss on Darwinism," and asserts that as such it "does not deserve a particularly large measure of publicity" (ibid.). Obviously upset that the matter has been receiving as much attention as it has, Dawkins asks why, and cites three plausible reasons for this 'unfortunate' phenomenon. First,

> There are those who, for religious reasons, want evolution itself to be untrue. *Second,* there are those who have no reason to deny that evolution has

happened but who, often for political or ideological reasons, find Darwin's theory of its *mechanism* distasteful. Of these, some find the idea of natural selection unacceptably harsh and ruthless; others confuse natural selection with randomness, and hence "meaninglessness," which offends their dignity; yet others confuse Darwinism with Social Darwinism, which has racist and other disagreeable overtones. *Third,* there are people, including many working in what they call (often as a singular noun) "the media," who just like seeing applecarts upset, perhaps because it makes good journalistic copy; and Darwinism has become sufficiently established and respectable to be a tempting applecart. (Pp. 250–51; italics added)

Dawkins is persuaded that the 'punctuated equilibria' theory is merely a 'minor gloss' on Darwinism because, in his view, Gould, Eldredge, and others interpret the 'gradualism' of which Darwin spoke as synonymous with development at a constant speed and thus as excluding altogether the possibility of saltations or irregular and unpredictable leaps. Dawkins rejects this interpretation of Darwinian gradualism, insisting, "In the sense in which Eldredge and Gould are opposed to gradualism, there is no particular reason to doubt that Darwin would have agreed with them" (p. 250). Surely Dawkins leaves no room for doubt but that he considers the punctuated equilibria theory as lying wholly within the confines of orthodox Darwinism. He concludes:

> What needs to be said now, loud and clear, is the truth: that the theory of punctuated equilibrium lies firmly within the neo-Darwinian synthesis. . . . The theory of punctuated equilibrium will come to be seen in proportion, as an interesting but minor wrinkle on the surface of neo-Darwinian theory. (P. 251)

Nevertheless there can be no denying that the punctuated equilibria theory has succeeded in striking a tender nerve of this doughty Darwinian. This is evidenced by his ringing peroration:

> It [the theory of punctuated equilibria] certainly provides no basis for any "lapse in new Darwinian theory." It certainly provides no basis for any "lapse in neo-Darwinian morale," and no basis whatever for Gould to claim that the synthetic theory (another name for neo-Darwinism) "is effectively dead." It is as if the discovery that the Earth is not a perfect sphere but a slightly flattened spheroid were given banner treatment under the

headline: COPERNICUS WRONG, FLAT EARTH THEORY VINDICAT-
ED. (Pp. 251–52)

The emotional overtones in Dawkins's critique of Gould clearly suggest that behind his opposition to the punctuated equilibria theory lies more than mere love of scientific rigor. At any rate, Dawkins is clearly persuaded that, when properly understood, the punctuated equilibria theory is fully consistent with Darwin's original insight into the key role sexual selection plays as the ultimate explanation for the diversity of living species, both extant and extinct, that are and have been identified either through direct experience or through a critical examination of the fossil record.

Moreover, the radical nature of the neo-Darwinian claim needs to be carefully attended to. Richard Dawkins, one of the leading and most outspoken contemporary advocates of Darwinian evolutionary theory, makes it perfectly clear in the *The Blind Watchmaker* that classical Darwnian evolution provides the explanation not only for the differentiation of living things, but for their primal origin as well. That is, the Darwinian view of evolution is not merely biological in nature, but cosmological as well. It purports to provide a consistent and scientific explanation for the origin of life itself, and it thereby becomes in effect a stand-in for philosophy as well as for religion.

With evolutionary theory the need for a pre-existing intelligence as designer of the universe and as the ultimate source of all living things vanishes. Life is simply the spontaneous result of the interplay of chemical forces on the molecular level. As Dawkins himself writes, this is the purpose of his book *The Blind Watchmaker*. "The basic idea of *The Blind Watchmaker*," he writes, "is that we don't need to postulate a designer in order to understand life or anything else in the universe" (p. 147). For Dawkins the cumulative selection process through which living things have evolved into more and more complex entities has *no purpose*. Evolution has, he says, "no long distance target" (p. 50). Additionally, "It has no vision, no foresight, no sight at all. It can be said to play the role of watchmaker in nature, it is the *blind* watchmaker" (p. 5). Dawkins therefore, like Darwin himself and numerous of the contemporary protagonists of the evolutionary position, is an atheist.

He credits Darwin with being the thinker who "made it possible to be an intellectually fulfilled atheist" (p. 6). This point is worth noting, because it discloses the manner in which evolution is understood by its more resolute defenders. It is intended as an explanation not only of how things change and evolve but of the very origin of life; it alleges to afford not an intermediate but an ultimate explanation of living things. Dawkins is unambivalent regarding this point, as he expressly denies that the evolutionary process itself has received its direction from a higher source. "Maybe, it is argued, the Creator does not control the day-to-day succession of evolutionary events; maybe he did not frame the tiger and the lamb, maybe he did not make a tree, but he *did* set up the original machinery of replication and replicator power, the original machinery of DNA and protein that made cumulative selection, and hence all of evolution possible." Dawkins's response to this line of reasoning offers little reassurance to those theists who are comfortable in similarly espousing Darwinian evolutionary theory. Dawkins eschews outright any such compromise. "This is a transparently feeble argument," he counters, "indeed it is obviously self-defeating. Organized complexity is the thing that we are having difficulty in explaining" (p. 141). Shortly thereafter he adds, quite naively, "To explain the origin of the DNA/protein machine by invoking a supernatural Designer is to explain precisely nothing, for it leaves unexplained the origin of the Designer. You have to say something like 'God was always there,' and if you allow yourself that kind of lazy way out, you might as well just say 'DNA was always there,' or 'Life was always there' and be done with it" (ibid.).

It is evident, then, that for Dawkins the evolutionary theory is not a subsidiary explanation of the origin of life; it is not complemented or even supplemented by other explanations; it is pure and simple *the* explanation for the appearance of life on planet earth. Evolutionary theory is, if properly understood, wholly autonomous, in Dawkins's view, explaining the rise not only of animal life but of human life as well. "Cumulative selection," he writes, "once it has begun, seems to me powerful enough to make the evolution of intelligence probable, if not inevitable" (pp. 146–47).

Dawkins professes wonderment at the "hostility" many feel toward Darwinian evolutionism, and the seeming inability of so many to grasp the concept of "cumulative selection" as the explanation of how life-forms originated and how the more complex life-forms developed from the simpler (ibid.).

The renowned molecular biologist of DNA fame, Francis Crick, shares Dawkins's perplexity. In *What Mad Pursuit? A Personal View of Scientific Discovery* (1988), Crick asks the question: "Why exactly is it that so many people find natural selection so hard to accept?" (p. 30). Responding to his own question, Crick presents three reasons: "Part of the difficulty," he writes, "is that the process is very slow, by our every-day standards, and so we rarely have any direct experience of it operating." Shortly thereafter he adds a further reason: "A second difficulty is the striking contrast between the highly organized and intricate results of the process—all the living organisms we see around us—and the randomness at the heart of it." He adds, however, that the contrast is misleading because of the "selective pressure of the environment." Crick fully agrees, then, with Dawkins's assessment that the environment, through cumulative selection, is the actual designer of the evolutionary thrust even though it does not consciously control or direct it. The "blind designer" is more accurately an unwitting one. Dawkins's choice of title, *The Blind Watchmaker,* is not only inaccurate but misleading, for what Dawkins is in fact advocating, and Crick along with him, *is a watchmaker who is comatose.* This in fact is what he himself actually implies when he remarks: "In the case of living machinery, the 'design-er' is *unconscious natural selection,* the blind watchmaker" (Dawkins, *Blind Watchmaker,* p. 37; italics added).

Thus, in Crick's words it is "the environment" which provides the direction, "and over the long haul its effects are largely unpredictable in detail." Crick grants that the human mind finds it hard to accept that there is no Designer, since the product of natural selection appears to be the result of conscious design (Crick, *Mad Pursuit,* p. 30). He does admit, nonetheless, that there are two "fair criticisms" of natural selection. One is that "we cannot as yet calculate, from first principles, the *rate* of natural selection, except in a very approximate way" (ibid.).

The other is that "we may not yet know all the gadgetry that has been evolved to make natural selection work more efficiently. There may still be surprises for us in the tricks that are used to make for smoother and more rapid evolution" (p. 31).

Does it not seem odd, however, that Crick is quite unaware of any inconsistency in his speaking of the evolutionary process in terms of greater or less efficiency, and of smoother or less smooth, more rapid or less rapid development? To characterize evolution in these terms entails viewing it from the perspective of achievement, and this would be to assign to evolution some sort of purpose or design. To say that such design is not truly inherent in the evolutionary process itself, but is merely overlaid upon the process in analogously Kantian fashion by intelligent observers, like ourselves, seems inadmissible if one takes seriously the oft-reiterated claim that evolution is blind and directionless. To apply the terms *efficiency* and *smoother* and *more rapid* to evolution is surely to introduce clandestinely a teleological dimension to natural selection theory, which one had been assured was quite alien to its 'nature'. To this point we shall presently return. Meanwhile, suffice it to say that Crick registers great surprise that not everyone shares his enthusiasm for the Darwinian philosophy of life, and its manner of accounting for life's origin, since for himself and many others it provides an explanatory process that is "powerful, versatile, and very important." He adds, "It is astonishing that in our modern culture so few people really understand it" (ibid.). Crick leaves little doubt that, for him, to understand Darwinism is inevitably to accept it.

EVOLUTION AND SPECIES

In the work to which we have already alluded, Dawkins emphasizes that Darwinian selection is not merely random (*Blind Watchmaker*, p. 49). In fact, this is probably his central thesis, and it is really but another way of his saying that natural selection is not step-by-step selection, but rather *cumulative selection*. Dawkins understands by *cumulative selection* the tendency of matter and, *a fortiori*, living things to tend toward ever greater complexification. In other words the changes that take place in nature are developmental by nature, vertically as well as horizontally. Despite the fact that Dawkins sees no significant difference between 'punctuated equilibria' and the nonpunctuated theory, i.e., orthodox Darwinian evolution, he does grant that there are differences between them that in reality would seem to be far from trivial.

The fundamental difference lies, allegedly, in the manner in which both of these positions view the term 'species'. This is an altogether central issue, indeed, the controlling issue, not only for the differences between punctuated and nonpunctuated equilibria theory, but also for the probing questions pertaining to the nature of living things and their origin, as well as the grounds for speaking of individuals as members of a group. Admittedly, the distinction between the individual and the class to which the individual pertains is one of the more vexing of all philosophical problems. The manner in which one responds to it controls the kind and number of options one has in responding to whole clusters of other related questions. One of these questions—indeed one of the major concerns of our present study—is the distinction between the human and the nonhuman animal. Correlative questions are the nature of a right and whether humans alone possess rights.

But let us return to our immediate concern—Dawkins's discussion of the differences between the 'punctuationist' (Dawkins's term for one advocating the punctuated equilibria theory) and the nonpunctuationist. Each views 'species' from a different perspective and each emphasizes a different detail of the fossil record. The punctuationist focuses on the "unchanging status" of living things, with the consequence that his attention is turned away from the individual organism and riveted on the class or species that survives. The punctuationist, Dawkins charges, makes "a big point of treating 'the species' as a real 'entity'" (p. 264). On the other hand the nonpunctuationist—and Dawkins numbers himself as one—"cannot see 'the species' as a discrete entity at all." This is because what exists are merely individuals succeeding each other, so that all one really 'sees' is "a smeary continuum."

For the nonpunctuationist "a species never has a clearly defined beginning, and it only sometimes has a clearly defined end (extinction)" (ibid.). Consequently, Dawkins continues, "The nonpunctuationist would not see a species as having a 'life span' like an individual organism." On the other hand, the punctuationist "sees a species as having a definite, or at least rapidly accomplished, end, not a gradual fading into a new species." From this it follows that "to a punctuationist, a species can be said to have a definite, measurable 'life span'." A species is then, considered "a discrete entity that really deserves its own name" (ibid.). The nonpunctuationist, on the other hand, views "the species as an arbitrary stretch of a continuously flowing river, with no particular reason to draw lines delimiting its beginning and end" (ibid.). When, then, the nonpunctuationist employs the name 'species' he appears always to be using it to designate nothing more than an arbitrary grouping of separate particles. That is, he uses species names "only as a vague convenience" (p. 265).

When the nonpunctuationist, or orthodox Darwinian, therefore, takes the long view of things, seeing only "individual groupings" following one upon the other, the name 'species' simply blends in with these groupings and is all but lost from view. As Dawkins states quite boldly, "When he [the nonpunctuationist] looks longitudinally through time, he ceases to see species as discrete entities" (ibid.). This view meshes well with Darwin's own firm conviction that "man and the lower animals do not differ in kind, although immensely in degree. A difference in degree, however great, does not justify us in placing man in a different kingdom" (*Descent of Man,* vol. I, p. 186, quoted by Howard, *Darwin,* p. 63).

We needn't pursue here whether Dawkins's characterization of the punctuationist position is altogether accurate, but suffice it to say that Gould himself continues to see a difference between the two positions. Further, he appears to have turned the tables on Dawkins, who had accused the punctuationists of an extreme Platonic view in looking upon species as independent entities. Gould counters by accusing Dawkins of considering the genes, through whose modification the individual undergoes change, as entities within themselves, distinct from the bodies they inhabit, the latter seeming to serve as nothing more than *loci operandi:* "Dawkins goes a step beyond reduction to parts and views genes themselves as the focus of natural selection. Bodies become mere survival machines, temporary homes for genes engaged in a more than metaphorical struggle to make more copies of themselves in future generations" (*An Urchin in the Storm: Essays about Books and Ideas,* 1987, p. 66). Gould concludes his assessment of Dawkins's position: "I find little defensible in this view."

CONSEQUENCES OF EVOLUTIONARY THEORY

Deserving of our special notice is Dawkins's view regarding 'species'. The inevitable consequences which follow upon his contention that 'species' lacks any grounding in our experience, as Dawkins himself expressly acknowledges, are sobering in the extreme. Since he views the term 'species' in strict consequentialist terms as nothing more than a vague convenience which the human mind employs as a useful descriptive tool, he is logically led to consider all living organisms as descending from "a single common ancestor." Dawkins allows that he is led to this conclusion by the recent biological discovery of the radical similarity in the genetic structure of all living things. "It is a fact of great significance," he argues, "that every living thing, no matter how different from others in external appearance it may be, 'speaks' exactly the same language at the level of the genes. The genetic code is universal. I regard this as near-conclusive proof that all organisms are descended from a single common ancestor" (*Blind Watchmaker,* p. 270).

Because of this genetic continuity, Dawkins doubts whether humans actually possess rights that are exclusively theirs. He questions, therefore, whether there is a defensible rationale for treating chimpanzees, and other primates, any differently than humans. Dawkins alleges that we humans employ a double standard whereby we attribute rights to humans which we do not similarly accord to chimpanzees. This arises, he suggests, from the fact that the original ancestral intermediates between ourselves and nonhuman primates are now extinct. Thus the differences between us are magnified, and we are provided a pretext for looking upon chimpanzees as ancestrally unrelated, when in truth they are our cousins.

Dawkins makes much of the fact that we and the chimpanzees share more than 99 percent of our genes (p. 263). He also claims that 'speciesism', whereby the human is held to be significantly superior to the animal, provides the ground for a one-sided human ethics and law. He implies that the rights we attribute to the human ought either to be applied to animals [chimpanzees?] as well or be radically revised so that the human is not awarded preferential treatment.

Dawkins also comes close to revealing an anti-Christian bias by finding it incongruous that "the breathtaking speciesism of our Christian-inspired attitudes" lead some to oppose the abortion of a human zygote [embryo, fetus?] while allowing the vivisection of adult chimpanzees. The passage merits being quoted in full, lest the reader conclude that Dawkins has been misinterpreted.

> Our legal and moral systems are deeply species-bound. The director of a zoo is legally entitled to 'put down' a chimpanzee that is surplus to requirements, while any suggestion that he might 'put down' a redundant keeper or ticket-seller would be greeted with howls of incredulous outrage. The chimpanzee is the property of the zoo. Humans are nowadays not supposed to be anybody's property, yet the rationale for discriminating against chimpanzees in this way is seldom spelled out, and I doubt if there is a defensible rationale at all. Such is the breathtaking speciesism of our Christian-inspired attitudes, the abortion of a single human zygote (most of them are destined to be spontaneously aborted anyway) can arouse more moral solicitude and moral righteous indignation *than the vivisection of any number of intelligent adult chimpanzees!* . . . The only reason we are comfortable with such a double standard is that the intermediates between humans and chimps are all dead. (Pp. 262–63; italics added)

It becomes painfully evident how profound and extensive are the implications of the neo-Darwinist view that all living things are descended from a single primitive ancestor. In this scenario the very existence of human rights becomes altogether problematical and the viability of human ethics equally questionable. If biology, especially molecular biology, became the supreme court for adjudicating what it means to be human, then most of what we identify as the fruits of human civilization could in time be swept away.

If 'species' is merely an ephemeral term of *vague convenience* lacking in significance then there is no longer any reliable gauge by which living things can be differentiated. All classification of living things becomes arbitrary, and, ultimately, meaningless. Indeed, the distinction between living and nonliving also dissolves in the solvent of *'molecularism'*, as it becomes more and more a challenge to distinguish physical and chemical biology from physics and chemistry.

Dawkins identifies himself as a hierarchical reductionist (p. 13), by which appellation one might justifiably conclude he intends simply to affirm that the evolutionary position denies all order to the universe. There appears, indeed, to be no standard available to him by which one thing could be measured as superior or inferior to another—or how one part of an organism could be marked off as superior to another part. One is left with little more than a very thin gene-soup in which one particle is indistinguishable from another. By denying any true meaning to species, Dawkins, and other committed Darwinists sharing his views, are denying that living things have shared essences; they are simply individuals spacially distinct from one another *as individuals,* and remain, therefore, inherently unclassifiable save in the most superficial way. From this openly nominalist perspective irresolvable tensions arise which Dawkins gives little evidence of recognizing, but which serve effectively to undercut any serious effort to study living things scientifically.

EVOLUTION AND THE UNITY OF LIVING THINGS

While Dawkins differentiates the living from the nonliving by ascribing to the former a quality of "adaptive complexity" (p. 304) whereby through natural selection the living organism is able to adapt itself to its environment in a manner more favorable to its well-being, he fails seriously to address the crucial question as to the origin of the organism's adaptive ability. Although Dawkins is critical of the punctuationist's focus on 'species selection' to explain the sudden appearance of new life-forms, insisting that there exists no dimension of reality which is not in and of itself individual, he logically must view 'species' as nothing more than a pure construct of the mind, while he himself is extremely reticent to tell us just what, then, an *individual* is. In his view it is the gene that plays the primary role conferring stability to the organism. How the various parts of that organism interrelate, and how it is possible for them to interrelate seem not to be questions that have aroused his interest.

The biologist (and seemingly scientists generally) often encounter difficulty in accepting the reality of a first life principle in living things, since one is unable sensibly to experience it. Yet scientists often postulate theories involving highly abstract principles in an effort to explain complex data. The universalization of their experience in the form of laws and principles that cannot be reached by sensory perception is, for scientists, an altogether common and acceptable practice.

Dawkins on several occasions speaks rapturously of the unbelievable complexity and beauty of various organs of living creatures and of their cellular and molecular structure. Perhaps the most striking instance of this occurs where he minutely and wonderfully describes the biological structure of the human eye (*Blind Watchmaker*, esp. pp. 80 and 288). Yet he stoutly maintains that the breathtakingly intricate and enormously complex organ of vision, which is the eye, *is* the 'unintended' result of natural selection. And yet it is obvious that the eye is but one of the organs employed by the individual who sees; the eye sees not just to accommodate the eye, but rather to facilitate those actions essential to the normal functioning of the individual animal or human possessing sight. One cannot, then, reasonably deny that there is a hierarchical, functional arrangement present within the individual organism itself. The eye ought not properly be said to see, but rather it is the individual who sees by employing the eye as its organ of vision. Indeed, it is to the hierarchical unity of the sensing individual that one refers in the very employment of the term 'organism'. Inexplicably, this obvious unity of the total functioning organism is mutely accepted but left unexplained. The source of the unity in an organism, both in being and in function, remains a crucial yet neglected question. Neither the eye nor any other organ within the organism is 'selfish' in the sense that its primal function is self-promotion; rather its specialized function aims at promoting and securing the health and well-being of the entire organism.

The organism is thus not a loose confederation of parts, each with its own constitution and center of government. It is, rather, one 'nation' or 'state' with a highly centralized governing system, which oversees and controls the activities of every part within its borders. Even on the cellular level the kinds of proteins produced by the individual cell are not under the full control of the cell itself, but these 'decisions' are made at the very apex of the hierarchical order of the organism of which the single cell is no more than a miniscule part.

Indeed, in the mature human organism there are estimated to be approximately 100 trillion cells all working to promote, not their own well-being, but that of the total living body in which they are incorporated. Thus, when some cells become 'disoriented' and no longer look to the promotion of the health of the organism of which they are a part, and their behavior turns roguish, as in the case of the cancer cell, the entire organism is debilitated. And if the revolt of the disaffected cells is allowed to go unchecked and to spread, it of course results in the demise of the organism itself, as its normal dynamic state of equilibrium is upset. Instead of a well-knit living being displaying activities of exquisite harmonious orchestration, its unity eventually collapses in upon itself and even the simplest collaborative activity ceases, never to be regained. The organism dies.

What, then, accounts for the unity of the living thing? What assures that all of the parts function not on their own behalf but on the behalf of the whole of which they are, in fact, merely a part? And how does one account for the sudden shattering of that harmonic unity when the myriad upon myriad parts of the simple organism rise up in anarchical protest and commence to function now not as coordinated parts but as themselves wholes, as independent entities? The phenomenon of death clearly reveals the occurrence of an irreversible macromutation of what was once a singular, vital entity; the hierarchical control that was formerly present has now suddenly vanished; the harmony previously noted has disintegrated, and none of the living activities the organism previously performed remain. Death is a sudden and total transformation of the organism. It is a phenomenon of which account must be taken in any comprehensive theory of living beings.

But Dawkins has taken no pains to explain either death's nature or its possibility. Whereas he has sought to explain the existence of the millions of varieties of life-forms through what he terms 'cumulative natural selection' and he has placed them on an ascending scale according to their adaptive complexity, he has kept silent on the subject of death. It is accepted as fact, an event occurring as a mere matter of course, a plain given, needing no explanation. But is it? Why does death occur? Why do all living things die? Why are the life spans of living things so extremely brief when contrasted with the 'age' of nonliving things, of planet earth, of heavenly bodies? These may not be the kinds of questions a biologist or paleontologist might ordinarily address, but the answers to them cannot but have far reaching implications for scientists as well as for all humankind.

ARISTOTLE AND THE PHENOMENON OF LIFE

Is there any way of explaining the living thing in terms of its parts? If there were, would it not follow that all bodies would be living? And would there then be any explanation whatever for the universal phenomenon of the death of living things? I submit that there would not. It was precisely to incorporate the common experience of living things as operative unities and the obvious phenomenon of limited life spans of all organisms that Aristotle postulated the existence of soul as the internal life source of living things. This empirically minded philosopher, who in his youth studied medicine, was, before turning to philosophy, an avid student of nature in all its forms and thus became the first biologist and taxonomist in the modern sense. Aristotle viewed the soul as the first principle of life within living things; that is, as the exclusive internal source of an organism's dynamic unity, of its ability to move and replicate itself and to perdure. By saying that the soul is the first principle of life inherent within the living thing, Aristotle is not saying that it is the organism's only principle of life, but that it is its first, most basic, source of unity, and is thus directive of all of the organism's functions and activities.

Aristotle did not come lightly to his postulation of a life principle or soul as the only way of explaining living things. His position was grounded on his continuing and assiduous observation and reflective study of animals of all kinds, including countless forms of marine life.

The Phenomenon of Things Living

One cannot come away from his biological treatises on the histo
generation of animals with anything less than a profound resp
the painstaking care with which he and his apprenticed students and
co-workers carefully described and classified the anatomy and living
habits of thousands of different fauna extant in his day.

As a philosopher imbued with an empirical sense of the real, Aristotle recognized that every event necessarily has a cause, *pace* Hume, and
must, therefore, have an explanation. The death of living things does
not just occur. Things do not, without rhyme or reason, simply either
fade away or come to be. Providing an explanation consists in nothing
more than a tracing out the cause or causes, i.e. reason or reasons, behind an occurrence. To allow for the possibility, let alone probability,
of a world without cause and without design would entail wholesale
abandonment of the scientific enterprise itself—an option no thinking
person could seriously entertain.

Our discussion of Darwin's evolutionary theory, though lengthy,
has been necessary. The Darwinian explanation of the ontogenesis as
well as orthogenesis of the many different life-forms has provided the
basis for the denial of a fundamental difference between the human
and nonhuman animal. Almost all advocates of such a position are, to
the best of my knowledge, ardent adherents of the classic or the neo-
classic Darwinian theory of the 'origin of species'. Hence it is important to recognize the Darwinian presuppositions to the view that there
is in fact no essential difference between human and nonhuman animals and thus there are no justifiable grounds for affirming that the
nonhuman animal is incapable of intellective reasoning. In the following chapter we will turn our attention to this allegation, namely, that
the nonhuman animal as well as the human is in fact endowed, to some
extent at least, with powers of intellective reasoning.

2 Intelligence in the Human and in the Nonhuman Animal

In the previous chapter we discussed the Darwinian theory at some length and came to recognize several of its fundamental assumptions. Perhaps the most important is the explicit acknowledgment on the part of Darwin himself that the term 'species' "is arbitrarily given for the sake of convenience to a set of individuals closely resembling each other" (as quoted by Philip Lieberman, *On the Origins of Language: An Introduction to the Evolution of Human Speech*, 1975, p. 173). This working definition of 'species' makes it clear that Darwin's fundamental philosophical convictions are thoroughly nominalist; as Lieberman notes, "The term *species* is simply a labeling device" (ibid.). In this chapter we intend to challenge the adequacy of this nominalist definition of 'species'. We begin by attempting to pinpoint, first of all, the reasons why Darwin and Darwinists alike opt for a nominal definition of 'species', and what the hidden implications of this view are. We next examine with some thoroughness the intellective act of the human, comparing it with the 'intelligent' behavior of the nonhuman animal.

We begin with an analysis of intelligence, because it is precisely here that we ground our definition of the human as an organism differing from the animal, not merely in degree but in kind as well. If the difference in cognitive powers between the human and nonhuman animal is merely one of degree, the foundation upon which a viable argument for the distinction of 'species' in the non-Darwinian sense must rest is dissolved and the argument collapses in upon itself. Hence the high importance of this issue.

The classical Darwinists see clearly enough that 'evolutionary' development through natural selection becomes internally incoherent if, by 'development', one understands an ascent from specifically lower life forms to higher. If discontinuity of species is initially granted, it becomes impossible, they allow, to account for the appearance of newer, more complex life forms on the basis of natural selection alone. Consequently, they argue that there is an underlying continuity in all life forms.

Both Mary Midgley (*Beast and Man*, 1978) and Stephen Jay Gould (*Wonderful Life: The Burgess Shale and the Nature of History*, 1989) strenuously object to viewing man as the upper end-product of a lengthy evolutionary process. The human is not the pinnacle of an imaginary evolutionary tree, and his coming to be was, they believe, altogether fortuitous. Writes Gould: "We must assume that consciousness would not have evolved on our planet if a cosmic catastrophe had not claimed the dinosaurs as victims. In an entirely literal sense, we owe our existence, as large and reasoning mammals, to our lucky stars" (*Wonderful Life*, p. 318). In Gould's reckoning, indeed, were the human to become obliterated on the entire surface of planet Earth by a mutated virus, for example, there would be no real possibility of its ever emerging again. He argues that the fragile calculus of random events which first accounted for the emergence of man could never again fall into place (p. 320). He seems to be striving to have it both ways. On the one hand there is the appeal to 'continuity' between the human and other mammals, since the human had other animal life forms as ancestors, and, on the other, there is at least the implicit appeal to discontinuity, for the emergence of man was the unforeseen result of a high risk, unrepeatable calculus of events. One might note that this same ambiguity pervades the Darwinian enterprise itself, and will be seen to surface once again in our examinations of human and animal intelligence as well as in questions of language.

We now turn our attention directly to the question of intelligence in the human and the animal. We shall first clarify the meaning behind the word 'intelligence' and in what sense human intelligence might be either similar to or different from the 'intelligence' of the nonhuman animal. As should be obvious, this is a matter of singular importance in comparing the human and nonhuman animal, although, regrettably, it is a point not infrequently passed over in haste when such studies are conducted. Eugene Linden, for example, states with regard to the human: "Man is different because that is the way we look at him" ("Talk to the Animals," *Omni,* no. 1 [January 1980], p. 109). Midgley, for her part, objects to the continual emphasis on instincts in referring to animal behavior, and firmly insists on the nonhuman animals being viewed as 'rational' creatures, affirming that they, too, act out of 'reason' (*Beast and Man,* p. 281). A similar view is expressed by Darwin himself, though he disavows any attempt to nuance his claim. "I will not attempt any definition of instinct," he says. "It would be easy to show that several distinct mental actions are commonly embraced by this term. . . . A little dose, as Pierre Huber expresses it, of judgement or reason, often comes into play, even in animals very low in the scale of nature" (*The Origin of Species,* pp. 207–8, as quoted by Howard, p. 67).

DARWIN AND HUMAN REASONING

As we have seen, Darwin disdains placing the human "in a different kingdom" from that of the animal, since he holds the differences between them to be merely differences of degree. His unwillingness to spell out the meaning of the phrase "a little dose of judgement or reason" seems surprising, coming as it does from one whose meticulousness in recording the results of his highly discriminating observations of nature is so generally characteristic of the man. His attention to detail and his standards of precision in recording findings remain to this day a source of wonderment and even despair for those who seek to follow in his scientific footsteps. How odd, then, that his quest for precision and exactness should suddenly slacken before the very threshold of seemingly the most important point of all of his investigations. So uncharacteristic does this appear that one is led to speculate as to the reason why.

One reason might be that 'instinctively' Darwin felt that by allowing a clear distinction to be made between the abilities of the human and of the animal, evolutionary theory—in which all animals were assigned a common ancestor—would be seriously compromised. It would be difficult to explain how, by natural selection, such an extensive rift between the human and nonhuman animal species could have been successfully bridged. Though one might still have been able to explain the origin of all animal life forms other than the human, the theory would no longer hold its appeal as embracive of all life forms, and the logic of its supportive infrastructure would have been fatally flawed.

Indeed, this seems to have been the scenario that flashed before Darwin's mind. Darwin himself informs us in his autobiography that as early as 1837 or 1838, thirty-some years before the publication of his *The Descent of Man,* he became quite convinced that the human species was not to be excluded from the general law of natural selection. "As soon as I had become," he writes, "in the year 1837 or 1838, convinced that species were mutable productions, I could not avoid the belief that man must come under the same law" (*The Autobiography of Charles Darwin, 1809–1882,* ed. Nora Barlow, 1969, p. 130). Yet, until the year 1871, when *Descent of Man* appeared, he sedulously avoided stating this view in his published works. He was aware that it would be ill-received in religious circles, especially by the Church of England. Also, he at least suggests in his autobiography that he deemed it advisable to heed the admonition given him by his father: it is best for a married man to keep doubts about the existence of God and immortality to himself, lest he unnecessarily cause his wife considerable grief. There is a most revealing paragraph in Darwin's autobiography, representative of his thinking just three years before his death, for it is dated 1879.

> Nothing is more remarkable than the spread of skepticism or rationalism during the latter half of my life. Before I was engaged to be married, my father advised me to conceal carefully my doubts, for he said that he had known extreme misery thus caused with married persons. Things went on pretty well until the wife or husband became out of health, and then some

women suffered miserably by doubting about the salvation of their husbands, thus making them likewise to suffer. (*Autobiography*, ibid.)

Thus it is clear from Darwin's own express admission where natural selection has led him, and how radically he understands its role as the unique source of all living things. By focusing his attention on the similarities between the many life-forms, Darwin gradually came to all but lose sight of their very significant differences, and his explanation of the origin of species through natural selection became in his eyes more and more plausible. With that conviction waxing, there was, then, a corresponding waning of his belief in a Divine being behind the emergence and proliferation of millions of different life-forms.

There are two parallel strategies one can employ in order to minimize the difference between life-forms, and here we have in mind those living animals termed human on the one hand and all other animal life forms on the other. It is here that the real test of natural selection occurs, and if the distance between the human and the nonhuman primate can be shown to be a matter merely of degree, then there could hardly be any substantive objections raised with regard to the origin of other animal life-forms. The first of these strategies consists in granting that there is a difference between instinctual and reasoned behavior, and then insisting that the 'higher' animals, at least, possess a true reasoning power process, with the result that no difference in kind exists between the mental abilities of the human and the nonhuman primate. The second strategy consists in refusing to acknowledge that there is indeed a true distinction between reasoning and sensation, for both are restricted to the level of experience which is always of singular things. In this case, it is argued, the continuity required for the coherence of the evolutionary theory is secured, since the human is not granted a capacity for knowledge that is essentially superior to that of other primates.

Now it would seem that Darwin at different times employs both strategies, but it is likely an inconsistency which actually matters little in the long run, for the end result is the same. The human and the nonhuman animal are, at bottom, kindred spirits, so there is nothing illogical in insisting that the explanation of the origin of the one can also serve to explain the origin of the other.

RICHARD E. LEAKEY AND ANIMAL KNOWING

The well-known paleontologist Richard E. Leakey opts for the first strategy, for he feels that recent studies of the higher primates establish their ability to reason. It is, he argues, for purely biological reasons that the nonhuman primates have made no progress toward the attainment of speech. "Biologically," he writes, "it is the *communication of thought* rather than the *thought itself,* that separates humans from the rest of the animal kingdom" (Richard E. Leakey and Rodger Levin, *Origins,* 1977, p. 204a; italics added). From this it seems clear that Leakey is attributing mental, that is, intellective, processes to the nonhuman primate as well as to the human, concluding that, while abstract mental thinking is necessary for the development of language, it is not, apparently, a sufficient condition, for he grants that the nonhuman primate is incapable of giving expression to its 'inner world' through speech. Leakey writes:

> A vital leap in the evolution of intellectual capacity would have been the ability to form concepts, to conceive of individual objects as belonging to distinct classes, and thus do away with the otherwise almost intolerable burden of relating one experience to another. Concepts, moreover, can be manipulated and this is the root of abstract thought and of invention. The formation of concepts is also a necessary, but apparently not sufficient, condition for the emergence of language. (*Origins,* p. 188a)

Thus Leakey seems to attribute to the nonhuman primate the power of abstractive thought along with the concomitant power of "manipulation of concepts" while at the same time grudgingly acknowledging that, despite these special mental powers, the nonhuman primate is, for whatever 'evolutionary reasons', biologically incapable of speech. Leakey seems to maintain this without intellectual discomfort. Moreover, he grants quite readily that, although the nonhuman primates possess 'intellective' powers, they do not possess the enormous capacity for learning which so strikingly characterizes the human. Yet he shows no qualms in attributing intelligence to the animal even though he does not explain why it fails to match the linguistic ability of the human. To say that the animal does not speak because it lacks the biological equipment to do so merely prompts the further question—"Why so?"

In seeking to defend Darwin from his critics, Leakey comments rather incongruously:

> Ever since Darwin tied knots between human beings and the rest of the animal world, many people have frantically attempted to untie them again, declaring that even though our roots are in the animal world we have left them so far behind as to make any comparisons utterly meaningless. To some extent this is true, because the quality that makes us unique in the biological kingdom is the enormous capacity to learn. (P. 208b)

Leaving untouched the crucial issue as to why the human possesses "the enormous capacity to learn," as he also does with regard to language, Leakey then rather weakly adds: "Humans can learn virtually anything, as the rich variety of cultures throughout the world testifies" (p. 208b). Yet he does grant that "the quality that accompanies the emergence of learning in the evolution of higher animals, namely intelligence, is surprisingly difficult to define" (p. 185a). So difficult indeed that Leakey seems not to have tried. Still, he does allow that there is a twofold level of self-awareness, which he refers to as the 'extreme' form and the 'simple' form. The latter is ascribable to the nonhuman animal, but the former is exclusively the prerogative of the human. "In its extreme form self-awareness manifests itself in notions such as that of soul, but in simple form it merely means to be aware of oneself as an individual among others" (p. 189b).

Again this is an unsubstantiated assertion since Leakey makes no effort to uncover the reason or reasons or to acknowledge that such reasons are in fact "uncoverable." He seems, however, to offer justification for his not doing so by pointing out that neither in the human nor in the chimpanzee is one able to determine with any assurance what one or the other might be thinking (ibid.). But of course the human can easily indicate to others much of what is going on inside his or her head, not only through language in the more restricted sense, but also by giving expression to their thought in action, e.g., by making or doing something. This the chimpanzee never really succeeds in accomplishing, other than to impress us with the fact that little or nothing is actually going on in its head.

Earlier it was pointed out that Leakey does not provide an argument to ground the obvious phenomenon that the human has indeed an enormous capacity for learning, howsoever that is to be explained. Now Leakey does appeal to the constraints of social living to explain, in part, why intelligence has arisen to such a marked degree in the human. He sees a definite correlation between "living in a stable social milieu" (ibid.) and the enlarging of the capacity to learn, i.e., to grasp intelligently.

> So it appears that the evolutionary process promotes its own progress: learning about the environment (which demands a certain intelligence) means living in a stable social milieu (which demands at least an equal and possibly a greater intelligence); as social intelligence increases, so too will the ability to learn; this in turn encourages an even longer social apprenticeship; and longer group living leads to more social intelligence. (Ibid.)

Leakey concludes with the disarming concession: "This is not to suggest that social life was the prime mover in the evolution of human intelligence, but it would be difficult to argue that it did not play a very important role" (ibid.).

What the prime mover in the evolution of human intelligence might be, Leakey does not disclose. His reduction of the socialization principle to a secondary role would undoubtedly have incurred Darwin's displeasure, since for Darwin socialization was of overwhelming importance in human evolution. As Jonathan Howard puts it, "The most distinctive feature of Darwin's human biology, and the consideration which enabled him to *avoid the enfeebling invocation of a mysterious power to which Wallace and others felt obliged to resort,* was his recognition of the overwhelming importance of socialization in human evolution" (*Darwin,* p. 68; italics added). Indeed, in Howard's estimation it constitutes the supreme integrating factor in Darwin's *The Descent of Man,* serving as "the thread which unites the two seemingly disparate themes—the origin and nature of man, and sexual selection—since sexual organization of a species was a necessary component of its social organization" (ibid.).

Leakey concedes that chimpanzees have likely not yet attained the intellectual level of humans, though perhaps they have. "It used to be said," he states, "that many animals know, but only humans know they know. For chimpanzees, at least, this is probably an injustice" (*Origins*, p. 189a). It will be precisely to the pursuit of this question— whether it is an "injustice" to deny to the chimpanzee that level of self-knowledge which Leakey is inclined to attribute to it—that the remainder of this chapter will be dedicated. Only a close look at the nature of knowing itself will enable one to conclude with any measure of scientific rigor whether or not the chimpanzees, and perhaps other primates of the animal world as well, actually possess the intelligence so prominently evident in at least every healthy, mature member of the human species.

HUMAN INTELLIGENCE AND SENSATION

While the fact of human intelligence is not a matter of dispute (for its very questioning would be self-negating), its nature is. It is one of the sublime paradoxes of human existence that something as closely related to us and so intimately a part of us as intellective awareness should remain one of the most controversial of all human issues. The fact that we humans are capable of stupendous achievements—as well in the realm of enhancing and transforming our environment, as in the pursuit and development of the arts and sciences—no one would dream of denying; but when it comes to explaining how such achievements ever came to pass, and of laying bare the hidden assumptions behind those explanations, one commonly encounters either silence or widespread perduring disagreement.

Darwin never developed or attended to an epistemological theory to accompany his convictions on the origin of species and the descent of the human. In fact, he never entertained much respect for philosophy as a discipline. Some tendency to minimalize the differences between the human and other animals is not an uncommon phenomenon among paleontologists and naturalists. As scientists, as seen, they are committed to what one often terms 'empirical evidence' and, though they must continually generalize upon the 'facts' which they as scientists observe, they retain an inherent distrust of theorizing itself. Since to them it seems undeniable that theorizing goes beyond the facts as actually experienced. The scientist by profession feels fully secure only when his or her theory can be directly corroborated by sensible observation. While there is obviously something wholesome and commendable about this attitude, and we will return to this later, there is also the likelihood that it can nurture a certain disdain for pure intellectual theorizing. It is also often matched by an exuberant exaltation of the place and worth of sensory experience.

A negative outcome can be the sharp dichotomization of intellective and sensory awareness, so that either the two are opposed as resolute competitors or—what is worse—one is said to treat of the factual and the other of the mythical. Now human understanding plainly involves a sensory as well as an intellective component. It is precisely the union of the two in one complex act of human knowing that leads to and constitutes the mysterious and seemingly paradoxical nature of human knowing. What further contributes to the problem is the pervading influence in Western thought of the skepticism embraced by Descartes, Hume, and Kant in particular, regarding sensory experience, which attempts to explicate the phenomenon of human intelligence either by eliminating one or the other of these components or by bringing them together in a quite awkward and artificial manner.

That the speculative and the practical must in some way be intimately united is clearly recognized by Howard: "A fact is only of interest in so far as it is included within or excluded from an argument" (*Darwin*, p. 91). That is, the human mind is continually given over to the classification of experienced data that is of prime importance. The amassing of data is mere prelude. The mind never rests content with what is initially plainly experienced, but pushes on to explore what lies beneath and beyond the *prima facie* experience. By comparing and contrasting one experience with another the mind seeks to uncover between them similarities and dissimilarities. Such discovery permits their assimilation into a class, for a class, while composed of many distinct individuals, is made up of individuals sharing something in common. The class as such, therefore, exists in the mind, yet it emerges from the experience of many individual things that actually share something in common, whatever that might be. Individual things are first sensed, but that they are seen as belonging to a class of things is not owing to mere sensory experience alone, but rather to something as intellectively grasped.

Classifying is thus one and the same as the act of generalizing, and it is through the power of generalization that the human is enabled to distinguish himself or herself from the nonhuman animal. For this reason, simple observation or sensing cannot be considered an end in itself; rather, it always leads further to the all-important uncovering of generalizations, viz., theories, hypotheses, classifications, latent within the experiences themselves. It is this power of generalization which sets the human animal apart, and which makes possible the explosive proliferation of technological miracles creating and undergirding contemporary civilization. Only humans manufacture and employ machines, invent computers and build airplanes, because only humans theorize.

The ability to classify entails the phenomenon of human consciousness. Specifically, the subject must be able to stand back in some manner, i.e. transcend, that which is being experienced. Unless the individual experience is somehow transcended, the one experiencing is unable to move beyond its own individuality, and consequently is blocked from entering into a concious and enabling union with others. This self-transcending awareness is essential if the experience manifold is to be grasped as a single whole. There must first be a distance between the subject knowing and the object experienced, if the object is to be grasped by an individual subject under the modality of universality. Without such distancing, or 'distanciation', there can be no distinct awareness of the self knowing in one act both itself and something not itself, i.e., of the self united with and at the same time distinct from the object known. In a word, unless the knowing subject transcends the object known, no self-conscious awareness is possible.

Further, if the object is to be recognized as a member of a class or group, it must be seen as somehow incomplete within itself, as not fully that which by its very being it asserts itself to be, i.e., as not exclusively or fully sharing in that which it is. To put it another way, what it is that makes the object experienced to be an individual cannot be that which renders it at the same time the member of a class, for otherwise the singular object would by itself alone constitute the class, since it would then belong to a class precisely as *this individual*. In that case the meaning of the term 'class' simply loses claim to intelligibility.

At issue here is one of the more basic problematics in philosophy, for it strikes at the very heart of the epistemological problem as it has been framed since the time of Descartes. If an individual is regarded as so unique that what it is is fully identified with itself, then all talk of its belonging to a class becomes straightway artificial and arbitrary. Though it may be 'said' to belong to a class, it really does not. The 'classification' of the experienced object has its genesis, in this view, not from the object but rather from the subject experiencing it. In short, the object known does not, according to the nominalist perspective, really belong to a class; it is simply perceived as so belonging. Classification or generalization is imposed upon the thing experienced by the one doing the experiencing. This nominalist solution to the problem of classification of the individual and the universal, proposed by Immanuel Kant, is his response to the phenomenalism advocated by David Hume.

While Kant intended to address the position advanced by Hume, who considered all forms of generalization to be mere mental constructs resulting from the frequent association of similar individual experiences, the solution does not differ all that significantly from that of Hume. This is because the Kantian generalization is found to be imbedded within the subconscious of the experiencing subject prior to any sensory experience. This renders it impossible to affirm, as Kant plainly recognizes, that the individual object experienced contains within it a reality transcending its own individuality. Hence for Kant the individual object belongs to a class only as a creation of mind. Classification thus becomes an 'epistemological entity' having no *bona fide* metaphysical or objective basis. The object experienced may conveniently be viewed as belonging to a species or class but actually it does not; it is simply a singular, isolated entity. Or, stated somewhat paradoxically, it is a class unto itself. Unless there is, in short, a sense in which the individual lies beyond itself, there is no *bona fide* factual basis for viewing it as belonging to a class or species. Now it is just this epistemological view that commonly prevailed in Europe during the eighteenth and the early part of the nineteenth century. The view was destined to provide a modicum of philosophical respectability to the Darwinian theory regarding the 'descent of man'. The very notion of the distinction of species had by the nineteenth century become so blurred and so out of fashion within the thought salons of Europe that the genesis of man or the human from 'lower' species did not appear so very implausible or to offer insuperable philosophic difficulties. (For a more detailed account of the intellectual climate of nineteenth-century Europe and how it facilitated the spread of Darwinism cf. Michael Landmann's *Philosophical Anthropology,* trans. David J. Parent, 1974, pp. 165–70.)

It was this nominalist-inspired philosophical atmosphere that prepared the soil for the emergence of evolutionary theory. And, as we have seen, this is precisely the vantage point from which Darwin himself was proceeding, even as early as his undergraduate years at Cambridge, for he seems always to have eschewed any firm philosophical distinction between species.

But the nominalist view is decidedly impoverished philosophically, and in no way consistent either with the totality of human experience or with the universally recognized accomplishments of the human mind, which have permitted the human literally to 'remake' the face of the earth. In a recent work, Derek Bickerton has eloquently contrasted the two distinct worlds of the human and the nonhuman primates:

> Yet if apes look around them, what can they see that their own species has made? At most, the beds of broken boughs that they built last night, already abandoned, soon indistinguishable from the surrounding forest. The contrast is no less striking if we look at how much of the world each species controls. The chimpanzee has a few patches of jungle, while we have the whole globe, from poles to equator, and are already dreaming of new worlds. Most species are locked into their own niches, ringed by unbreachable barriers of climate, vegetation, terrain. We alone seem magically exempt from such bounds. (*Language and Species,* 1990, p. 1)

What, then, differentiates the human from the other primates? What is this striking contrast to which Bickerton refers? Quite simply, the human's ability to generalize and classify: to uncover within the singular things experienced a dimension that is not in every sense singular but transcends the individual *in its very act of being singular,* thus making classification and new orderings *objectively* possible. Such classification entails an act of discovery, but not one of pure invention; that is, the individual is subsumed under a general classification because it is, though an individual, already more than an individual. The individual thing is implicitly or potentially universal, and it is the human who has the power to penetrate within it, to elevate it and so enter into dialogue with it at the level of universality.

Through this unique access to the universal domain of the singular, the human is able to join together the seemingly disjointed, immutable individuals it experiences, classifying them into species and then reclassifying these in turn into super-species or genera and beyond. In this fashion, everything is experienced as somehow interrelated and sharing diversely in a common reality, though without being that reality. The name given this highest or most general classification, which is all-inclusive simply because it transcends all species or kinds of things, is *being*. 'Being' thus constitutes the ambient world horizon of the philosopher, whose vision extends to the outermost limits of experienced reality, integrating all individual things to the extent that they somehow share in the actuality of being, and are thus existent things, no matter *how* they exist, or *when*, or *where*, or *why*.

The supremely integrating power possessed by the human we call intellect. It has the entire world as its purview because it, and it alone, is endemically ordered to what *is*. Only this unique ordering to all things can account for the limitless variety of human interests, which in turn leads to the wondrous achievements of the human on every imaginable level of doing and making. It explains, further, why the human is never satisfied with his or her present lot, but is continually seeking to better it. This, again, is achievable only because the human is capable of uncovering the universal within the rich manifold of its singular experiences. The inner world of universalized experience is not just one series of temporally related events whose association is altogether ephemeral and without enduring meaning, as would be the case were it exclusively wedded to a world of singulars. Rather, such experience is constituted by a grasping of the unity among these divers experiences. This is precisely what is meant by the process of coming to the aware-

ness of the universal, non-individual dimension of each singular experience, and, consequently, of grasping these singular things not exclusively as separate and unrelated but as one. This is insight or true understanding, and is an activity that finds no exact counterpart in any of the
other animal species. In fact, the very term "intellect" derives from this
act of gathering and unifying, for its etymological roots are the two
Latin words *inter* and *legere,* 'to gather up' or 'bring together'. For reasons of euphony the 'r' of *inter* is transposed to 'l', thus forming the
Latin word *'intellegere'* from which of course the English word 'intellect' derives.

By penetrating to the level of the universal, the human is liberated
from the narrow constraints of that singular event, and is permitted to
move about freely, not in a world of his or her own making entirely,
but in a world uncovered among the shadows of a time-conditioned
world. It is in this sense that the human is able to transcend the limits
of its own self and to 'observe' timeless and spaceless relationships
among 'events of the day' which survive the ephemeral risings and set-
tings of a fickle sun. The intellective experience enjoyed by the human
is not dissimilar to the astronauts' experience of weightlessness in space
travel—an experience, by the way, that was accurately "foreseen" by
the scientists before it was actually 'experienced'. What made the experience of space travel possible was its anticipation, before the fact, by
the human mind, which already, while still firmly planted on planet
earth, coursed through the starry expanse at a speed far superior to
that of light. It is this built-in capacity for 'mind travel' which has given
humans the gift of being able wondrously to transform their world
through science and technology; to maintain in a permanent state of
presence events of the past; to visualize that which is not yet but which
can be; to express through the work of their hands the inner workings
of their own minds; to enter intimately into communion with the world
in which they live; to explore the very nature of that communion; to
muse about the meaning of their own selves; and to give expression to
and exchange with others through the medium of language the inner,
timeless world of their own consciousness. Clearly, the human is not
just another biological entity, but is simultaneously a citizen of two

worlds, the biological and the meta-biological. All efforts *
understand the phenomenon of the human in reductiv
founder on the shoals of the human's continuing achievements, w.
at every turn surpass those of all other known life-forms. It is a
supreme irony of the human condition that it is the human alone who
has ventured the claim that human intelligence is merely a 'biological'
function. This very affirmation, itself unique to the human, is time-
transcending and hence, quite obviously, simultaneously self-refuting.

ANIMAL INTELLIGENCE

Earlier it was stated that no other animal is capable of replicating
the human's intellective art. It is now opportune that we examine this
claim in some depth and attempt to clarify in a systematic way why this
is so, and precisely to what extent the human state of consciousness
differs from nonhuman consciousness. This is a deceptively difficult
problem for, although it seems plain enough to the unbiased observer
that the nonhuman animal lacks the intellective capacity described
above, there is a lingering reluctance by many to grant that humans
have a corner on intelligence. Others there are who would argue that,
as yet, not all the pertinent facts are in. There is still the possibility, they
feel, that the higher primates at least may well possess intelligence in
the human sense, even though we are presently unable to detect it. This
is, for example, the position we have seen taken by, among others,
Richard Leakey who suggests that chimpanzees possess intelligence
but, owing to physiological limitations, they are unable to give expression to it in language. Others would extend this line of reasoning to the
apes (Jane Goodall) and even to the baboons (Shirley Strum). Even
Derek Bickerton, who clearly acknowledges the superiority of human
intelligence to any form of consciousness thus far observed in the nonhuman animal, still has some doubts as to whether the higher primates
might not possess similar powers after all. He enumerates the wide-
ranging, extraordinary mental capacities of the human:

> Each of us has a lively and persistent sense that we are able not only to act
> in the world, but also to stand back, so to speak, and see ourselves acting;
> review our own actions and those of others, and deliberately weigh and

judge them; seek in ourselves for the motives that inspire those actions; catalogue our hopes, our fears, our dreams, and perform countless other overactions that we subsume under the head of 'mental activities' or 'consciousness'. (*Language and Species*, pp. 1–2)

Hard on the heels of this description of the proven qualities of highly developed human consciousness, Bickerton places everything in provisional brackets by adding: "We do not know whether any other species has *these particular capacities*" (ibid.; italics added). But one may ask what further evidence is required? What indeed would pass for further evidence? If it is not evident from their performance over these many years, even millennia, that the nonhuman animal does not possess the intellective ability of the human, then is it possible that it ever could be?

Observing that "no one has shown convincing evidence that any other species has a consciousness that resembles ours" (p. 2), Bickerton goes on to argue that "failure to find something is no proof of its non existence." He fails, however, to recognize that this principle needs to be contextualized, for its validity depends mightily on the nature of the problem to which it is applied. One cannot conclude that gold does not exist in a certain area because it has not yet been found there, but that is a far different matter than concluding that animals lack an intellective capacity because they have never displayed it. To argue so is to apply scientific methodology to a non-scientific problem. Yet no scientist would deny that gold is denser than silver, even though his refusal to do so is based on his failure to find an instance of silver being denser than gold.

Bickerton himself seems to recognize this, for he grants that there is a 'likelihood' of a causal linkage between man's dominance of nature and his highly developed intellect. Clearly, it is illogical to argue that the nonhuman animal is or might be intelligent (in the way humans are), even though it never is observed to make use of such an intelligence. If it were intelligent in the same way, then it would in unmistakable fashion display it, just as the human does, to survive. To maintain otherwise is implicitly to argue that the activities of nonhuman life-forms bear no real relation to those life-forms themselves, but emanate from them in complete randomness. The advocacy of such a premise would be utterly destructive of the life sciences themselves, since it would render all classifications meaningless and without foundation.

If activities of the nonhuman animal were variable and unstable, the animals themselves would be doomed to a hasty and 'premature' death. They would be unable to fend for themselves; they would have no way of discerning who their 'natural' enemies or prey were, and they would soon fall victim to the vagaries of a sometimes chaotic world. Not only are the activities of any organism related to the species of which an animal is a member, but they are directly and strictly proportioned to it. It is this close relation alone that permits the activities of individual species of animals to be studied and analyzed. In short, the science of ethnology is predicated on the uniformity of the activities of animals within a certain species, for it is primarily from this behavioral uniformity and permanency of 'life-style', and not simply from their outward appearance, that their species—that is, the kind of organism they are—is revealed.

Indeed, the sole justification for the study of animal behavior lies in the perceived reward of being able to uncover the specific natures of the animals through a carefully monitored and prolonged observation of their activities. In no other way could a determination then be made by the naturalist as to whether this particular animal was a member of this or that species. In consequence, animal husbandry would soon be gone with the wind. There is no possibility of ascertaining a priori what kind of an animal this happens to be without observing its activity, for activity follows upon nature, and not the inverse. Since nature is the permanent underlying capacity for action in the category of substance, it is revealed only in the activity flowing from it. There is, certainly, a direct correlation between behavior and bodily structure, but the purpose or raison d'être of the structure becomes known only when it is actually deployed in an activity of some kind. Of course much of the problem here stems from an implicit nominalist view of nature so commonly shared by naturalists and scientists which causes them to bridle at the least suggestion that natures are the controlling factor regarding the activity of organic beings.

Even if, in the nominalist view, the claim denying the reality of any such thing as 'nature' was asserted, the naturalist's meticulous classification of individual life-forms as belonging to this or that group, itself belies the ground of such a denial. Otherwise, the various classifications are merely arbitrary affirmations without meaning or purpose. Yet, obviously, taxonomists take these classifications seriously, as even a casual perusal of *Nature* magazine clearly attests.

Therefore, if it is true, as Richard Leakey and others theorize, that the higher primates are intelligent in the 'human' sense, but are prohibited from expressing that intelligence because of their inability to speak, then they have unduly restricted the meaning of speech, or language. As the well-known adage has it, actions speak louder than words. Behavior itself is a form of language, taken in the broader sense, manifesting as it does the inner structure of things. If the nonhuman primate actually possesses intelligence, there is no way this could be hidden. Though lacking the power of speech, they would still possess the power to act, and through their activities of making and doing their alleged intellective power would unerringly manifest itself. The behavior of animals constitutes a form of language, body language, which the naturalist can read and interpret. The animal, if gifted with intelligence, would no more be incapable of performing 'simple' intelligent actions, such as small children perform as a matter of course, than would a human, mute from birth, be prevented from displaying his or her intelligence through the performance of everyday tasks. We find no animal capable of such commonplace but, absolutely speaking, complex human activities as preparing a meal, lighting a fire, tidying a room, or reading street signs.

These activities, and thousands of others like them that humans perform routinely, are activities that have never, even under the most favorable of circumstances been observed in the nonhumman primate. Even with the most painstaking effort, no human has ever succeeded in teaching animals to perform them. The reason for this is ultimately simple: the nonhuman primate lacks the power of classification as earlier described. It cannot stand back from its actions; it is incapable of intelligent generalization. The only environment in which its activity unfolds is that of the rigidly and starkly singular; that is, the individual act of a nonhuman primate aims at obtaining an immediate, rightly limited, sensory goal.

Instinctual activity, which will be discussed more fully later, is capable of immediate, spontaneous association whereby diverse singular objects become linked together to form a descriptive manifold whose unity depends upon spacial juxtaposition and temporal sequence. Thus, when Jane Goodall, Richard Leakey, and others refer to instances in which chimpanzees have been seen to employ small twigs to "fish for termites" as clear signs of the chimps' ability to reason intellectively, they are altogether misguided. Such instances do not parallel either the human's ability to fashion tools for use in performing a great number of tasks or the human's foresight in storing them safely for future use. The chimpanzee indeed is able to grasp a connection of usefulness between this particular twig and these particular termites, but many animals make similar use of claws, pincers, or beaks to accomplish like goals. This is quite different from 'grasping' the connection between twigs and termites in general. Were the chimpanzee able to apprehend this universal connection intellectively, that is, as detached from the world of the singular, it would be logical to expect the chimpanzee to store twigs for future use and even to negotiate an exchange in such things as food, sexual favors, a place to sleep, etc. But the chimpanzee shows little interest in the past or the future. This can only be because it does not grasp them; its conscious state does not transcend the near, immediate present.

It is, however, important to recognize that we are not claiming that the activities of nonhuman primates do not 'aim' at a definite goal; through its acts the animal seeks its own protection or the defense of its mate or offspring, or the procurement of food. But though, as in the example above, the chimpanzee acts intelligently, it does not follow that these actions are actions of intelligent beings, for they do not emanate from a self-conscious center of reflection that grasps the singular as silhouetted against a general, universal horizon.

Yet the real crux of the matter is, again, just what one understands by intelligence. If, like Jane Goodall, one views human intelligence as but one of various levels of intelligence, providing but a limited view of our world through a conspicuously narrow window, then one has brought the intellective act down to the level of the merely sensory. By employing the narrow window model of intelligence, Goodall is able to argue with some superficial plausibility that the chimpanzees, and perhaps other primates, are fundamentally as intelligent as the human, even though they peer out upon the world through a quite different window frame. She writes:

> Most of us, when we ponder on the mystery of our existence, peer through but one of these windows onto the world. And even that one is misted over by the breath of our finite humanity. We clear a tiny peephole and stare though. No wonder we are confused by the tiny fraction of a whole that we see. It is, after all, like trying to comprehend the panorama of the desert or the sea through a rolled-up newspaper. (*Through a Window: My Thirty Years with the Chimpanzees of Gombe*, 1990, p. 10)

Since Goodall has minimized the wondrous quality of human intelligence, she has narrowed the knowledge gap between the human and the nonhuman primate. By doing so she has conditioned both herself and her reader to accept her position that the chimpanzee is as truly 'intelligent' as is the human, only in a different way. Because the human intellect is the power of a finite being, she insinuates, its vision is not only restricted to a limited portion of the world but blurred as well. Hence, since it looks through a different window altogether, the chimpanzee sees a quite different world than does the human, with the result that very likely we have as much to learn from this most intelligent of animal primates as it has to learn from us. "If only we could," she opines, "however briefly, see the world through the eyes of a chimpanzee, what a lot we should learn" (p. 11). Strangely, Goodall does not recognize that the human is fully aware that *what is 'seen'* at any given time is but *a small portion* of what is 'seeable'.

Yet Goodall is heartened by the fact that "once again, as in Darwin's time, it is fashionable to speak of and study the animal mind" (p. 18). She is convinced that the time is past when the nonhuman primate is any longer describable as a non-intelligent being, where intelligence is viewed as a uniquely human prerogative. She states very plainly:

Gradually it was realized that parsimonious explanations of apparently intelligent behaviors were often misleading. This led to a succession of experiments that, taken together, clearly prove that many intellectual abilities
that had been thought unique to humans were actually present, though in a
less highly developed form, in other, nonhuman beings. Particularly, of
course, in the nonhuman primates and especially in chimpanzees. (Ibid.)

It is evident that Goodall conceives of intelligence among animal
forms as being of several different levels so that one animal can be coherently viewed as having more or less intelligence than another. Yet
this is like saying that one woman can be more or less pregnant than
another. One is either intelligent or one is not, and, as discussed earlier,
intelligence opens out onto the general, the non-particular, and hence is
not restricted to a partial 'window view' of the world. Its vision, to
continue that metaphor, is as one from the top of a hill extending in all
directions, and not as the restricted view one has in standing on the
side of that same hill. An unrestricted 'view' is endemic to human intelligence, for that is how it differs in kind from any sensory act, which
invariantly operates in a limited field. For instance the sensory act of vision is restricted to those sensible 'things' reflecting color; the act of
hearing, to that which vibrates and hence gives off sound, and similarly
for the other senses, each of which possesses a highly restricted field of
activity. Sound cannot be seen nor can color be heard.

Yet the intellect, on the other hand, has an unrestricted field, for it
is capable of making contact with whatever *is* in some way, it making
no difference in *what way* its object happens to exist. Intellect can
know '*all* kinds' of things, and this, most literally. To express this notion in more precise philosophic terms, the field of intellective knowing
is 'being itself'. *To be understood* something does not have to exist in a
particular way, as it does for sensory knowledge; it merely has to exist
in some way. That is why all things can serve as the object of intellective knowing. As Aristotle stated the point, "The intellect is in a way all
things, for everything is a possible object of thought" (*On the Soul*, Bk.
III, ch. 4, 429b.).

Now, regarding the chimpanzees, whom Goodall considers to be
the most 'intelligent' of the nonhuman primates, it is clear that their

learning ability, while comparatively significant, is, in contrast with humans, very limited. They can improvise tools of sorts by preparing small twigs taken from branches to be used to flush out termites from logs, but their daily life patterns are hopelessly repetitive and monotonous. Their activities are invariant and directly bear upon their immediate environment. Most of the daytime activity of the adult chimpanzees is spent on the nurturing of the young, moving to new camp grounds, seeking food, grooming each other by removing insects, engaging in sexual activity, and preparing a safe haven for their nocturnal rest. There is no indication that they seek to better their lot—that they have the slightest inclination to work, to provide permanent housing for themselves, to raise their own food, to make use of fire to warm themselves, or to prepare meals. From their actions it is reasonable to conclude that the chimpanzees live from day to day simply because they lack the capability of projecting into the future.

The chimpanzee society remains glued to everydayness and to its immediate survival concerns. Their behavior as a group is altogether predictable, for it is captive to the physical elements that environ it; this is their world. These primates are notably incapable of seeing beyond present needs, and this is just as true of the highly select chimpanzees who have received the 'benefit' of living in a human's world, of being educated and of being pampered by foster human parents, as it is of those chimpanzees living in the wild. Thus, 'privileged' chimpanzees retain, at bottom, essentially the same behavioral patterns as their cousins in the distant jungles of Africa. Their behavior toward the human remains as self-centered as ever. They are always on the receiving end, never reciprocating through any kind of nonselfish actions.

Only the most naive form of anthropomorphism can induce one to fail to see the gulf that remains between the 'educated' or 'urbanized' chimpanzee and the average human, regardless of the latter's educational background. Whether or not the DNA molecule of the chimpanzee is 99 percent similar to that of the human, it is undeniable that the one percent difference makes all the difference in the world (cf. Goodall, *Through a Window,* p. 13). The only truly viable criteria available to us for ascertaining whether or not chimpanzees are intellective as are humans is whether they act as humans act. Whether or not they use twigs to fish for termites, or show emotions of affection, fear, anger or sadness in ways similar to humans, the only true test of their being intelligent is whether their behavior is nature directed and controlled, that is, whether it is on the whole invariant and unchanging. If, and only if, like humans, they share in a consciousness that is unrestricted in scope and transcends the exigencies and needs of the moment, can they be said to be intelligent.

It is all too clear that the chimpanzees and, a fortiori, all other non-human primates, fail to measure up in their behavioral patterns to the unrestrictedness so manifestly a part of every normal human. If their repertoire of activities is severely limited, then they cannot be judged to share in self-directed intelligence, and they are, consequently, basically incapable of forging new ways and new methods to deal with the succession of crises confronting every finite being in its struggle for continued existence. A being that shows no interest in improving its lot cannot by definition be possessed of intelligence, for it manifestly operates within a highly restricted horizon. "By their fruits you shall know them," is an oft-quoted proverb and it finds cogent application here (Matt. 7:16).

Actions do literally speak louder than words. It is through its activity that the inner core of a being is brought to light. This also applies to the case of the human. Intelligence only becomes manifest to us by its incorporation into the physical world through language, doing, and making. While not physical directly, and precisely for that reason, intellection is only obliquely recognizable. It cannot be sensed in itself, but it can be detected through the manner in which it organizes and orders the physical world, i.e. through its effects. Organization and order is the tell-tale trace intelligence leaves behind: the signature of its presence. It is precisely upon this foundational principle that the science of archeology, for example, is based. Human artifacts such as shards of pottery, arrowheads, ruins of temples, tombs, or human dwellings are identifiable only as infallible indications of human insight and intelligence. They unquestionably are not the fortuitous result of random interplay on the part of nature.

The chimpanzee, however much it may physically and genetically resemble the human, does not produce artifacts. The troop moves through the jungle consuming leaves and berries, attacking monkeys, bushpigs, baboons and eating them alive, and, when successful, fashions an impromptu nest of sorts in the trees where it may safely pass the night. But the next day the troop moves on to fresh feeding grounds (cf. Lieberman, *Origins of Language*, pp. 153–54). Viewing the cycle of activity of the chimpanzee from the human perspective, we cannot help but find it to be exceedingly limited. The actions of these primates are suited to their day to day needs, but do not go beyond them. This is why one can affirm with confidence that they do not share in the power of intelligence.

But, most important of all, the spectrum of activities of these primates, as already indicated, would not remain as limited as they in fact have, were they intelligent, but would gradually have expanded as insight was added to insight, and the acquisition of more and more leisure time would have provided them additional opportunities for applying their 'minds' to new and more complex problems.

When Lieberman concludes that "the tools that people use do not reflect their innate cognitive or linguistic ability in any direct manner" (*Uniquely Human: The Evolution of Speech, Thought, and Selfless Behavior,* 1991, p. 161), the terms 'innate' and 'direct' must be understood in a very restricted way if one is to concur. The authentic use of tools surely provides valid grounds for inferring that the tool maker possesses an inherent intellective ability. At the same time, the level to which this native intelligence has been developed is indicated by the quality of tools people use, even though the employment of sophisticated tools does not of itself imply, as Lieberman correctly argues, that only the people who employ them are intelligent. That is, all intelligent beings are equal insofar as they have inborn ability to obtain insights and an inherent potential for unlimited development. From this it also follows, as Lieberman states, that "we [humans] have no inherent cognitive advantages with respect to first-century Romans and Gauls even though the TGV high-speed train traveling between Paris and Nice at 200 kilometers per hour makes for a smoother ride than a chariot" (ibid.). However, though we moderns have no 'inherent cognitive advantages' over our forbearers, we do nonetheless have advantages. The sophistication and speed of modern railway travel in France provides a clear indication of the creative adaptability of human intelligence which, over a period of many, many years, progressively devised ways to improve on the Roman chariot.

Because, then, the menu of activities of the chimpanzee is so firmly restricted as to be in no way reasonably comparable to that of the human primate, and further, since the chimpanzee in its behavior has shown no interest in 'learning' how to use, let alone fashion, such simple tools as knives, hoes, brooms, etc., it is hardly coherent to assume that the chimpanzee possesses an innate but as yet undeveloped cognitive ability. The unifying process through which the differences within the experiential manifold are transcended is an activity which no sensory power, however refined, can accomplish. As seen, understanding does not consist in mere observation or listening (though, granted, it always entails some form of accompanying sensory activity). If the nonhuman primate were intelligent, it would give evidence of forging new ways and new methods of dealing with the everyday succession of crises confronting every finite organism in its struggle to protect, maintain, and better its life, exploiting its inmost possibilities for development and progressively improving its lot as circumstances allow.

Still, though the nonhuman primate lacks the human prerogative of intelligence, that does not mean, as already mentioned, that the nonhuman animal does not act 'intelligently'. An organism is acting intelligently if it acts according to design and if its activity is goal oriented, even though not necessarily consciously so. An organism does not behave as it does merely to amuse itself, but rather to provide for its many needs, by food gathering, nest building, and self-defense; thus it safeguards its continued existence and promotes its development, as well as provides for the continuance of its own species through propagation and the rearing of its young.

One cannot help but marvel at the purposefulness of all animal activity, nor fail to note the ingenious manner in which the animal utilizes its special physical powers and anatomical endowments in order to provide for its own well-being and survival. Aristotle echoed this when he remarked, "In all of nature there is cause for wonder" (*Parts of Animals* I, v. 645a). Equally deserving of our admiration and wonder is the austere economy of an organism's activity, for even though the repertoire of activities of each animal species varies greatly according to its singular needs, the number and variety of those activities is, as already stressed, palpably restricted. Further, what is especially noteworthy, these activities are invariant, not only with respect to the individual but as regards all adult members of the same sex and species as well. It is this very invariancy of behavior that provides the basis for viewing zoology as a science and not an art.

Once a particular species of animal has been painstakingly studied, and its behavioral patterns traced in detail, one knows in a very accurate way the behavior of other animals of that same species living in a comparable kind of habitat, under similar climatic and environmental conditions. One can without reservation join in Lieberman's tribute to Jane Goodall and others who, by their heroic commitment, have literally opened up for us the hidden world of the chimpanzee and other primates: "Thanks to the dedicated work of Jane Goodall (1986) and other patient observers who followed her example, we now have some idea of chimpanzee culture" (*Uniquely Human,* p. 150).

It is unfortunate that Goodall herself has not been more discriminating in her appraisal of what she has observed during her thirty some years with the chimpanzees. She expressly attributes intellective abilities as well as human emotions to the chimpanzees. In her recent work, *Through a Window,* she remarks: "People sometimes ask *why chimpanzees have evolved such complex intellectual powers* when their lives in the wild are so simple. The answer is, of course, that their lives in the wild are not so simple!" (p. 23; italics added). But her descriptions of these primates' lives in the wild would indicate that these animals do indeed live lives that are rigidly structured and highly repetitive, immensely so when compared to the behavioral pattern of the human. Goodall further claims that recent experiments "clearly prove that many intellectual abilities that had been thought unique to humans were actually present, though in a less highly developed form, in other, nonhuman beings. Particularly, of course, in the nonhuman primates and especially in chimpanzees" (p. 18). Comments such as these lead Lord Zuckerman, in a recent review of *Through a Window*, to comment that Goodall is "overwhelmingly anthropomorphic! She not only knows what chimpanzees have in their minds when they do the things they do, but also what their emotions are when they do them" (*New York Review*, May 30, 1991, p. 44). His point, I believe, is well taken.

Unquestionably, Goodall, relying confidently on her Darwinian assumptions, shows little inclination to differentiate between intellective and sensory activity, seemingly concluding that any purposeful activity performed by the nonhuman primates takes its rise from an intellective power of the primate. Further illustrative of her anthropomorphizing tendency is her following assessment of chimpanzee behavior:

> Indeed, the study of chimpanzees in the wild suggests that their intellectual abilities evolved, over the millennia, to help them cope with daily life. And now, the solid core of data concerning chimpanzee intellect collected so carefully in the lab setting provides a background against which to evaluate *the many examples of intelligent, rational behavior* that we see in the wild." (*Through a Window*, p. 23; italics added)

Toward the end of this same work, after emphasizing the physical similarities between the chimpanzee and the human, Goodall provides us with her most definitive assessment of the chimpanzee's cognitive abilities:

> There are, of course, equally striking similarities between humans and chimpanzees in the anatomy of the brain and nervous system, and—although many have been reluctant to admit to these—in social behavior, cognition and emotionality. Because chimpanzees show intellectual abilities once thought unique to our own species, the line between humans and the rest of the animal kingdom, once thought to be so clear, has become blurred. Chimpanzees bridge the gap between 'us' and 'them'. (P. 249)

This account of chimpanzee behavior illustrates the need for carefully distinguishing between intellective and sensory acts, lest the line of demarcation between the human and the nonhuman indeed not only become blurred but be dissolved. Unfortunately, neither Goodall nor others of similar persuasion have shown any lingering concern for this dimension of the problem. Lord Zuckerman, commenting on Goodall's musing that "If only we could however briefly see the world through the eyes of a chimpanzee, what a lot we should learn" (*New York Review,* p. 11), evinces little or no enthusiasm for what the results might be. "But even if it were possible," he writes, "to see the world in this way, Ms. Goodall does not tell us what we should expect to learn. My response to this particular suggestion would be that *we would learn nothing meaningful to us as human beings"* (p. 44; italics added).

Precisely. From her admittedly lengthy and painstaking observation of chimpanzees in the wild, Goodall is drawing inferences that go far beyond the domain of the naturalist. She is wading into the deeper waters of philosophy, and she needs to adjust to the change in her aquatic environment by directly acknowledging as much. The conclusions from Goodall's experiences with primates, being by nature philosophical, are by that same token subject to philosophical critique and evaluation. Yet she seems inclined to find dialogue at the philosophic level unnecessary, even unhelpful—a point to which Mary Midgley draws our attention even though her own immediate concern is with a quite different group of scientists. In referring to the contribution made by neurobiologists, regarding the nature of the human, she states: "It is philosophy that they are doing. All important and original biological thinking involves some philosophy" (*Beast and Man,* p. 174). Surely Midgley's sage comment can be applied as well to anthropologists and naturalists, especially when it comes to drawing comparisons between the human and the nonhuman animal. Midgley's further observation that "philosophy and biology are not in competition; they are different aspects of one inquiry" (ibid.) is also well worth heeding, for knowledge cannot work against itself. All inquiry, after all, must take its rise from the same commonly shared experiences. Were it otherwise, then either scientific knowledge itself would not be possible or there would need to be as many sciences as there are inquirers.

DARWIN AND INSTINCTUAL BEHAVIOR

Earlier, in discussing intelligence in the nonhuman animal, we had occasion to refer to instinctual behavior, indicating that we would later return to this subject for a more detailed comment. Instinct is, to borrow Darwin's terse definition, (*Descent of Man,* Pt. 1, ch. 3, p. 288b) an "inherited habit." Habit may be defined as an ability to perform a complex activity effortlessly, with reasonable skill and proficiency. A habit thus differs from a trait or mannerism. Habit understood as mere mannerism is a purely descriptive characteristic which does not in any way facilitate the performance of a task. Smoking, scratching one's head, speaking with nervous rapidity, are inconsequential activities as far as achievement is concerned. They are thus not habits, properly so-called. A true habit is purposeful and is either task- or action-oriented. Ordinarily, unless the term is further qualified, a habit is an acquired characteristic. Thus, any acquired ability one possesses which permits one to perform some specific task with comparative ease and proficiency, is a habit in the proper or philosophic sense. It can be seen, therefore, as a relatively permanent quality which empowers a subject to act in some way. For example, the ability to play the piano, the ability to write or to speak a foreign language, are acquired, not inborn, characteristics and are true habits. They permit one possessing them to act without hesitation and with little conscious effort. The accomplished pianist needs only a piano in order to perform; one who has learned to write needs nothing more than paper and a pen or pencil to be able quickly to transfer thought to paper. Once acquired, a habit is a relatively permanent acquisition which one can make use of whenever one wishes.

An instinct differs from a habit in that it is inherited rather than acquired, as Darwin has remarked. Darwin further distinguishes between instincts that are simple and those that are complex. The former inherited tendencies are ordered to the performing of comparatively simple actions. Among these are included the sucking instinct of infants, the sexual drive, the instinct to ward off or flee from danger, and to react to pain. The first of these the human shares with all nonhuman mammals, and the latter three are shared with all other species of animals. The complex instinct, as the term would indicate, refers to the inherited ability to perform tasks of much greater intricacy whose completion will often require the development of a skill over a rather extended period of time. Examples of such instinctual habits abound especially in the world of insects. Darwin singles out two such 'marvelous instincts', those of "sterile worker-ants and bees" (ibid., p. 288b). To these might be added the instinct of some birds to build nests; of some spiders to weave webs; and of beavers to build dams. Darwin expressly alludes to all of these skills. The following passage, in which he contrasts the ability of the human to fashion a canoe with the animal's ability to perform a complex but specific task, is of particular interest:

> . . . but there is this great difference between his [man's] actions and many of those performed by the lower animals, namely, that man cannot, on his first trial, make, for instance, a stone hatchet or a canoe through his power of imitation. He has to learn his work by practice; a beaver, on the other hand, can make its dam or canal, and a bird its nest, as well, or nearly as well, and a spider its wonderful web, quite as well, the first time it tries as when old and experienced. (P. 289a)

Yet, when it comes to explaining the relationship between instinctual knowledge and intelligence, Darwin remains tied to his theory of natural selection, and, as a consequence, lacks a viable criterion by which the two might be distinguished one from the other. His assumption, simply put, is that insects and other higher animals, such as the beaver, are *intelligent* because they possess complex instincts of a remarkable nature. In making this assumption he has simply equated acting in a purposeful, intelligent manner with 'being intelligent'. As the following passage discloses, this opens the flood gates to unrestricted ambiguity in the employment of the term 'intelligent':

> The fewness and the comparative simplicity of the instincts in the higher animals are remarkable in contrast with those of the lower animals. Cuvier maintained that instinct and intelligence stand in an inverse relation to each other; and some have thought that the intellectual faculties of the higher animals have been gradually developed from their instincts. But Pouchet, in an interesting essay, has shown that no such inverse relation really exists. *Those insects which possess the most wonderful instincts are certainly the most intelligent.* In the vertebrate series, the least intelligent members, namely fishes and amphibians, do not possess complex instincts; and *amongst mammals the animal most remarkable for its instincts, namely the beaver, is highly intelligent.* (P. 288a; italics added)

For animals to perform feats of remarkable complexity, as many of them do, is certainly *a sign* of intelligence, for such tasks could never be accomplished randomly. But on Darwin's own admission these abilities were not learned but inherited, so they were not acquired through the intelligent effort of the individual animal. Not only that, but the invariant character of these instincts within a given animal population or species is clear indication that the animals themselves have no effective control either over their possession or of their use. They do not act the way they do because they have chosen or choose to do so—otherwise they would exhibit marked variations in their task performances. Furthermore, they would not, as obviously they do, spontaneously follow their instincts on cue. For example, it would be impossible a priori to identify the building of dams or canals with a particular species of animal. Yet not only do beavers perform identically the same tasks generation after generation, but the manner in which those tasks are performed is likewise invariant. This can only be because the animal itself has no control over what we refer to as its instinctive behavior.

In addition, and this is of great importance, the anatomy of individual species of animals is closely proportioned to their species-related tasks. The beaver could not build its dams if it were unable to swim and if it did not have large, sharp teeth to bring down trees to be used in the dam's construction. To *be intelligent,* as is the human, is one thing; *to act intelligently* through instinct, as do the animals, is quite another. Humans, precisely as intelligent beings, are in command of their activity, and their repertoire for immediate action is limited only by the number and kinds of previously acquired habits and skills. Besides, the kinds of additional long-term human activities possible as a result of newly developed habits is without limit. The human can learn to play not only one specific musical instrument but any instrument; can learn not only one specific language but any of some 5000 different languages humans actually speak. Which explains why some humans speak French and others Russian, for example, and why some individuals play the violin, while others play the flute or bassoon, and still others can do all of these.

Darwin is at a loss to explain the origin of instincts, which he is constrained by dint of logical consistency to attribute to unknown causes acting on the 'cerebral organization'. He is equally perplexed by the fact that the higher animals possess so few inherited habits, and the human even fewer. He is, nonetheless, handsomely confident that natural selection holds the key to the origin of the vastly intriguing phenomena of instincts. He writes:

> But the greater number of the more complex instincts appear to have been gained in a wholly different manner, through the natural selection of variations of simple instinctive actions. *Such variations appear to arise from the same unknown causes acting on the cerebral organization,* which induce slight variations or individual differences in other parts of the body; and these variations, owing to our ignorance, are often said to arise spontaneously. We can, I think, come to no other conclusion with respect to the origin of the more complex instincts. (*Descent,* Pt. 1, ch. 3, p. 288a & b; italics added)

To appreciate why the behavior of the higher animals, viz. of the primates especially, is less dependent on complex instincts (inherited habits), it behooves us to consider for a moment the *raison d'être* of sensory activity. As a form of knowledge, sensation puts the animal in direct contact with its environment. Yet this cognitive union with nature and the surrounding world is not for the animal a matter of simple luxury, something nice to have but without which the animal could still manage. Rather the animal, in sensing, becomes attuned to its 'world' for the simple reason that it is dependent upon its environment for food source, shelter, and protection. If we restrict our consideration for the moment to the external sensory activities alone, viz. seeing, hearing, smelling etc., we may state in general terms that all animals experience the same 'world' in the sense that they see, hear, and scent the self-same objects of nature. At the same time, however, because of differences between species of animals, some of which are considerable, what is one animal's delight might well prove to be another's poison. The particular needs of the various species also vary greatly, and the manners in which they interpret the world of natural things unfolding before them through their sensory perception also differ significantly. What one animal can digest, another cannot; what is prey to one species of animal is a predator to another. In order, therefore, to explain why one animal is singularly attracted to an object sensed while another is thoroughly frightened, angered, or repelled by it, one needs to take into account not only the nature and qualities of what is sensed but, most especially, the manner in which such objects relate to the well-being of the sensing organism itself.

Clearly then, animals must be equipped with powers of discrimination by which what is sensed is known or perceived by them as being suitable or unsuitable for their own particular needs. Such a discriminating power cannot of course be identified with the special senses, say vision or hearing, for, as mentioned, in general all animals possessing these powers visually see the objects in a similar way (if we may, for our present purpose, prescind from the clarity of the perception, its intensity, etc.). The eagle, for example, possesses incomparably better eyesight than does a horse, which enables the former to spot its prey while circling unseen high overhead, and thus to be able to swoop upon it and take it by surprise. Such extraordinary vision is of primary importance to the eagle, for otherwise, because of its imposing size and lack of quick, adaptive maneuverability, it would rarely succeed in capturing its prey. The horse, on the other hand, has far less need of being able to identify objects at some distance visually. It is not handicapped, therefore, by possessing a vision that is adapted to seeing objects close at hand. Moreover, the horse compensates for its poor vision at a distance by possessing a highly sensitive sense of smell, which enables it to detect the existence of bodies of water many miles distant.

Yet, as the Swiss biologist Adolph Portmann informs us, some insects such as bees do not perceive the color red, though their color spectrum is quite comparable in extent to that of the human. Yet bees do have the ability to perceive ultraviolet light "as a color in its own right," a color of which the human has no experience (Portmann, "Colour Sense and the Meaning of Colour," in *Color Symbolism: Excerpts Eranos*, 1977, pp. 5–7).

Thus, as already indicated, all animals must indeed effect a kind of judgment or evaluation regarding the sensed objects surrounding them. They must ascertain the unique value a sensed object has for them and for their species. In many instances there is no margin of error allowed them. A false or incorrect evaluation of a sensation in a given instance can readily result in the animal's death or its sustaining crippling injury. This evaluation must, then, consist in a form of sensory judgment, and it does, therefore, imply a kind of reasoning, since it involves a calculus of sorts by which the animal interprets a particular object experienced as suitable or unsuitable to its well-being.

What is important to note, however, is that, although animals of different species will evaluate the very same things in quite contrary ways, evaluations by animals of the same species are uniformly similar. The only feasible explanation is that the evaluations made are inherited or instinctual. This Darwin expressly acknowledges, writing: "It is, however, certain, as we shall presently see, that apes have an instinctive dread of serpents, and probably of other dangerous animals" (*Descent*, Pt. 1, ch. 3, p. 288a). Now of course these instinctive reactions to experience observed in animals reflect intelligence inasmuch as they serve to protect the animal from elements in their sensed world that could prove hurtful to them, but the fact that all apes have a dread of serpents indicates that the evaluation arrived at by the individual ape is not the result of a reasoned reflection but rather of an instinctual process. It is not, in short, a learned behavior, but behavior imprinted a priori in its nature.

INSTINCTUAL CONTRASTED WITH FREE BEHAVIOR

In order to identify more clearly the difference between the instinctive judgments of the nonhumans and the free evaluative judgment of the human, we turn our attention momentarily to the human act of willing. We have already considered the manner in which human intelligence attains or knows its object. The human experiences the world both through its senses and through the intellect that transcends the sensory powers. Although the human intellect knows the objects that are sensed, it knows them from an atemporal and aspacial perspective. This, as seen, is what permits the classification of known objects under general headings. More succinctly, the intellect attains to its object universally and immaterially. Whatever the intellect knows, it knows as an expression of being, and what it can know is anything at all that exists.

Hence, since its own knowing shares in being, it is able to know its knowing as well. This unique, transcending quality of the human intellect undergirds human freedom and the ability to choose.

Because the intellect is open as a knowing power to whatever is, it can view the same object or experience as diversely related to the individual who understands. Therefore, anything at all can be seen as being in some way advantageous, that is, as having some value. The self-same object may be evaluated as being advantageous in one way and disadvantageous in another. Thus, one individual may find a chocolate fudge sundae very appealing because of its wonderful taste, but the same individual might simultaneously view this same sundae as unwholesome and disadvantageous because of its high calorie and fat content. The same sundae can be viewed not only from the standpoint of its pleasant taste but also from that of its dietary value.

It is this unique ability of the human as an intellective being to translate its experience into the universal mode that grounds and renders possible an act of free choice. Consequently, the human's evaluations of its varying experience are not determined or controlled by inherited habits or instincts. Rather, the knowing subject itself controls and determines its own behavior. And this is the singular result of the mind's knowing things from a universal or general standpoint. Were its activity under the control of pre-determined rules, a denial of the unique nature of the intellective act itself would be entailed, and this would concomitantly deny the existence of a value freely chosen.

For the mind's act to be predetermined by an established program, its viewpoint would have to be restricted. But whatever it is that is grasped by the intellect can be viewed as in some ways advantageous and in some ways disadvantageous. Thus the human is free to select the value or goal it itself prefers. And hence what the human does is fully under the control of the individual. This is precisely, then, what is meant by saying that the human is free: its behavior is self-determined, in that it itself controls which of the behavioral options it will select. This explains why human behavior is unpredictable on the individual level. One cannot foretell unerringly what a human will do, precisely because the human subject alone is the ultimate controlling cause of its own action. At most one can guess what a human might choose, but no one can truly 'know' a decision before it is made. One knows the decision another makes only at or after the moment it is made.

But, with the nonhuman animal, the case is quite different. Since it does not possess an intellective power, and thus does not possess fully reflexive self-knowledge, its viewpoint is restricted to what it senses. Consequently, the animal is incapable of determining through a reasoning process what would be best for it. To compensate for this lack the nonhuman animal is provided with inherited habits which 'blindly' guide its behavior in a manner most consonant with its well-being.

It is Thomas Aquinas's view that, since the nonhuman animal is seen to lack the power to reflect fully on its own act of knowing, it is incapable on its own of assuming the responsibility of fully directing its own activity (*ST* I, q. 83, a. 1; the translation of this and subsequent quotations from Aquinas are mine). Accordingly, the nonhuman animal lacks the capability correctly to interpret its sensory experience in a way that could adequately safeguard and guarantee its own well-being. That is to say, the animal would be incapable of making sense out of its experience, unless nature itself provided it with the proper interpretive guidelines by which it could unerringly estimate the value relation obtaining between its experience and itself. Such interpretive guidelines are what is referred to by the term instincts. They consist in a priori sense estimates which the animal possesses by reason of its nature, and which are not, therefore, self-acquired through individualized experience.

The working of the modern computer provides an illustrative paradigm. The instinctual knowledge the animal possesses can be likened to the software in a computer, for instinct is equivalent to a built-in program. The computer functions with unerring accuracy provided the one employing it faithfully follows the informational format that has been programmed into it. Just as the computer did not provide its own software, but received it, so the animal did not draw up its own behavioral code, and can only resolve the experiential problems presented to it within the strict parameters of the specific program it has received. It is incapable of interpretively modifying its instinctual guidelines, and even less so of exchanging one set of guidelines for another. Nor is it capable of developing new guidelines from the old. Like the computer, it can function well within the limitations of the program it has received, but lacks the ability to create new programs on its own: In short, it cannot program itself.

INSTINCTS AND REASON

Nonetheless, Aquinas's understanding of instinctive animal behavior does leave room for an important estimative act on the part of the animal; in fact, it is the instinctual programming of the animal that renders such an estimation possible. For the animal does have to make an assessment of what it experiences sensibly. This assessment, however, is confined to a radically subjective evaluation. The evaluation the animal makes focuses exclusively on clarifying the correlation between whatever it is experiencing and its own individual well being. The animal is interested not in the qualities of the object sensed in themselves, but in the object and its qualities precisely as they have some bearing on its own special world. Michael Landmann, in discussing the relation between human and animal consciousness, states that the latter "relates everything to itself and sees everything only in its own perspective, a perspective that determines value or nonvalue, takes itself as center. . . . The animal has primarily and only the subjective world given to it by nature. . . . Within a more limited circle the animal knows from the first what each thing means and requires of it" (*Philosophical Anthropology*, translated by David J. Parent, 1974, pp. 196–97).

The important function of the a priori program, then, is to provide the nonhuman animal, which is lacking the intellective fullness of reasoning, with the ability unerringly and with benefit to itself and its species to interpret whatever it experiences. This can be achieved only through an estimation which the animal itself makes, although the quality of the estimation conforms to the a priori guidelines for acting with which nature has endowed it. Because of the need the animal has to evaluate its experiences, Thomas Aquinas clearly acknowledges that the nonhuman animal does in *some sense* reason, and such evaluation is indeed comparable to a kind of judging (*De Veritate*, q. 24, a. 2).

Still, such 'reasoning' is comparatively elemental, falling well short of the reasoning displayed by the human. Though first and foremost a biologist, Portmann also clearly recognizes that there is a significant difference between the human and the nonhuman animal. He traces that difference to the ability of the human to penetrate the barrier of naive, sensory experience, obtaining thereby insights permitting the construction of instruments which in turn allow for an investigation of a reality yet unknown ("Colour Sense," p. 9). This ability to go beyond sensory knowing is reasoning in the proper sense of the term, and it involves a conscious and creative joining together of separate objects or elements of experience. This is rendered possible only by first grasping the reality of the thing experienced independently of its relation to the subject experiencing it, and, further, by the subject's simultaneous grasping of its own nature as conformable to the object known. In this way the formulated judgment is itself evaluated. Through a reflective act the knowing subject is capable of evaluating its own evaluation, and of thus consciously ascertaining the truth of its original judgment. Such a conscious bringing together of disparate elements of experience is what Aquinas terms a *collatio;* and it is this conscious 'collation' that constitutes an act of reason properly so called (*De Veritate,* q. 4, a. 2). This act, then, differs markedly from the sense judgment made by the nonhuman animal which is instinctual rather than reflectively conscious, resulting as is does from a wholly 'nature-determined' estimation. The instinctual 'judgment' thus spontaneously results from the stimulation of the precoded sensory receptors. This explains why it is that the behavior of animals of the same species is predictably uniform.

Thus the nonhuman animal's response to an experiential stimulus can be viewed as analogous to pre-programmed telephone answering systems. Or again, in illustrating the instinctual phenomenon, Aquinas has recourse to the familiar example of how sheep, upon seeing a wolf, instantly flee. There are here two 'judgments' made by the sheep. The first judgment recognizes that the wolf's presence is threatening; the second, that the sheep should flee. This example could easily be replaced by Darwin's example of the ape's instinctive dread of serpents, for both illustrate the same programmed reaction to danger. Be it noted, however, that the object of the fear or dread in each instance differs. How an animal reacts instinctively to its environment and what it experiences clearly depends on the nature of the animal in question. If nature has provided it with the physical defenses needed to neutralize the aggressive capabilities of another species of animal, it exhibits no fear of it. Thus a wild boar does not fear a rattlesnake, however poisonous; nor does a cougar fear a solitary wolf, since both the boar and cougar are instinctively aware that neither is endangered by an encounter with such animals, no matter how threatening they may be to animals of other species.

A further nuancing is needed, however, which, though not found expressly in Aquinas's explanation of animal behavior, is an implicit corollary: The animal's judgment with respect to the suitability or unsuitability of an object does not imply that the animal always arrives at this sensory judgment instantaneously. On occasion the animal is sometimes deceived by what it senses, judging a situation to be harmless which is actually dangerous, or, contrarily, dangerous which is totally unthreatening. An example of the former might be an African impala approaching a river bank to drink, not realizing that what appears to be a harmless object is actually a totally motionless, half-submerged crocodile. Or again, Darwin relates an example of an animal perceiving danger where there is none. One day he brought a stuffed, coiled-up snake into the monkey-house of the Zoological Gardens. As he relates the incident, as soon as the monkeys saw the imitation snake they immediately became violently alarmed and agitated (*Descent,* Pt. 1, ch. 3, p. 290b). There are also occasions when the animal will display great curiosity about what it might see or scent, not immediately realizing what it is. Yet, straight upon recognizing what it is, a reaction directly follows, the natural consequence of a judgment having been formed that such and such an object is somehow suitable or unsuitable to it.

Lastly, there are instances when animals can be observed to hesitate as to what course of action to follow, as they exhibit indecision and a certain anxiety and tension before acting. Such might be the case, for example, if a dog's master, going for a swim, wades into the water and invites the dog to accompany him. The animal may be observed to come to the waters edge, wade in a short way, and then quickly retreat to the safety of the shore, half barking and half whimpering. It obviously wants to heed its master's call, but fears doing so. As long as the fear or distaste for getting wet outweighs the desire to heed the master's voice, the animal does not enter the water. A situation such as this can indeed give one the impression that the animal is 'reasoning' about what it should do. However, this form of behavior can readily be explained as an instance of the kind of 'reasoning' to which we saw Aquinas refer earlier, involving a spontaneous hesitation between two contrary options, neither of which is momentarily grasped as compelling one way or the other.

In such cases there is, indeed, a certain comparison the animal makes between options, but the comparison is always between concrete, singular options. Thus it is reasoning in some sense, but it is not the same kind of reasoning we find in the human. In the former case the animal is irrevocably restricted to the concrete or singular situation with which it is confronted. Moving within the world of the singular, it forms a particular judgment merely on the basis of which of the options strikes it as the more favorable on the sensory level. Once the experience is interpreted, i.e. registers, as clearly favorable or unfavorable, the instinctive reaction kicks in predictably and automatically.

Doubtless to some this explanation will seem contrived and self-serving. Aquinas of course was well aware of this. According to appearances, the animal does seem to reason much as we do and to make free judgments, and its actions do seem to reveal a sublime kind of prudence and even sagaciousness. Yet, he still firmly insists that the ordered activity notable in mammals does not and cannot ultimately proceed from themselves, but is the enactment of a received behavioral program. This is so, he argues, for the reason already alluded to, viz., that animals belonging to the same species invariably act in similar and hence predictable ways. If the orderliness and intelligence their actions clearly reflect originated from the individual animals themselves, their activity would reflect widespread diversity, just as individual humans react in unpredictably different ways even in situations that are highly similar (cf. *ST* I–II, q. 13, aa. 2 and 3).

This conclusion, of course, goes beyond mere appearances. It is a philosophical conclusion, emanating from the appearances but not confined to them. As earlier suggested, much of the difficulty one encounters with regard to the question of the intelligence of the nonhuman animal stems from a pervasive unwillingness on the part of ethologists and naturalists to recognize the validity of philosophical reasoning. Wishing to base their conclusions on "the sensed facts themselves," scientists often display a not uncommon antipathy to generalizing, and indeed a rather profound mistrust of what they negatively characterize as 'metaphysical thinking'.

Though Darwin grants as incontestably true that the human "is capable of incomparably greater and more rapid improvement than is any other animal" (*Descent*, Pt. 1, ch. 3, p. 294a), he still firmly rejects the view that "no animal during the course of ages has progressed in intellect or other mental faculties," affirming rather that to take this position is "to beg the question of the evolution of species" (p. 295b). The reason Darwin alleges in support of his own claim is purely anatomical. "We have seen", he says, "that according to Lartet, existing mammals belonging to several orders have larger brains than their ancient tertiary prototypes" (ibid.). In so arguing, Darwin is clearly affirming that there *is* a direct ratio between brain size and intellective acumen—a generalization that dangerously approaches the borders of a 'metaphysical' assertion. Darwin has, apparently, no qualms about admitting the superiority of human intelligence on the one hand, and affirming on the other that the nonhuman animal is *in principle* equally as intelligent, even though the latter displays nothing comparable to the human by way of reasoning or progressive development. The major difference in the unequal progress each has made is, according to Darwin, "mainly due to his power of speaking and handing down his acquired knowledge" (p. 295a). We have seen that Leakey, Bickerton, and Lieberman have made this argument, which seems peculiarly circular unless one assumes that language and intelligence command no strict correlation. At any rate, such a claim must remain unsubstantiated until such time as the hidden presuppositions of the nature of language are flushed out. Since the question of language will be taken up in a later chapter, we postpone further consideration of Darwin's claim that the human's singular intellective powers of reasoning are wholly owing to the 'fact' that somehow along the way the human learned how to talk.

That Darwin, in discussing the human and animal intelligence, is weaving together surreptitiously two quite different meanings of 'intelligence' can be fairly concluded from the following quote in which he approvingly cites a Mr. Leslie Stephen, whom he finds to be an "acute reasoner." The following quotation is appended in a footnote.

> I am glad to find that so acute a reasoner as Mr. Leslie Stephen ["Darwinism and Divinity," *Essays on Free Thinking* (1873), p. 80], in speaking of the supposed impassable barrier between the minds of man and the lower animals, says, "The distinctions, indeed, which have been drawn, seem to us to rest upon no better foundation *than a great many other metaphysical distinctions;* that is, the assumption that because you can give two things different names, they must therefore have different natures. It is difficult to understand how anybody who has ever kept a dog, or seen an elephant, can have any doubt as to an animal's power of performing the essential processes of reasoning." (*Descent,* Pt. 1, ch. 3, p. 294b; italics added)

As we have seen, however, it is not merely a matter of giving the same activity two different names; rather, what is involved is the legitimacy of dissolving the very real differences between two pronouncedly different kinds of knowing acts. It is hoped, therefore, that the foregoing serves as an effective rejoinder to Darwin's brag which he unsheathes at the beginning of his third chapter of Part 1 of *The Descent of Man:* "My object in this chapter is to show that there is no fundamental difference between Man and the higher mammals in their mental faculties."

3 Freedom, Human and Nonhuman

We now turn our attention to the issue of freedom as it pertains to the nonhuman animal. Though this problem is not isolated from the major questions hitherto addressed, it is indeed correctly seen as an altogether integral component of the very complex phenomenon of animal behavior. One is quite justified, it would seem, in concluding that it is precisely the freedom aspect of animal behavior that has allegedly provided the most convincing support for the view that the nonhuman animal does not differ in kind from the human. It appears, that is, that animals, particularly the 'higher' species such as mammals, exhibit an independence of activity generally associated with freedom, since at least on occasion, their behavior is apparently as unpredictable as that of the human. It is, then, to these questions and others such as relate to the quality of freedom in the nonhuman animal that we now attend.

Just as some argue in favor of the commensurability of the human and the nonhuman animal by affirming a distinct similarity between their respective acts of knowing, so there are those who in similar fashion argue for a parallel conclusion that the human and the nonhuman animal are both commensurably free in their behavior. Obviously, one may develop this line of reasoning either by concluding affirmatively that the animal is as truly free as the human, or by concluding negatively that the human is no more free than the animal. Both lines of argument lead to the conclusion that humans do not differ significantly from nonhuman animals in respect of freedom.

Let us begin by considering what freedom means for the human. What is there about human activity that first leads us to describe or refer to that activity as free, and what are the actual presuppositions of such free behavior? In its root sense the word 'free' means to be unhindered or unobstructed. To say that human behavior is free is simply to say that it is unhindered in some significant way. This last qualification is necessary because, when one claims to be free, one never truly intends that such freedom be understood as absolute. There are some things that, because of natural or physical restrictions, one simply cannot do. Nor do we, as we employ the term freedom in everyday language, mean that everything a human does is truly a free act. Some acts are seen as free and others are not. For example, we would never claim that we freely coughed or sneezed unless we deliberately so intended. That is why, if either should occur at an inopportune time, as when someone is speaking to us, we would normally excuse ourselves, the clear implication of our apology being that the abrupt interruption our cough caused was wholly unintentional and beyond our control. Briefly, we did not interrupt the conversation deliberately or freely.

A free act, then, is seen to be an action that truly flows from the individual person in a unique way; that is, not only in the sense that the individual caused the act, but in the further sense that the individual had it within his or her power to perform or not to perform the act. Thus, even though in ordinary language one would say that "a man was killed by a falling tree," the implicit canons of language would not allow one to say that "a tree murdered a man." The tree did not commit murder because it was not responsible for its toppling over. This could be attributed only to some other agent, as, e.g., a woodcutter who felled the tree, or a strong wind which blew it down. In the case of the free act, the individual is a true cause of the action, with the result that the action is rightfully ascribable to him or her. If one is not free, then no action can correctly be attributed to them in the same sense as when we say: "She told me she did it herself," or "He indicated they had decided not to buy the house." The free act is an unhindered act because it is within the power of the individual who performs it. As defined, then, an act freely performed is one that could just as well not have been performed.

Yet to say that an act is free because it is unhindered, is not the same as saying that it is not in some way dependent on some factor distinct from the self or individual performing it. Whenever one chooses, one chooses something. That is, one always has something in mind in acting freely. The free act always involves a reaching beyond oneself. There is always a purpose or reason for acting or choosing. Some, among whom could perhaps first of all be numbered the behaviorists, consider that it is precisely this factor of the act of choosing which rules out the possibility of there being such a thing as a 'free' act. They argue that if the decision is made for a reason, that this same reason is a cause of the choice itself, and, consequently that, since the act is caused, it cannot be said to be free. The dependency of the acting subject on a value which is independent of itself is incompatible, in this view, with the claim that the choice is free. This, at bottom, is the underlying reason supportive of the behaviorist's or determinist's contention there can be no such reality as the alleged act of free choice. In this view all human actions that are consciously performed are subject to elements which are not under the control of the choosing subject, and consequently are not fully free, though they may be 'perceived' to be such by the individual performing them.

This serious objection of the determinist directs our attention to the very aspect of the free act that differentiates it from the act of the sensory appetites. The point at issue here is precisely the sense in which the intellective appetite, which is attributed only to the human, differs from the sensory appetite, which the human and the nonhuman animal share in common. An understanding of the difference between the intellective and the sensory appetite is, then, crucially important as one confronts the philosophical question of freedom as it relates to the nonhuman animal.

Before clarifying the difference between the intellective and sensory appetites, a word should first be said about the nature of appetite in general. Etymologically, the word 'appetite' derives from the Latin *appetere (adpetere)*, meaning to seek after, or to incline toward. In its most fundamental sense the word refers to the natural tendency everything has to be itself, to retain its own identity, and, if referred to a living thing, to develop and unfold its singular capabilities that it might become *actually* what it already is *potentially* by reason of its nature. Physical growth, for example, is the direct outcome of something 'desiring' to mature or develop, and thus become actually what it is already but in an imperfect, inchoate way. Thus the puppy is already a member of the canine species, although not a fully grown one, and as a puppy it 'wants' or has a natural inclination to 'grow up' and become a 'big' dog.

When, however, we are speaking of appetite as a tendency or inclination toward an object, as we clearly are in applying it to the question of human and animal freedom, we do not employ it in precisely the sense as that just referred to. Rather than being a natural or inborn tendency, an appetitive act is an acquired act; it is dependent as well as consequent upon an act of knowing. Indeed, the appetitive act is correlative to a knowing act, for what is desired is a good presented to the appetite by a knowing power, which latter is, as seen earlier, itself dependent for its act of knowing on its having assimilated an added actuality acquired through experience. Appetition, therefore, as an act of an appetitive power is an act that inclines or tends initially toward an object that lies outside the subject doing the experiencing. In short, the appetitive power is attracted to a good newly acquired by a knowing power, which good has an existence independent of that of the knowing, desiring subject.

Now appetitive acts can be seen to be of two kinds, since there are two distinct levels of knowing, and the appetitive act is, as seen, itself a response to an act of knowing. If the attraction is toward a good presented to an appetite by a sensory knowing power, the appetitive act is sensory, and hence of a sensory appetitive power. If, on the other hand, the attraction or desire is toward an intellective or nonsensory good or value, the act is an act of an intellective appetite commonly referred to as 'will'.

Finally, it should be noted that the act of desire belonging to either the sensory or intellective appetite is not 'natural' in the same way as is the puppy's 'desire' to develop into a mature member of its species. The intellective and sensory appetitive acts do not emanate solely from the natures of the subjects doing the desiring, but also from the natures of the particular desired objects. The acts of appetition deriving from the appetitive powers differ one from another, then, in that *what* is desired arises solely and uniquely from the particular objects desired. That is to say, the content of the desire, what it is that is desired, is wholly and exclusively measured by the object desired. This comes about because acts of the appetitive power follow upon a cognitive act, whereby a new formality is acquired by the experiencing subject, and this newly acquired form or value itself exerts a new, additional attraction upon the appetitive power beyond that of the natural inclination the subject has toward its fulfillment as a member of its particular species. In summary, a natural appetite differs from the desire exercised by an appetitive power in that the former is present from the start by reason of the nature which an organism possesses, while the latter is desire arising out of the acquisition of a new reality, which reality determines the content or value of the act of desire. What has just been said, however, is not to be understood as implying that, in the case of the will, it must necessarily choose whatever good is presented it, but rather that it can only choose *among the goods* that are presented it. The will is free to choose *what* it chooses, but it is not free *not to choose*.

An everyday example may illustrate the above. When one scents the distinctive aroma of freshly baked bread or pastry, one immediately experiences being attracted to it, even perhaps to the point of sampling it. The attraction is clearly toward an object distinct from oneself, in this case bread recently taken from the oven. The attraction experienced only occurred as a result of one's having scented the wonderful aroma. Upon reflection, one will find this to be the case in every instance of desiring something. I find, that is, that I desire only what I have somehow first come to experience through some kind of knowing. Desire is, as an act of experiencing, always an attraction toward something that is already united with the subject in a nonphysical way through knowledge. But it is most important to emphasize again that all desire involves an urge to go out of oneself in some way—to be drawn toward another.

This brings us, then, to the central question relating to the nature of human willing. Do I necessarily tend to or desire the things I do desire, or am I free to pick and choose what I desire? In other words, is it within my power to control those things I will, or is my willing firmly controlled by situations and circumstances which in no way are of my own making? If the former, I am of course free; if the latter, I am not free. The response to this question rests entirely on how one construes human intellection, and hence defines the nature of the knowing power. This is why the analysis of the human act of understanding, which we undertook earlier, and the manner in which it differs from the sensory act of cognition, is so important for decoding the nature of human volition. For, as should now be clear, since the appetitive act follows upon a knowing act tending toward what is cognized, if the intellective and sensory acts of knowing are similar in kind, there could be no basis for distinguishing intellective and sensory appetition in any significant way.

But since, as seen, the human intellective act extends to whatever is, i.e., to anything whatsoever, its horizon of 'vision' is, accordingly, panoramic, or unlimited. Consequently, the object of desire to which the intellective appetite or will tends is the wholly good, i.e., the good as unrestricted. This is because the intellect as an unrestricted power presents to the will a limited good against a backdrop of unrestricted good. Thus the will by its very nature has the capability of being attracted to anything whatsoever, provided it be intellectively known, however imperfectly. The particular kind of being something is, is, then, truly irrelevant as to whether will can desire it or not. Provided it *is* in some way, it can be viewed as possessing some perfection or good, and hence it can be understood as desirable or good and positively willed. The will is therefore necessarily free to choose any particular good presented to it by the intellect, since every particular good, no matter how perfect it might be and hence how desirable, will always be lacking in some goodness that *other things* possess. In other words, everything the will confronts is understood by the intellect to be both at once desirable and undesirable, depending on how viewed.

The will, then, *can* choose any of the individual goods or values presented to it, precisely because it can choose any good whatsoever; but it *need not* choose this or that particular good because that good can simultaneously be viewed as lacking some other good. It can thus be viewed as suitable and not suitable, though in different respects, at one and the same time. The end result, then, is that the will is radically unhindered in any individual act of choosing a particular good. There are no built-in constraints that point it inflexibly in the direction of any particular value or good. Rather, the will possesses the option of moving in any direction as regards the whole field of limited goods offered it, because the only necessitating object which can command its consent is one which is unlimitedly good, and such a being it of course does not encounter in its present state. This is because the unique source of its objects is a cognoscitive power (the intellect) which derives its knowledge exclusively from sensory experience. As long as such a restriction obtains, the human intellect can have no *direct* knowledge of unlimited being and hence of unlimited good.

THE FREE ACT AS SELF-DETERMINING

What yet remains to be considered is the seemingly contradictory contention that, although the human will is free or unhindered in its pursuit of particular goods or values, in choosing a good the will is nonetheless caused, because it receives specification or determination from the very value it has chosen. Many have rejected the freedom of the will precisely because they understood—correctly—that according to this view the will would be caused (determined) by the very value it chooses. The apparent enigma of a 'caused' act being also 'free' is resolvable by having recourse to the Aristotelian distinction between formal and efficient causality. The value of the object chosen in the act of choice does indeed cause the will, inasmuch as it provides the will the content or specification of its act. 'What' I choose, as distinct from 'why' I choose it, actually specifies, and hence formally determines, the will in the latter's very act of choosing. The will cannot simply just 'choose'; it must choose *something,* that is, some good or value. In this sense, in choosing, the will is truly dependent on that which it chooses.

Yet at the same time the will, in being formally determined by what it chooses, is free in the very important sense that the will itself has singled out and accepted this particular value among all those available to it. The will's root tendency is to incline to the unrestricted good, since it is an intellective appetite receiving its object from intellect; hence, if it does actually incline toward this or that restricted or particular good, it does so ultimately for *one reason* only, viz., not because it has to but because it chooses to; that is, the will itself controls and chooses the very value which specifies and, in the sense of a formal cause, determines it. In brief, the will functions as the efficient or agent cause of its own act of choosing, while what it chooses is the formal cause of that selfsame act. This or that particular value determines the will in the act of choosing *only because* the will wants and hence allows it to. For this reason, the act of choosing can most aptly be defined or characterized as an act of *self-determination*. In this manner the paradox—how there can be an act that is both free and caused—is resolved.

The act of free choice remains enigmatic only to those who find Aristotle's distinction between efficient and formal cause a mere hairsplitting logical exercise, and who thus are left without the requisite cognitive tools to provide a reasonable explanation to one of the most central and important questions touching upon the nature of the human person.

This question concerns not only the philosopher but all thinking persons. An eminent ethologist, W. H. Thorpe, who taught for many years at Cambridge and has been a visiting professor at three prestigious universities in the United States, recognizes the centrality of this question for all students of the human. After remarking that "a large number of otherwise mechanistically minded scientists" when questioned closely, grant that the phenomenon of self-reflective mind in man requires laws beyond those governing physics, chemistry and biology, he concludes:

> After all, it [a theory of freedom] seems to be the simplest theory capable of validating, of making sense of, man's personal activities, including of course those mental activities which have produced the whole scientific world picture and on which it solely rests. It also raises another very important issue, or rather the same issue in another form; that is the question of the actuality of 'free will'. ("Vitalism and Organicism," in *The Uniqueness of Man*, ed. John D. Roslansky, 1969, p. 74)

Emphasizing the centrality of the question of free will and its particular relevance for the young, Thorpe comments further:

> This question of 'free will' is one which often worries people immensely; and I find that students particularly tend to be troubled by the question "Is

my will really free?" "Can I choose?" "How is it possible that I can choose in the kind of mechanistic materialist world in which we live?" Now, I think there are some very important things to say about this and I am going to try and say them—albeit very quickly and superficially. (Ibid.)

Following the lead of Prof. D. M. MacKay, Thorpe agrees that physical analysis alone will not bring us to the right understanding of mind. Rather, what will be needed, he suggests, is a new understanding of the intertie between the brain and the activity of the body. In his own words:

> It seems that what we call 'mind' has a working contact with matter more intimate then one form of energy upon another. They seem to be truly complimentary in some very mysterious way and this mysterious unity is what we know as 'personal agency'. I believe it is absolutely essential to stress this idea of a personal agent if we are to begin to get to grips with the problem of the mind and its place in nature. (P. 79; italics added)

The rather detailed explanation of 'personal agency' or of self-determination, which we have taken some pains to outline above, seems fully consistent both with our internal awareness that we are indeed in control of the decisions we make, and with the scientific demand, at the same time, that it is nonsense to speak of a happening that is uncaused. The act of choice is indeed caused both by the object chosen and by the agent choosing, but the causality exercised by the object is formal only. It is hierarchically subordinate to the causality of the willing agent in that it is the agent itself that fully controls the content or value that, in the act of choosing, determines and specifies the act. To express this somewhat elliptically, it is the agent choosing who causes the manner in which he or she is caused. In this sense the act of free choice is, as we shall now strive to establish, unique in the world of nature. In concluding this section we wish to acknowledge that the substantive portion of the position we have attempted to outline regarding human agency is a development of the analysis of free human behavior made by Thomas Aquinas. The reader wishing to consult directly the latter's treatment of this matter will find the following citations helpful: *Summa Theologiae* I, question 82; *Summa Theologiae* I–II, question 10; *De Veritate*, questions 22 and 24.

HUMAN AND NONHUMAN APPETITION

We are now prepared to undertake a comparison of the appetitive activity of the human with that of the nonhuman animal. The question is: Do the higher animals, at least, exhibit in their behavior an activity in any way comparable to the free agency of the human? The task we undertake is formidable, and we do not approach it without a certain sense of apprehension.

The investigation of the phenomenon of human freedom has enabled us to understand in some measure, as well as to appreciate more fully, the close intertie between knowing and freedom. The advantage knowing provides to those individuals possessing it is simply that by broadening the scope of their activity, a whole new range of options for acting is made accessible to them. That is why it becomes so much easier to predict the 'activity' of either non-sensing or non-living creatures of nature, once we acquire an elementary acquaintance of their behavioral patterns. Thus, we know that iron is an excellent conductor of heat and, when heated, first turns red, then white. We exploit this property when we use it as a filament in our electric stoves. Whenever we turn on the oven we are confident that it will heat up. We also know that metals contract when cooled and expand when heated. Consequently, when they are used in construction, these properties must be taken into account, if structures such as bridges are to remain safe for use during the various seasonal temperature changes. In like manner, on a windy day in autumn we know that the fallen leaves will occasionally move about tossed by the wind, collecting against fences and shrubs, which impede further movement. Natural objects are completely controlled by their environment, so that, to the extent that we are aware of the physical conditions of their environment, we are able accurately to predict the way in which they will react. In this way we have learned how to control and use to our advantage the presence of the natural objects around us. By modifying their environment, we superimpose a new order upon them, by exploiting and redirecting the natural tendencies they are known to possess. Clearly, this is how humans have come to change and transform the world in which they live, and it is this very order in nature which provides the basic assumptions themselves underlying all levels of scientific inquiry.

Without a presumed uniformity within nature, there could be no such thing as science. The inanimate objects that constitute the world of nature simply are what they are, and, though they undergo many modifications at the hands of their environment, these modifications themselves are rigidly controlled by the laws governing the interaction of bodies, thus rendering such interaction accurately predictable. If the environmental changes are known beforehand, those modifications which will follow can be 'known' or anticipated well in advance of the event itself. Indeed, all tools and machines are merely planned extensions of this basic phenomenon. The car keys I carry in my pocket are incontrovertible evidence that I fully accept the existential reality of my car as the concrete embodiment of a manifold of ideas possessing the wondrous ability to transport me quickly and comfortably from one locale to another.

But when we turn our attention to those beings that possess knowledge in some form, the situation is, to be sure, considerably altered. As we have already considered, knowing beings are very differently related to their surroundings than are non-knowing or inorganic ones. Through knowledge the perceiving organism imbibes something of its surroundings, but in a way quite different from the manner in which non-cognoscitive beings do. The modification brought about by knowledge in the knowing subject is of an altogether different order than that whereby, for instance, water becomes heated. Knowing entails a true union between the knower and the object known. The knower undergoes modification in knowing, though the object that it knows remains unchanged. What is seen or heard, for example, is in no way altered by its being sensed, but there is a remarkable change in the one seeing or hearing. (An obvious exception to this is to be found in the sense of taste: something tasted cannot be sensed unless it is at least partially dissolved in the mouth, for only in this way can flavors be released and the sense of taste activated.)

If asked "What is the purpose of knowing?" "What particular value does it bring to the knowing subject?" one might first point out that there is a decidedly liberating dimension involved in knowing. Being united through knowledge with portions of the environment, the knower is enabled to interact with what it knows, with the result that the knower can act in new and different ways. The environment truly becomes for it a world of unlimited potential, a rich storehouse from which can be drawn out what is useful for the knower's growth, development, and overall enhancement. In knowing, the knower has truly become what he or she knows. Through knowing, the knower acquires qualities and perfections possessed by the objects known and is enabled to act in new ways that transcend his or her own original limited capabilities. Knowledge, then, provides the knower with an additional dynamism for action that non-knowing beings not only lack initially, but simply have no way of acquiring. It is in this sense that all forms of knowledge serve to liberate the perceiver from its original limitations by opening up to it new vistas, and permitting it to search out and pursue those particular items of value it uniquely requires to maintain itself in being, as well as to grow and develop in accord with the demands of its own nature. In a word, the cognizing or knowing being is in a position to take the initiative in supporting and providing for itself. In this way it is less subject than, say, shrubs or trees, to the unpredictable moods of nature to supply for its various needs such as favorable climatic conditions, a sufficient water supply, an adequate food source, etc. The tree is unable to move to more fertile land or seek out a more abundant supply of water, if either or both of these is of poor quality or lacking; the horse, on the other hand, can search out other areas better suited for grazing and can locate a water source even many miles distant. In this sense the animal possesses an independence utterly inaccessible to the plant and inorganic world of nature.

In any discussion of how the nonhuman animal contrasts with the human, there are decidedly different views regarding the question of freedom. Some there are who, not acknowledging any distinct line of difference between the behavior of the nonhuman animal and the human, incline to the view that the former is effectively free in the same fundamental sense as is the human. This is clearly the position taken by Darwin, and it is shared by many contemporary naturalists. For them, one convincing argument supporting the equality of the human and the nonhuman animal (the higher primates particularly) is the often-observed tentativeness of the animal's behavior, particularly the element of unpredictability. This they take to be a clear sign that the animal's behavior is not wholly controlled by instinct, but that these animals, like the human, are empowered to give direction to their own actions. On the other hand, there are others who see animal behavior as strictly controlled by instinctual patterns which are inherited, thus denying to them any real form of freedom, and regarding all of their actions as unqualifiedly determined. There is a third, not widely recognized position which occupies a place somewhere between the two positions just indicated. This view, taken by Aristotle and later refined and developed by Thomas Aquinas, best satisfies the findings of scholars who have studied the higher primates closely in their native habitats. As we shall see, it seems as well to meet the objections raised by supporters of the other two positions briefly outlined above.

Now whether or not the nonhuman animal acts out of freedom is truly a matter that can be determined only by a careful observation of its behavior. We humans have no other way of entering their inner world. Yet we cannot grasp the nature or depth of animal consciousness through sensory observation alone, for such knowledge must transcend the limited capacity of the sensory powers. It can only be acquired through the intellective medium, to which the nonhuman animal is not privy.

As humans we do not consciously recall what it was like to experience on the sensory level alone without our sensory acts being accompanied by the transforming dimension of the intellective awareness. And even if we could, there would be no manner in which that experience could find intelligible expression, for to do so goes beyond the merely sensory. This point is decisively made by George Klubertanz in his informative work *The Discursive Power.* In discussing the estimative or highest cognitive sensory power of the nonhuman animal, he comments:

> But we are in the difficult position that we are men [humans], not animals. Our estimative [power], in adult life, works precisely as discursive. Properly

to understand the operation of the estimative, we would have to remember why and how, as little children, our sense appetites were aroused. If we were able to remember this, it would be most probably impossible to give any intelligible account of it. It seems therefore inevitable that we must rest content with a merely analogical knowledge of how the estimative works. (1952, p. 237)

Realizing, then, the difficulty of coming face to face with the problem we are investigating, namely that of animal freedom, we must content ourselves with an indirect analysis of the animal's behavior. We simply have no direct conscious experience as humans of operating strictly at the sensory level alone.

Let us begin with a commonplace observation: animals are attracted to objects they sense. Perhaps less obvious is that, what on one particular occasion may attract an animal, at another may not. As a consequence, their behavior may seem to be unpredictable, and some have concluded that the animal's behavior must be interpreted as simply being free—that the animal, is, in a sense, not unlike the human, fully in control of its actions. Otherwise, its behavior would be altogether invariant and hence predictable. This line of argument is reacting, and, I believe, rightly so, to a rigid and seemingly Cartesian understanding of instinctual behavior, which would not allow any appreciable level of behavioral variability among animals of the same species or even by the same individual animal at different times and places.

However, a fully mechanistic interpretation of 'instinct' in the animal is neither necessary nor called for. Instinct, which guides the animal in its behavior, is an inherited characteristic, and instincts found in one member of a species are found unerringly to exist in other bonafide members of the same species. The instinct is hence a priori and thus frees the animal from the tedium and danger of lengthy trial and error experience before the expertise needed for meaningful activity is acquired.

This claim that instinctual knowledge is innate, or prior to experience, ought not to be seen as denying the facts Portmann refers to in his discussion of the nature of instincts. Young chicks of various species become bonded with animals even of a different species, including humans, some days shortly after their birth. This 'imprinting', which is observed to occur in various species of animals both invertebrate and vertebrate, can itself be seen as but a further determination of a yet more basic and more general orientation which the self-preservation instinct itself aims at. This bonding or imprintation is reasonably interpreted as a further specification of an original instinctual urge which has become activated through contact the new born animal has made with other creatures shortly after its birth.

In this way the young chicks of ducks and of song-birds, for example, normally become permanently bonded with the mother bird, but because of various circumstances of a wholly contingent nature, this attachment can be formed with other creatures with whom it makes contact during the early days of its existence. As Portmann specifically indicates, however, there is a statute of limitations of sorts operative here, for after a fairly short period, usually not exceeding twelve days, the chicks permanently lose the ability to become bonded in this manner to other creatures. One might with reason say, therefore, that the imprinting to which Portmann refers is truly an extension or application of the self-preservation instinct on the part of the animal experiencing it. Indeed, the occasional abnormal imprinting clearly supports the contention that this form of "deviant" behavior is possible only because of an antecedent for bonding (*Animals as Social Beings*, 1964, pp. 105–9).

In its own way, the instinct is a directive within the animal that is universal, inasmuch as it is uniform and shared by many individuals within a species. But precisely because it is universal, it must find its application in the world of singular things and individual activity. All sensory activity is of itself singular, occurring in a particular time and place. There must, therefore, be an element of contingency accompanying each application of the instinctual judgment to the singular event. The matter here is very similar to the application of general knowledge on the part of the intellect to a singular action. In the human, this involves what Aristotle terms the 'practical', as opposed to the speculative, intellect. The practical intellect is not understood to be a special power distinct from the speculative (which is always concerned with

universal knowledge), but is merely that same power as ordered to doing or making rather than to knowing for its own sake.

The process of connecting intellective knowledge, which is general
and universal, to action, which is always singular, is what Aristotle understands by the term practical reasoning. The practical syllogism consists of a general, universal judgment, constituting the first or major
premise of the reasoning process, and a singular judgment which is
subsumed under the major premise. From these two judgments, a third
judgment results, which is a synthesis of them both. An example will
serve to illustrate the point. As the first or general premise let us take
the statement: "All persons should be treated justly," and as the second
premise the statement: "Socrates is a human person." The conclusion
deriving from these two judgments is: "Therefore, Socrates should be
treated justly." Here one can see how the conclusion necessarily follows from the prior two premises. If the first two statements are found
to be true, the third statement, "Socrates should be treated justly,"
must follow from them and consequently must also be true.

Though the above example is illustrative of one type of practical
judgment, since it involves an activity and refers that activity to an individual person, Socrates, it is not practical in the fullest sense. The conclusion is not restricted to a singular event, but still remains on the level
of universality. The conclusion, "Socrates should be treated justly," although it refers to an individual person, and in this sense is a singular
proposition or statement, is not restricted in its application to a particular time or place. What the conclusion affirms is that Socrates should
always be treated justly regardless of the circumstances. It is in this
sense, then, that the statement is indeed universal, since it claims to be
true for a time which is coterminous with the lifespan of Socrates, and
perhaps even beyond, in that it might be interpreted to mean as well
that, even after his death, his memory should still be treated with respect. However that might be, this kind of practical syllogism can be
termed speculative-practical, since the conclusion, while referring to a
mode of activity aimed at a single person, still does so in general or universal terms.

A syllogism or reasoning process fully practical would involve the application of a general judgment exclusively to a singular event. One will readily note the difference between the syllogism just considered and the following: "It is wrong to condemn an innocent person to death." "In condemning Socrates to death you are condemning an innocent person." Therefore: "In condemning Socrates to death you (members of the jury) are acting wrongly." Here the major premise is, as was true of the prior example, a general, universal statement. However, the minor or second premise is not. Rather it is a statement referring to a uniquely singular happening, and hence is a practical judgment in the fullest sense. For want of a ready-made term, we may refer to this type of judgment as a practico-practical judgment, i.e., one that is practical to the highest degree possible. The synthesis effected here is that between a universal principle and an altogether singular event, whose truth rests completely on direct experience. The entire worth of this latter reasoning process rests upon the premise that Socrates is indeed innocent, and this can be established only through experience. His innocence can be known only by examining his actions. Knowing that Socrates is a person would not alone suffice to support the truth of the statement "Socrates' condemnation is unjust."

We have dwealt at some length on the nature of practical judgment since it plays a vital role in determining a measured and nuanced response to the question of the nature and scope of knowledge and freedom in the nonhuman animal. Since animals do and must make practical judgments or evaluations regarding what they individually sense, an understanding of the nature of those judgments is crucial toward that end.

The animal cannot be said to enjoy a universal awareness in the most proper sense, because it is incapable of representing to itself whatever it might be sensing as other than singular. Further, since sensory activity always requires the presence of its object, animal consciousness will always be a consciousness of the present. We will see shortly how this is true even with regard to those higher animals which possess a developed memorative power. Unable to interpret the object sensed in any other way than as suitable or unsuitable, the animal is spontaneously, i.e. involuntarily, attracted toward it or repelled by it accordingly as the object is judged as favorable or unfavorable (cf. Thomas Aquinas, *ST* I–II, q. 13, a. 2, ad 2). Consequently, the animal does not enjoy freedom of action in the same manner the human does.

Now in some species of animals the judgment itself appears to be instinctually made when certain of its natural predators are involved. As seen, Darwin refers to the natural dread apes have of serpents, and this seems to be true of many other species as well. Similarly, sheep have a natural fear of wolves, fleeing from them as soon as their presence is detected. Such instinctual judgments are commonplace in the animal world, and it can readily be understood why this is so, for, if the animal had no inherited awareness of who its predators were, it would have little chance of surviving in its natural habitat. This would prove detrimental not only to the individual, as is obvious, but to the species as well. It does not follow, of course, that an animal always immediately apprehends its predators. The latter may employ stealth or disguise to approach its prey undetected, with the result that the prey can, to its detriment, often be tricked into making an altogether incorrect assessment of its situation. This explains why decoys are successfully used by the duck hunter, and lures by the fisherman.

THE ANIMAL AND FREEDOM

It might appear that the nonhuman animal acts freely, since the manner in which it responds to a similar stimulus or experience can vary. If this be so, it is argued, then it clearly is making a free judgment about its experience. In other words, the higher animals display on occasion the ability to 'reason' in evaluating their situation, and hence they must indeed be capable of grasping universals. If such be the case, there then seems no grounds for denying them the capability of acting freely.

This objection did not escape the attention of Thomas Aquinas. That there is a certain likeness to free will in the behavior of some animals—that they are able to act one way at one time and another on a different occasion, even when the circumstances are similar—is granted by Aquinas, simply because in each separate instance they have made a *different assessment* of the situation. An experience previously estimated as suitable, may now be judged as unsuitable, and so the response will differ. To this extent, then, Aquinas acknowledges that there can be found in animal behavior a certain resemblance to authentic free acts. Yet he denies that this shows them to be free in the same sense that the human is, because the nonhuman animal always spontaneously reacts according to the particular singular judgment it makes. It cannot assess a situation as dangerous and fail to respond to it save in a determinate and prescribed way (*De Veritate*, q. 24, a. 2, resp.). But it can, of course, assess a similar situation differently on different occasions. The complex nature of the animal's experience permits this, since at one time the animal's attention may be focused on one aspect of the total experience and at another time on another.

This accounts for a certain amount of flexibility in the behavior of the animal, so that it is never possible to predict with certainty how some higher animal is going to react in a particular instance. If a human or another animal approaches it, for example, it can, depending on various contingencies, evaluate the situation as threatening or non-threatening. Its reaction will depend exclusively upon its assessment of the particular situation at hand, and its response will follow spontaneously upon this assessment it. Yet, since how this individual animal will assess its situation on this particular occasion is uncertain, there is no way one can unerringly predict how it will react. Aquinas refers to this level of "freedom" the animal exhibits as a "certain conditional or qualified freedom" (*De Veritate*, q. 24, a. 2, resp., & ad 3), but he denies outright that this manner of behavior is comparable to an authentically free judgment where the options for judgment are non-determining and where, consequently, no judgment is received by the will as an unqualified command.

It is because they fail to see a difference between conditioned and unconditioned freedom that authors such as Jane Goodall and others are led to speak of the freedom of the nonhuman animal as differing only in degree from the freedom experienced by the human. But the limited number of options available to the animal is an altogether telltale sign that they are lacking the human's grasp of universals, with the consequence that the animal's action following upon such a judgment cannot be an act of genuine self-determination. Self-determination is made possible only by a concomitant implicit awareness of good as unrestricted. This awareness serves as a backdrop against which the particular good is silhouetted. It is thus incompatible with a judgment that fails to transcend the singular everydayness of sensory perception.

In other words, the direction ultimately given to the activity of the animal derives from its own specific nature through the mediacy of inherited characteristics. Consequently, the animal does what it does without truly understanding what it does, and without knowing ultimately why it does it. It has subliminal, not overtly conscious, knowledge of the particular goal of its action, which is to satisfy an attraction toward something or positively to counter an immediate threat arising from a sensory experience. This in turn is conjoined with an instinctual assessment mandating the action which follows. In this sense the nonhuman animal's awareness of the goal and purpose of its activity is severely limited, if, that is, it be compared with that found in the human, who comprehends goals from a universal perspective (cf. Aquinas, ST I–II, q. 11, a. 2).

It is, then, important to recognize the difference between denying that the nonhuman animal has a fully conscious awareness of the goals of its activity, and saying that its activity is, therefore, directionless. This latter is far from the case, for nothing is more evident to the unbiased observer than that the activities of the animal follow recognizable patterns in accord with the species to which they belong. Furthermore, there is an ingenious correspondence between what animals do and the unfailing promotion of their own and as well as their species' welfare. This view is unconditionally opposed to the convictions of both Hume and Kant, for whom all recognition of an intelligible relation between events and achieved goals is attributable solely to anthropomorphic interpretation on the part of the human. That is, superimposed upon the blind events of nature is a goal-directed intent, which has no basis in experience itself, but rather is purely the work of the human mind. That such a view is an incoherent interpretation of the events themselves becomes evident from the uniform manner in which such 'interpretations' are ceaselessly made. No merely subjective hermeneutic can account for them. If the 'order in the universe' is but the product of the 'free-play' of human imagination, all scientific study of nature, both inorganic and organic, becomes inherently counterpositional, for the assumptions upon which any such inquiries would rest would entail a flagrant denial of the underlying assumption that there can be no such assumptions. Cassirer seems to have tripped over just such an inconsistency when he called for the introduction of a new standard of truth which is *logical,* not empirical, and which thus entails a new form of "intellectual interpretation" (cf. *Essay on Man,* 1944, p. 209).

The activity of nonhuman animals, is, assuredly, directed toward the achievement of clearly recognizable goals, even though these goals themselves are hidden from those seeking to achieve them. Were it otherwise—were the nonhuman animals in full control of their activities and, consequently, of the goals themselves toward which these activities are directed—there would not be latent within these activities the tell-tale thread of uniformity reflected not only in the lifetime patterns of behavior of individuals but within the behavioral patterns of their progenitors, as far back as human investigation can take us. Rigid, species-related, uniform behavioral patterns, spanning the generations and uniting them in a compact behavioral unit, are totally unexplainable within the thematic option that such behavior results not from internally inherited (and hence controlling) habits, but from a free interplay of the experiencing subject and the particularized experience which one encounters within the ever-expanding ambit of human experience.

FREEDOM AND TECHNOLOGICAL ADVANCEMENT

Perhaps nothing is so evident to the thinking human of the late twentieth century as the exponential rate of advancement in technological artifice within the astoundingly brief span of the past one hundred years. The contemporary human has at his or her disposal today household appliances such as the microwave oven, the self-defrosting refrigerator, automatic garbage disposal, as well as other household conveniences such as the television set, the radio, video-discs, desk-top computers, etc. which, at the turn of the century, were not even on the wish list of the average citizen, unless his or her imagination had been stirred by the creative fantasy of science fiction authors such as H. G. Wells. Yet, what but a few short years ago seemed wholly imaginary has now become, in most Western countries at least, fairly common household furnishings.

It is a part of the human condition and the intellective ability of the human that we are capable of ingeniously adjusting to new situations. Humans have the power of insightful discovery and are thus not chained to instinctual inflexibility, as is the nonhuman animal. The human alone is capable of such progressive behavior. The field studies of Jane Goodall, Dian Fossey, and other naturalists less celebrated have provided a close look at the behavior of the nonhuman primate in the wild, and there is always in their observations the implicit recognition that the activities of these animals have been constant and unchanging, and can be trusted so to remain. Contrarily, however, no such generalization of the human could be taken seriously by naturalists or any member of the scientific community. The behavioral patterns of the human are much too variable to allow one to treat them as static. Many of the activities performed by the human today would have been all but unthinkable as little as a hundred years ago. Insight alone, coupled with its correlative, freedom, can explain so great a disparity in the behavioral histories of the human and the nonhuman, for they play the key role in accounting for the radically creative shift in modes of behavior. Insight affords a breakthrough to new horizons; it makes possible the 'uniform discontinuity' between the varied activities of the human as they change with the succession of one historical epoch upon another.

It is knowledge that is the key to the human's behavioral advancement, and not change in the genetic code of the organism. Because the knowledge of the human transcends the barriers of sensory experience, it is not constricted to satisfying the immediate needs of a particular organism. As Aristotle remarked concerning the human intellect, "It has no nature of its own, other than that of having a certain capacity" (*On the Soul*, Bk. III, ch. 4, p. 429a).

At the same time, even though the activity of the nonhuman animal is not shaped by its grasping of ends to be achieved and the subsequent tailoring of activity to the requirements of those ends, there is, nonetheless, within the behavioral repertoire of some of the nonhuman animals a certain undeniable similarity with the human's mode of acting. Animals do *appear* in certain instances to employ a reasoning process not unlike that of the human who, through practical reason, brings together in a single practical judgment the universal and the particular, applying the former to the concrete, singular experience.

This recognition that the higher primates perform acts that *seem* to display a prudential assessment of singular situations—that is, they *seem* to be applying a general, nonsensible principle—has led Darwin, Leakey, Goodall, and others to conclude that the animals acting in this manner do indeed manifest a power of reasoning in no significant way unlike that possessed by the human animal. Goodall, for instance, lays great stress on the ability of chimpanzees to "fashion tools," observing them break off branches from trees, stripping them clean of smaller branches and leaves and then using them "to fish" for termites and ants (*Through a Window*, p. 59) Goodall sees this as strong reinforcement of the data assembled in the lab pointing to "intelligent, rational behavior" in the chimpanzee (p. 23). For Goodall there seems to be no doubt that various tests run on chimpanzees in labs have confirmed "again and again" that "their minds are uncannily like our own" (p. 21). One of these tests seeks to measure what psychologists call "cross-modal transfer of information." Goodall states that "it had long been held that only humans were capable" (ibid.) of this form of information transfer. What she is referring to here is the ability of the sensing organism to correlate the activities of different sensory powers. Thus, "if you shut your eyes and someone allows you to feel a strange shaped potato, you will subsequently be able to pick it out from other differently shaped potatoes simply by looking at them" (ibid.).

Now using precisely this line of argument, Aristotle concluded that there must be an *interior sense* that coordinates and differentiates the activities of the special senses (*On the Soul*, Bk. III, ch. 2, 425b). For Aristotle this coordinating sense was not a power possessed only by the human; it was a necessary capability of all sentient organisms possessing more than one sense. There is simply no other way of explaining this obvious given of experience, that sensing organisms do indeed bring together as well as differentiate between their various sensing activities. Thus for instance, the sensing organism, the human included, can put together its seeing and its hearing, identifying the colored object seen as the very one emitting this or that sound. It is the only manner in which the organism could make meaningful use of what it senses. In what is, then, an instance of a strange reversal of logic, Goodall, after stating, "It turned out that chimpanzees can 'know' with their eyes what they 'feel' with their fingers in just the same way" (i.e., as do humans), confesses with complete candor: "In fact, we now know that some other nonhuman primates can do the same thing. I expect all kinds of creatures have the same ability" (*Through a Window*, p. 21). In what sense, then, does the 'fact' that chimpanzees possess this ability indicate that they are in any *singular* way intelligent like the humans? The fact that all other multi-sentient animals possess this same coordinating power ought then to serve to elevate them as well to the level of intellectual beings. This line of argumentation thus logically leads to the conclusion that all multi-sentient animals are intelligent—a conclusion one suspects few humans are prepared to accept.

Richard Leakey has made a similar rediscovery of Aristotelian reasoning, without, apparently, realizing its origin. Unlike Aristotle, however, Leakey appears to identify the highest intellective act of the human with the act of coordinating the various sensory activities. For the Greek philosopher, of course, such coordination is but the beginning of a much more complex activity, which terminates in the act of understanding and which in turn makes possible what Leakey himself recognizes as the distinguishing factor of human intelligence, namely its flexibility. "The secret," Leakey states, "of the human mind is that rather than having the ability to learn variants of *specific* tasks or behavior patterns, it simply has the *ability* to learn, to be adaptive to practically anything that the environment has to offer" (*Origins*, pp. 191b–192a). This is, doubtless, a remark admirable for its insight and clarity. Yet it loses much of its luster when read in light of the comment that follows, wherein Leakey seems to infer that this coordinating ability is the seat of the intellective act of the human, evolution's crowning "biological success."

> But the signals from the ears, nose and eyes do not remain separate: they
> are integrated to form a more complete picture, and this integration is per-
> formed by the outermost crust of the brain, the cerebral cortex. It is this
> part of the brain that shows the most dramatic structural advances through
> evolution, and in the human brain it becomes the crown of biological suc-
> cess. (*Origins,* p. 192a)

GOODALL AND THE APES OF GOMBE

The observations Goodall records of the chimpanzees of Gombe in the earlier chapters of her recent work tend to emphasize the 'positive' features of the chimpanzee's talents and character, thus placing them in a more favorable light and making it tempting to view them as display-ing human traits. The later chapters present, by contrast, not only a less flattering picture of chimpanzee society but a shocking one as well. Goodall openly admits that the violence she experienced among the chimpanzees after having tracked them for many years and observed them as a nomadic community shattered her previously formed con-ceptions of them as a genuinely peaceful society. It is best to let her re-late her reactions and to express her own inner feelings of revulsion as the violent side of the nonhuman chimpanzees' nature came to light:

> The intercommunity violence and the cannibalism that took place at
> Gombe, however, were newly recorded and those events changed for ever
> my view of chimpanzee nature. For so many years I had believed that chim-
> panzees, while showing uncanny similarities to humans in many ways were,
> by and large, rather 'nicer' than us. Suddenly I found that under certain cir-
> cumstances they could be just as brutal, that they also had a dark side to
> their nature. And it hurt. Of course I had known that chimpanzees fight
> and wound one another from time to time. I had watched with horror
> when adult males, all inhibitions lost during the frenzy of a charging dis-
> play, attacked females, youngsters—even tiny infants who got in their way.
> But those outbursts, shocking though they were to watch, had almost never
> resulted in serious injuries. The intercommunity attacks and the cannibal-
> ism were a different kind of violence altogether. (*Through a Window,* pp.
> 108–9)

In the immediately subsequent passage Goodall presents a grue-some first-hand account of what she observed that so revulsed her. All of the proper names occurring here are names given to individual chim-panzees by Goodall herself:

> For several years I struggled to come to terms with this new knowledge. Of-ten when I woke in the night, horrific pictures sprang unbidden to my mind—Satan, cupping his hand below Sniff's chin to drink the blood that welled from a great wound on his face; old Rodolf, usually so benign, standing upright to hurl a four-pound rock at Godi's prostrate body; Jomeo tearing a strip of skin from De's thigh; Figan, charging and hitting, again and again, the stricken, quivering body of Goliath, one of his childhood he-roes. And, perhaps worst of all, Passion gorging on the flesh of Gilka's baby, her mouth smeared with blood like some grotesque vampire from the legends of childhood. (P. 109)

It must have been painful for Goodall to pen these passages. They unquestionably weaken her case for a more benign judgment and hu-mane treatment of the nonhuman primates, which constitutes the ma-jor thrust of the concluding chapters of her book. Yet, with another cu-rious twist of logic, Goodall seeks to compensate for this revelation of the dark side of the nature of the chimpanzee. To exonerate these pri-mates she claims that "though the basic aggressive patterns of the chimpanzees are remarkably similar to some of our own, their compre-hension of the suffering they inflict on their victims is very different from ours" (p. 109). And how does their comprehension differ from that of the human? In a statement highly critical of the human but prej-udicial to her prior claim that the chimpanzee is intelligent in the hu-man sense, Goodall explains the difference between the two by appeal-ing to the moral innocence of the chimpanzee: "Chimpanzees, it is true, are able to empathize, to understand at least to some extent the wants and needs of their companions. But only humans, I believe, are capable of *deliberate* cruelty—acting with the intention of causing pain and suf-fering" (ibid.).

While the human is, admittedly, capable of the most hideous cruel-ty, since the human can employ his intelligence to inflict unspeakable pain both mental and physical upon his victims, we generally recognize this as an aberration, and it is certainly not tolerated by public senti-

documentary e.g.?

ment. When those perpetrating such crimes are apprehended, they are, in most societies, appropriately brought to justice and punished. To imply, as Goodall seems to, that humans are, on the whole, cruel and sadistic, is unfair and one-sided, though it is undeniable that humans are capable of committing unspeakable crimes. Such individuals, however, are censured and punished, because we recognize them as responsible. Goodall is quite right in arguing that animals are innocent of their "crimes," but precisely because they act neither with premeditation nor out of freedom.

In an earlier book (*The Chimpanzees of Gombe: Patterns of Behavior,* 1986, pp. 291–92) Goodall gave a minute, altogether revolting description of the hunting practices of the chimpanzees she herself observed. As Lieberman, who has quoted the passage, comments, "Whereas human hunters kill their prey before they start to eat its flesh, chimpanzees do not seem to care whether the victim is dead or not" (*Uniquely Human,* 1991, pp. 153–54). From the description of their hunting practices it seems that Lieberman has been kind to a fault in his assessment of the chimpanzees, for what seems more likely is that chimpanzees 'prefer' to consume their victims alive. Omitting most of the gruesome details, which do not make pleasant reading, suffice it to say that, according to Goodall's own description of the live dissection by a group of chimpanzees of a male colobus monkey, it took nine minutes for the monkey to expire. More nauseating yet was the case of a nearly year-old baboon who was consumed by only one adult male chimpanzee, and who was "still alive and calling feebly for forty minutes after his capture." Three large bushpig young "took between eleven and twenty-three minutes to die as they were slowly torn apart" (Goodall, *Patterns of Behavior,* p. 292). While such scenes as these may not be everyday occurrences in the life of the chimpanzee in the wild, they do seem common enough to constitute a certain pattern of behavior. On this point Lieberman comments:

> Typically they [chimpanzees] begin to eat small prey by biting open the skull, which causes death, but their purpose appears to be limited to keeping the victim immobile to facilitate the process of tearing into its flesh. The victim may scream and thrash about, but the chimpanzees' chief interest

seems to be that the meal proceeds in an orderly manner, whether the victim is dead or alive. (P. 153)

Goodall cannot have it both ways. If the chimpanzees are as intelligent as she and others seem to think they are, then their acts are cruel and criminal; if they are to be exonerated on the grounds of not intending to inflict pain, but merely following their instincts, then they cannot be viewed as endowed with reason in the sense in which the human is.

Despite the dark side to the nature of the chimpanzee, which Goodall has carefully witnessed and described, there can be no question but that these animals rank as among the most 'technologically advanced' of the primates. In addition to fashioning sticks from branches to use as an instrument to fish out termites and ants, some groups among them have also been observed in the wild to have "devised a method" of cracking open nuts by striking them with rocks against a hard piece of wood, which is used as an anvil. Indeed one troop of chimpanzees was found to have invented a system for cracking a particular type of hard nut by striking it on three different sides. It was also noted that this method of cracking open the nuts was taught to the younger female members, whose exclusive task it was to provide food for the troop (Lieberman, *Uniquely Human*, p. 151). Further, in the laboratory chimpanzees have been observed to identify themselves in a mirror held in their hand (Goodall, *Through a Window*, p. 21), and some have been taught sign language at the linguistic level equivalent to that of a two-and-a-half-year-old human infant (Lieberman, *Uniquely Human*, p. 155).

These are notable achievements and need to be viewed seriously, but they also need to be taken in context. When at Gombe, Jane Goodall first observed chimpanzees using twigs to root out termites, she tells how she wrote to Louis Leakey (the father of Richard Leakey), who had originally suggested to her that she take up the work of observing primates in the wild. "I well remember writing to Louis about my first observations," she writes, "describing how David Greybeard not only used bits of straw to fish for termites but actually stripped leaves from a stem and thus *made* a tool. And I remember too receiving the now oft-quoted telegram he sent in response to my letter: 'Now we must redefine *tool,* redefine *Man,* or accept chimpanzees as humans'" (*Through a Window,* p. 19). In his review of Goodall's book in which the above statements occur, Lord Zuckerman, a scientist himself—who, as he tells us, spent the first five years of his career as prosector to the Zoological Society of London—comments on Louis Leakey's reply: "That statement could only have been made by someone who was both unaware of the fact that there are several animal species that normally use stones and other objects for a variety of natural purposes, and that there are other characteristics besides using objects as tools that apply to the definition of Man" (*New York Review,* May 30, 1991).

Near the end of *Through a Window,* Goodall informs her readers of the true purpose of her work in Gombe. "Louis Leakey," she writes, "sent me to Gombe in the hope that *a better understanding of the behaviour of our closest relatives* would provide a new window onto our own past" (p. 206; italics added). She then provides her own assessment of how her work has opened a window on the nature of Man.

> The opening of this window onto the way of life of our closest living relatives gives us a better understanding not only of the chimpanzee's place in nature, but also of *man's* place in nature. *Knowing that chimpanzees possess cognitive abilities once thought unique to humans, knowing that they (along with other 'dumb' animals) can reason, feel emotion and pain and fear, we are humbled. We are not, as once we believed, separated from the rest of the animal kingdom by an unbridgeable chasm.* Nevertheless, we must not forget, not for an instant, that even if we do not differ from the apes in kind, but only in degree, that degree is still overwhelmingly large. An understanding of chimpanzee behaviour helps to highlight certain aspects of human behaviour that *are* unique and that *do* differentiate us from the other living primates. Above all, *we have developed intellectual abilities which dwarf those of even the most gifted chimpanzees.* (P. 206; italics added)

The juxtaposition of two statements found in this passage—whereby apes and humans are said to differ only in degree, on the one hand, and that human behavior possesses aspects that are unique and do differentiate us from other primates, on the other—is curiously perplexing. There seems to be here a high level of inconsistency. The differences in "intelligence" which she admits to be compellingly large, "dwarfing those of the most gifted chimpanzees," seem to require no explanation. We have here an instantiation of the illogical claim that the more things differ the more they are the same.

Though Goodall seems not to have a very clear notion as to what the words 'kind' and 'degree' actually intend, the grounds for her contention seem quite clearly to be what she perceives as the "intelligent" activities of the chimpanzees, especially their ability to "fashion and use tools." It is to this claim that we finally direct our attention before concluding our reflections on the behavior of the human and nonhuman animal.

That there is something altogether remarkable in the activities of certain of the nonhuman primates is undeniable. What we should question, however, is whether these activities proceed from the animal's nature, and hence are an extension of instinctive habits or directives, or whether they emanate from a free association of separate components of experience and are, consequently, genuinely intellective and free.

AQUINAS AND ANIMAL BEHAVIOR

Aristotle and Thomas Aquinas made several observations regarding animal intelligence that can be of help in sorting out the salient features of this elusive and complex phenomenon. Aquinas saw most clearly that an adequate distinction between human nature and animal nature must find its basis not only in the differentiation between intellective and sensory knowledge, but in the simultaneous recognition of what they both possess in common as well. Perhaps Aquinas's most informative treatment of this question is found in his *Summa Theologiae,* where he raises the question whether the nonhuman animal is endowed with the ability to choose freely. His basic response is that the sensory appetite is determined (limited) to the sensible good or value by the very nature of the animal itself, and that in its response to the options experience offers, the nonhuman animal is not choosing, properly speaking, but is acting according to the demands of its particular nature. As is frequently his practice in addressing a question, Aquinas first provides a list of objections that have or could conceivably be raised against the position he is defending. An objection he raises against his own conclusion—that the nonhuman animal does not possess free choice—directly addresses the contemporary problem discussed above. The objection, in full, is as follows:

> Moreover, as it is argued in the 6th book of the Ethics [Aristotle]: "It is a mark of prudence to choose well those things which are ordered to an end." But the nonhuman animal displays prudence. . . . And this seems applicable as well to sensory activities: for there appear to be in the activities of animals wondrous instances of sagacity, as, for example, among the bees, spiders, and canines. For if a dog, in tracking a deer, should come upon a three-way break in the trail, it will seek to determine with its sense of smell whether the deer took the first or second trail, and, if it discovers that it did not, confidently moves along the third fork of the trail. This is as though by exploring the first two trails it is making use of a disjunctive syllogism whereby it concludes that the deer must have taken the third trail, since it did not take either of the other two. Consequently, it seems that the nonhuman animal is capable of choosing. (ST I–II, q. 13, a. 2, obj. 3)

The imagined objector, then, is arguing that the hound has exercised an act of practical reason in sniffing out the route the deer has followed when the former is confronted with a tripartite break in the trail.

In responding to this argument Aquinas readily grants that the activities of the nonhuman animals reflect a high degree of order, and hence of intelligence, but sees this as contained in their respective natures: "In the activities of nonhuman animals there appears a certain wisdom [sagacitas] to the extent that they have a natural inclination to certain highly ordered operations [ordinatissimos processus], as directed by the highest art" (ST I–II, q. 13, a. 2, ad 3). He then concludes that it is because of this wisdom or intelligence which is reflected in the activities of the nonhuman animal that the latter are said to act with prudence and insightfully. His clinching argument in support of this view is the same as that seen earlier, namely, that "all animals of the same species act in a similar way" (ibid.). This they assuredly would not do if, like humans, they were freely following their own individual appraisal of a situation.

The principle underlying Aquinas's argument is metaphysical in nature, and is given at the beginning of his response to the objection cited above. Again quoting Aristotle (Physics, Bk. III), he states that "motion is the actuality of the moveable as received from the mover." From this Aquinas concludes: "The power of the mover becomes manifest in the motion of that which is moved." He argues further:

> Consequently, in all things moved by reason, there appears the order of reason of the mover, even though those things moved by reason do not possess reason. In this way an arrow is directed toward its target by the action of the archer as if it possessed the reason directing it. And the same thing appears in the movements of clocks and all human inventions which are *the products of practical human reason* [*quae arte fiunt*]. (Ibid.; italics added)

There is, then, a certain parallel between artifacts, which depend upon human ingenuity, and the activities of the animal, which include the constructing of such things as nests, dams, hives, etc. The former are products whose intelligibility derives from intellect, while the latter are products whose intelligibility derives from nature. Just as the arrow is not the cause of the directed motion it has, neither is the nonhuman animal the cause, as an individual, of the activities it pursues. This is why we do not hold the animal responsible for its acts.

Since, then, for Aquinas prudence is a practical habit of intellect which facilitates the latter's application of general rules of conduct to particular instances, and which, therefore, aids in the formulation of correct judgments regarding actions to be taken or things to be made, he grants that in an extended sense of the term, the animal can be said to act 'prudently' in the exercise of some of its activities. These 'prudential judgments' of the nonhuman animal are, however, formally grounded in instinct. The dog is hunting a scent and follows its trail (*ST* I–II, q. 3, a. 6). It is only through a certain likeness which this action and others like it have to the human prudential act that they may be said in a secondary sense to be prudential (cf. *In VI Ethicorum*, lect. 7, 1214).

Aquinas employs basically the same argument in addressing the question of whether the animal has any knowledge of the future. "They do not know the future," he states, "but by natural instinct are moved to something in the future as though they foresaw the future," adding, "This instinct is given to them by the divine intellect which foresees the future" (*ST* I–II, q. 40, a. 3, ad 1). The direction and goal of the nonhuman animal's activity are not provided or determined by the individual animal itself, but rather are provided by nature (*ST* I, q. 18, a. 3). "The nonhuman animal prefers one thing to another because its appetite is naturally ordered to it. Consequently, immediately upon either sensing or imagining something to which its appetite is naturally inclined, it is moved toward it without choice" (*ST* I–II, q. 13, a. 2, ad 2). It should be noted again, however, that *the manner in which a particular experience will be apprehended by the animal,* is simply unpredictable. What Aquinas is not saying is that the animal will always act in the same way given this or that experience, but that the animal will always act in the same way if the experience is similarly apprehended by it, either as favorable or unfavorable, and to the same degree.

Is the view of Aquinas as outlined above outmoded? Have recent findings by anthropologists and naturalists, especially the discovery that some primates fashion and make use of primitive tools, shown Aquinas's position to be too restrictive of the nonhuman animal's capabilities? Many thinkers, Richard Leakey among them, claim that it is no longer realistic to speak of a distinction between the human and nonhuman in traditional terms. Acknowledging that though instincts play a very important role in the behavior of animals, he feels that these should be understood in a less inflexible manner. "One of the most efficient pieces of biological machinery," he writes, "is instinct, an innately programmed response to a specific stimulus," adding, "The notion of instincts as powerful forces guiding animals' behavior patterns has been overestimated; and the flexibility of responses depending on prevailing environmental forces has until recently been largely ignored" (*Origins,* pp. 208b–209b).

Yet, as we have sought to make clear, instinctual behavior properly understood does not eliminate the phenomenon of contingency. Cassirer, too, clearly recognizes an element of contingency in animal behavior, for he notes that the reactions of animals can vary appreciably, but he attributes this variability to an inborn stimulus of sorts that does not depend on a direct sensory experience. Perhaps he understands by this non-immediate stimulus precisely what we have just described as a diverse inner evaluation of the direct experience or stimulus, though he does not expressly say so (*Essay on Man*, p. 33). There will thus always remain an element of uncertainty in animal behavior stemming from the fact that conditions vary according to time and place, and these are integral components of the experience influencing the manner in which a situation will be apprehended and hence evaluated by the individual animal. Although Leakey seems unaware that this is the manner in which instinctual behavior has been more traditionally understood— and certainly so by Aquinas—the main thrust of his comment is nonetheless well taken: instinctual behavior need not entail a rigid inflexibility. But is it possible to maintain that all animal behavior is in some sense instinctual, or, to phrase the question somewhat differently, can one explain satisfactorily the 'tool making ability' of the chimpanzees without attributing to them a power of reasoning that does not differ in kind from the reasoning powers of the human? I believe that one can, but this will require a closer look at the sensory act as found both in the human and in the higher nonhuman animals, e.g., the primates.

As seen earlier, according to Aquinas, the sensory act reaches only the singular; it must leave it to the intellect to uncover the universal dimension in that same singular experience. At the same time, however, Aquinas does acknowledge a kind of awareness of the universal on the part of sense. How does this occur? Aquinas is less explicit regarding the manner in which this takes place than he is in providing a reason why it must be so, which is that such "universal" knowledge is essential to justifying the phenomenon of intellective knowledge. He states this plainly in his commentary on Aristotle's *Posterior Analytics,* where he comments, "If, however, the sense were to apprehend the singular thing and in no way should apprehend the universal nature in the particular, *it would not be possible that sensory apprehension could cause in us a knowledge of the universal*" (*In II Librum Anal. Post.*, lect. 20; italics added).

One will readily understand why Aquinas and Aristotle insist on this when one recalls that, for both, the universal is 'discovered' in the singular experience and made known to the intellect through the illuminating power of an active intellect, which does not '*create*' the universal but simply *uncovers* it as already virtually, that is, potentially, present in the sense experience itself. If one were to insist that the universal knowledge of the intellect in no way had its ground in the singular thing, one would of course have no other option than to grant, along with Kant and others, that the universal is an a priori form of the intellect. Cassirer, it seems, seeks to devise a variation on the position of Kant by claiming that true, that is, scientific, knowledge is not a knowledge of *reality* but of *"reality symbolized"* (*Philosophy of Symbolic Forms, III,* 545) and leaving rather ambivalent, as Wilbur M. Urban has commented, the role of sensory experience and the source of truth itself. Urban notes further that there is "a gradual shift [on the part of Cassirer] of the locus of verification from the intuible to the meaningful" ("Cassirer's Philosophy of Language," in *The Philosophy of Ernst Cassirer,* ed. Arthur Schilpp, 1949, p. 428). What precisely is for Cassirer the relation between "intellective symbol" and sensory reality? Urban's response is highly critical. "To say that physical science, in the later stages of its development, is no longer concerned with the actual, but solely with formal principles and structure, is seemingly to enunciate a paradox of the most astounding sort" (ibid.). Cassirer seems to appropriate as expressive of his own view the position he attributes to science, viz., that "truth is not to be attained so long as man confines himself within the narrow circle of his immediate experience, of observable facts" (*Essay on Man,* p. 208). Similar remarks lead Urban to conclude that Casirrer's philosophy is properly termed a phenomenology and not a metaphysics ("Cassirer's Philosophy of Language," pp. 436–37). And if these Kantian options are rejected, then the universal becomes nothing more than a pure mental construct formed from a frequent association of singular impressions, as Hume had contended. The point at issue, then, is essential to an integral explanation of the origin of intellective knowledge as emanating from the sensory experience of particular things.

As just indicated, Aquinas does not attempt to elucidate precisely how the sense actually 'grasps' the universal in the singular; he simply points to the fact that it must be so, providing a particular instance in which this takes place. He states his fundamental claim as follows: "It is manifest that it is the singular thing that is sensed properly and of it-self, but nonetheless the sense in some way apprehends the universal" *("sed tamen sensus est quodammodo et ipsius universalis") (In II Lib. Anal. Post.*, lect. 20). Aquinas then provides an example intended as il-lustrative of his claim: "For the sense knows Callias not only as Callias, but also inasmuch as he is this man [human], and similarly Socrates inasmuch as he is this particular man." He then concludes: "And from this it follows that by this sensory apprehension the intellective soul is able to consider 'man' in them both," i.e., Callias and Socrates (ibid.).

From this it is apparent that Aquinas is willing to grant that there is some sense at least in which one can say that the sensory power does apprehend the universal. But let us press this point further. If the senses are able somehow to grasp the universal, how might this differ from an intellective grasp of it? While what follows is not explicitly stated by Aquinas, it is offered as an extension of his views expressed above, and it will show some coherence with contemporary research findings on the activity of some of the more highly developed primates. How might the intellective awareness of the universal be definitely distinct from this sensory grasp of the universal to which we have seen Aquinas re-fer? The intellectual awareness is a knowledge that is immaterial. This of course is expressly the position of Aquinas. In the *De Veritate* he says: "The intellect has a certain operation which is of those things known by the sense but in a higher manner, since it knows universally and immaterially what the senses know materially and particularly" (q. 25, a. 3). Because the intellective awareness is a grasp of a singular thing but universally and immaterially, the intellect is capable of know-ing its own act and the nature of that act, with the consequence that it is able to know the universal as universal. Thus the intellective act is by nature 'reflexive', self-present. This means that, in knowing Callias or Socrates, the intellect knows individual humans but not as individuals. Rather it knows them universally and immaterially, that is, as human.

Contrarily, the highest of the sensory powers—the cogitative sense—although capable of knowing Callias and Socrates as human, only knows them as singular humans but not as members of a common species. For this reason the knowledge the nonhuman animal would have of Callias and Socrates, since they lack the immaterializing and universalizing power of intellect, would be restricted. Such knowledge would not grant the animal an awareness of the individual human as related to human in the abstract. The nonhuman animal cannot formulate judgments such as "The human as a human is a living being." Accordingly, then, the nonhuman animal would have an inchoate knowledge of the individual human as a human but a very imperfect knowledge withal, since the universal could be recognized only as under singular conditions and never as universal in the unrestricted sense. A nonhuman animal might be capable of having an intimation of the universal so that it could experience similarities between singular things it sensed. This could explain how it is possible for such an animal to profit from past experience; how it might avoid the repetition of experiences that caused pain in the past; or how it could welcome experiences that previously provided pleasurable sensations.

Indeed, in his commentary on Aristotle's *Metaphysics* Aquinas appears to affirm just this. In commenting on how knowledge takes its rise in the human from the world of singular things which the individual initially contacts through the exterior senses, Aquinas states that from sense experience an *'experimentum'* is formed, which arises from the comparison of many singulars received in the memory. He goes on to say that this kind of comparison is proper to the human and is effected by the latter's 'discursive power' (*In I Meta.*, lect. 1, nos. 14–16). Now for Aquinas the discursive power is a sensory power whose function it is to compare experiences *(intentiones)* received from the proper or special senses and conserved in the memory. The discursive or cogitative power *(vis cogitativa)* is the highest *sensory* power found in the human—sensory because it compares particulars; it can also be referred to as the 'particular reason' *(ratio particularis)*. Its counterpart in the nonhuman animal Aquinas calls the 'estimative power' *(vis aestimativa)* (*ST* I, q. 78, a. 4). Its function, according to Aquinas, is to make use of instinct to evaluate experiences the animal's special senses have provided.

How does the estimative power of the animal differ from the discursive power of the human? As sensory powers, qualitatively they differ not at all. How then, do these powers perform decidedly different functions? In the human the sensory discursive power is not the highest knowing power, since in the human there is the added power of intellect, which operates on the immaterial, universal level. Because of the affinity and closeness of the intellective to the sensory discursive power, the latter is able to rise above its sensory limitations and to imitate, after a fashion, the reasoning power of the intellect itself. As the intellect is able to reason on the level of the universal, the discursive power of the human is able to reason on the level of the singular. That is why it is sometimes referred to (for instance, by Aristotle) as the "particular reason." It will be recalled that earlier we discussed the vital role of the cogitative or discursive power in our reflections on the practical syllogism whereby the intellect combines and funnels back universal ideas to the sensible world of the singular.

Because of the centrality of the relation between the discursive power in the human and the estimative power in the animal as it relates to reason in the animal, we cite again Aquinas's own words as he describes this relationship:

> The excellence which the discursive and memorative powers have in the human is *not owing to their being sensitive powers* but to a certain affinity and nearness to the universal reason which in a way overflows into them. *They are not, therefore, different powers than those found in other animals only they are more perfect.* (ST I, q. 78, a. 4, ad 5; italics added)

Precisely because the highest sensory cognitive power in the human is accompanied by the illuminating power of intellect, it is called cogitative or discursive, rather than estimative, as it is referred to in the other animals. As a sensory power considered in itself, the discursive power does not differ from the estimative. This is an important point to retain for the purposes of our present investigation, and especially for the interpretation of the passage from Aquinas's commentary on the *Metaphysics* recently cited.

KNOWING AND FREEDOM IN THE NONHUMAN ANIMAL

It will be recalled that in Aquinas's commentary on the *Metaphysics* he affirmed that an '*experimentum*' arises from the comparison of many singulars received in the memory, that this comparison is performed by the discursive power, and, moreover, that it is an act proper to the human. This act of comparing carried out by the discursive power parallels the intellective act of reasoning in that, while the intellect compares universal ideas, the discursive power compares individual images and impressions retained in the sense memory.

Now it is Aquinas's subsequent comment immediately following that is most relevant to the point we are presently considering, for here Aquinas is speaking directly of the estimative power in the animal. He states, "And, because from many sensations, and memories, animals become accustomed to seek or avoid something, hence it is that they seem to share in something of the *experimentum; even though but little.*" (*In I Meta.*, lect. 1, no. 11; italics added). Here Aquinas grants that in some way the animal is able through an accumulation of experiences to make a kind of comparison between them, and thus in some way to go beyond the singular experience as such. Yet again, the final qualifying clause, "even though but little", would appear to put into question the prior comment that the nonhuman animal is capable of an act that in some way entails reasoning. Unfortunately, there are to my knowledge no other references in Aquinas's writings which indicate precisely how he intends, "seem to share in something of the '*experimentum*'," to be taken.

What, then, can we conclude regarding the level of knowledge Aquinas considers the nonhuman animals possess? It would appear that what Aquinas intends is that the nonhuman animal in its actions can give the 'appearance' of acting out of reason because it does act 'intelligently'—that is, its actions are ordered toward an end. The order found in these actions, however, does not arise from a reasoning process within the animal itself, but is rather the result of inborn attractions or instinct. The activity of the animal is directed not by a free judgment, then, as in the case of the human, but by a spontaneous judgment of the estimative power, which assesses the experience as advantageous or disadvantageous. For the human the judgment follows upon deliberation; thus one and the same experience can be evaluated as either pleasant, useful, or befitting or as some combination of these. In the animal the judgment follows directly upon the experience. That the animal does not deliberate in the authentic sense of searching for the appropriate activity against a background of all but unlimited choices is evident to Aquinas because those animals belonging to the same species act habitually in the same way. Such uniformity of behavior occurs not just for a short period of time only, nor is it limited to the life span of the individual but is seen to even continue generation after generation. It is only to be explained, in Aquinas's view, if it is *the nature* of the animal as this particular species of animal and *not that of the individual animal itself* which grounds and primarily controls the judgments made apropos of the manifold of experiences originating from the external senses. There is a further passage from the *De Veritate* which seems fully to corroborate the above summary of Aquinas's views on animal behavior:

> But of these things which have movement of themselves, for some the movement comes from a judgment of reason, for others it comes from a natural judgment. The human acts and is moved by a judgment of reason; for they deliberate about what is to be done. But the nonhuman animal acts and is moved by a natural judgment. This is clear from the fact that all animals belonging to the same species *act in a similar way*. Thus it is that all swallows make their nest in similar fashion. It is also clear from the fact the judgment of the animal is restricted to a determinate activity and does not extend to all. Thus it is that the bee does not work at any other project than the making of hives of honey, and the same is true of other animals. (Q. 24, a. 1; italics added)

In further elaborating his views relating to the activity of the nonhuman animal, Aquinas makes the broad claim that motion and action in them is not dissimilar from the motion and action found in inanimate things, in the sense that neither controls by a free judgment how they react to a given singular situation. The inanimate object is clearly totally dependent for its motion on an agent extrinsic to it; the nonhuman animal, on the other hand, is dependent upon its own nature for the judgments it makes and, consequently, for the actions that follow.

> Just as heavy and light bodies do not move themselves so that they are the cause of their own motion, similarly, neither does the nonhuman animal judge of its own judgment, but follows the judgment instilled within it by God. Consequently, the nonhuman animal is not a cause of its own choosing, nor does it possess freedom of choice. (Ibid.)

An animal does not control its choices, because it does not enjoy the vision of unrestricted knowing as does the human animal, and it is upon this kind of knowledge that the full meaning of liberty depends *("tota ratio libertatis ex modo cognitionis dependet")* (*De Veritate,* q. 24, a. 2). Yet as Aquinas states, the situation is significantly different in the case of the human.

> On the other hand it is the human who through his power of reasoning is able to judge about things to be done and about his choices as well, since he knows the meaning of the end of an action and of those things that are means toward an end, as well as the relationship and order of one thing toward another. The human is thus *the cause of himself not only in acting but also in judging,* and that is why he is said to have free choice as though one were to say that he has free judgment as regards acting or not acting (Q. 24, a. 1; italics added).

We return to the question earlier raised regarding the level of knowledge in the higher primates, occasioned by the recent 'findings' Jane Goodall and others have recorded regarding certain activities of the chimpanzees. When the chimpanzee employs twigs to ferret out termites, or small branches of trees to probe for ants in ant hills, are they not really making use of tools, and does this not, then, entail an authentic reasoning process? And if this is so, would it not further seem that the gap between the human and nonhuman animal's reasoning power has all but been closed, and the continuity between the human and nonhuman primates established as an unassailable given? This is the conclusion at which Goodall, as well as others sympathetic to her interpretation of the behavior of chimpanzees and baboons in the wild, arrive. The late Dian Fossey's parallel study of the gorilla in its native habitat near Uganda (*Gorillas in the Mist,* 1983) led her to a similar conclusion, even though the gorilla proves to be a primate considerably less gifted than the chimpanzee.

Yet, assuredly, the most remarkable account of the chimpanzee's use of tools is found in Lieberman's recent book, *Uniquely Human.* Relying on the observations made by Christophe and Hedwige Boesch a decade ago in Tai National Park in the Ivory Coast in West Africa, he describes how chimpanzees use stones to crack open nuts:

During the nut season they spend on average two hours each day systematically gathering and cracking nuts, a rich source of food. . . . The nut-crack-
ing technique is not mastered until adulthood and at least four years of
practice are necessary before any benefits are obtained. To open soft-shelled
nuts they use thick sticks as hammers, with wood anvils. They crack harder
shelled nuts with stone hammers and wood anvils. These nuts have three
segments, and the chimpanzee must rotate the nut on the anvil between
each successive hammer blow to extract the whole kernel. Mothers overtly
correct and instruct their infants from the time they first attempt to pound
nuts, at age three years. (P. 151)

Utilizing the same informational source, Lieberman adds that "the
stone anvils are stored in particular locations to which the chimpanzees
continually return, and the wear patterns on the stones indicate that
they have been used for generations" (ibid.). Further, Lieberman mentions that Goodall also observed chimpanzees making use of leaves as
sponges to soak up water, as recorded in a film made in 1986.

Putting aside for the moment the question as to whether the activities of chimpanzees described above are comparable in kind to the human's use of tools and their manufacture, it will first be of interest to
note that most of these activities are not restricted to chimpanzees.
Moreover, they had also been observed by others many years before
Goodall, Fossey, Boesch, and others had occasion to observe them. We
have already alluded to the comment of Lord Zuckerman who, in referring to Goodall's first-hand experiences in Gombe, wryly remarks
that "there are several animal species that normally use stones and other objects for a variety of natural purposes" (*New York Review,* May
30, 1991, p. 44). But surely most interesting of all is the citing of numerous instances of tool-using behavior on the part of chimpanzees
and other animals by Darwin himself in *The Descent of Man,* written
over one hundred years ago. The following passage indicates some of
the tool using behavior he himself observed; for other behaviors he re-
ports what he has learned from his readings.

It has often been said that no animal uses any tool; but the chimpanzee in a
state of nature cracks a native fruit, somewhat like a walnut, with a stone.
Renegger easily taught an American monkey thus to break open hard palm-
nuts; and afterwards of its own accord, it used stones to open other kinds
of nuts, as well as boxes. It thus also removed the soft rind of fruit that had

a disagreeable flavour. Another monkey was taught to open the lid of a large box with a stick, and afterwards it used the stick as a lever to move heavy bodies; and I have myself seen a young orangutan put a stick into a crevice, slip his hand to the other end, and use it in the proper manner as a lever. The tamed elephants in India are well known to break off branches of trees and use them to drive away the flies: and this same act has been observed in an elephant in a state of nature. . . . As I have repeatedly seen, a chimpanzee will throw any object at hand at a person who offends him; and the before-mentioned baboon at the Cape of Good Hope prepared mud for the purpose. In the Zoological Gardens, a monkey, which had weak teeth, used to break open nuts with a stone; and I was assured by the keepers that after using the stone, he hid it in the straw, and would not let any other monkey touch it. (Pt. 1, ch. 3, pp. 295b and 296a)

Surely these are remarkable feats that can only arouse our admiration and wonder. That they involve knowledge of some kind is undeniable, although this may well have been a position earlier subscribed to by Descartes and his disciples. The real point at issue on the contemporary scene, however, is whether or not the nonhuman primates and at least other of the 'higher' animals are able to 'think' in a way comparable to that in which humans think; not whether they possess any kind of knowledge whatsoever. Thus Lieberman's contention that his dog and mollusks 'think' in some sense is unlikely to meet with serious opposition today, though it is certainly susceptible of further nuancing. Lieberman states:

I am not claiming that my dog thought as well as I do, but he clearly did think. The mollusk experiments show that they also think, insofar as some of the neural mechanisms involved in associative learning are present in these simple animals. Either we must accept the proposition that they "think" to some degree, or we must arbitrarily decide that associative learning is not a cognitive act. (*Uniquely Human*, p. 125)

Even Aquinas acknowledged that all forms of sensation among animals possessing two or more senses entail reasoning, if by reasoning one understands some manner of 'associative learning', to use Lieberman's apt phrase. The animal depends upon its senses in reading the perils and advantages of any situation in which it finds itself. Were it incapable of making any association between its experience and its own well-being, its sensory activity would serve no purpose whatever.

The animal would be utterly incapable of putting itself in touch with its surrounding world, and without such communication it would inevitably and very quickly perish.

But do the examples cited above indicate that the animal is reasoning as the human reasons? In interpreting its sensory experience, does the animal exhibit full control of the judgment it obviously makes before acting? There is, I submit, no way of settling this question if we confine ourselves merely to that which we are able directly to observe. Unquestionably, the activity of the nonhuman animal is, as already acknowledged, replete with intelligence. Its actions do 'aim' at an end; its activity is nothing if not truly functional. These various activities reveal a purposefulness that clearly betrays intelligence, but simple observation alone is incapable of settling the question as to *the source* of that intelligence. Does the animal make its judgment freely so that in acting this way or that it could at the same time have acted otherwise? If the intelligence reflected in the action of the primate, or other nonhuman animal, is a conscious possession of the animal, it must be in a position to evaluate its situation as does the human in any number of different ways.

If such be the case, each separate evaluation can then provide a sufficient, although *not necessitating*, impetus for action. The upshot of this must be that there will then be no rigid pattern of uniformity remarked in the animal's behavior, an unmistakable pattern of randomness will be noted. But such randomness is not observed in animal behavior. On the contrary, so patterned is its behavior that it becomes, within the parameters of a given species, highly predictable by one well acquainted with the animal's habitat and particular species. It is this uniformity of animal behavior that makes the science of zoology as well as animal husbandry possible, according to which not only the physiological characteristics of animals of different species are meticulously listed but the behavioral traits and patterns utilized as well. It is the predictability of animal behavior that has, since time immemorial, been exploited by the hunter and fisherman, who, through learning the special feeding habits and other traits characteristic of individual species of animals, are more readily able to lure and entrap them. Yet such behavioral uniformity would be inexplicable if the hunted animal were intelligent and free in the sense that the human animal is. We are left, then, to conclude that the source of the direction of the species-specific behavior of the nonhuman animal is not the individual animal itself, but rather the nature which it shares with other members of its species. This is simply the only conclusion one can legitimately draw from the available evidence.

One who operates out of a Humean epistemology—according to which all reasoning consists only in a contingent association of events according to temporal and spacial relations—understandably will confound animal 'reasoning' with the human's ability to reason. Hume's knowledge theory, which is a sophisticated restatement of the nominalist position espoused earlier by William of Ockham, allows for no systematic discrimination between sensory and intellective levels of knowing. Hence, once it is granted that the animal is 'in some way' capable of performing 'reasonable' acts, the way lies open to admitting the animal to the intellective world of the human. For Hume's disciples, including Charles Darwin, the task of differentiating between the simple association of singular empirical events and the *conjoining* of the singular with the universal as humans constantly do, is not merely Herculean but impossible.

But if the animal is incapable of reasoning in the human sense, in what sense can it be viewed as capable of reasoning? This is the question to which we now turn. It is a question touching on the crucial point involving the animal's use of tools. What kind of knowledge does the animal employ in utilizing certain objects as tools? If for the animal, tool using is a learned experience, as it fully appears to be, what renders such knowledge possible? Is true universal knowledge of some kind involved?

To respond to these questions we will need to recall what was said earlier regarding the nature of knowledge and the distinction between sensory and intellective knowing.

1. First it should be borne in mind that in animate beings all knowledge, whether sensory or intellective, entails in its very nature as knowledge a special kind of union between the one knowing and the

known. Knowledge is a union which in itself is non-physical or immaterial, even though, as is the case with sensory knowing, it occurs under material conditions.

2. Second, knowing as it is found in animate beings entails reasoning *of some sort*, whereby there is an act of association, a proceeding
from one thing to another.

The human mode of knowing requires a reasoning process because
of the intellect's dependence on sensation to supply it with the content
of its knowing. The enormous number and variety of sensory perceptions which the human experiences on a daily basis must be classified
and ordered. The human reasoning process is precisely the activity of
the intellect inquiring into the meaning of what it has obtained through
the medium of the senses. The purpose of the reasoning process, then,
is to search out the true meaning and fuller significance of what the intellect has acquired and what the individual is experiencing. Otherwise
the information obtained could not be utilized for the benefit of the
knowing subject. This can only be accomplished, however, if the store
of information can be further unified and integrated. Although the reasoning process just described is proper to the human alone, it is noteworthy that precisely because the human is, as human, a kind, i.e.
species, of animal, that its knowing process includes reasoning. Were
the human intellect altogether independent of the sensory world for the
source of its intellective knowledge, its knowing would not be discursive but merely intuitional or intellective, coming in sudden bursts
without the labor of a thinking process either preceding or following
the acquisition of new knowledge.

Now the foregoing finds an analogous application in the case of the
nonhuman animal. Unless the animal was able in some way to coordi-
nate the almost unlimited number of different sensory perceptions and
pass an evaluative judgment on them, relating them in some way to its
own utility, they would be of no benefit to it whatsoever. Without an
evaluative assessment there would be no way in which the animal
could distinguish the significant from the insignificant experiences it
has, giving singular attention to some while wholly ignoring others.
Take away an animal's ability to adapt its activity to the situation and

[margin annotation: Human Intellect is dependent on sensory exp.]

it has lost its ability either to feed or defend itself. It would lack the ability to survive.

In this way the nonhuman animal's performance does bear a resemblance to the act of reason in the human. The judgments it forms are based on an experience and an inborn template which enables the animal to discern the suitability or unsuitability of a particular response to what it is experiencing. As already emphasized, the nonhuman animal is not the agent determining for itself what in fact is suitable or unsuitable. It is the persistent uniformity of its responses which alone justifies our making this inference. The underlying criteria by which the animal 'reasons' are inborn or instinctual. Consequently, then, the behavior of the nonhuman animal can be said to bear at least a faint likeness to the human act of intellective reasoning. The animal's judgments always refer directly to singular actions occurring at a particular time and situated in a particular place, which explains why the attention span of the animal is limited to a stimulus actually physically present, and why the object of its interests is so fickle—often shifting, quickly, even instantaneously, without the benefit of reflection.

Another manner in which the nonhuman animal can be said to share in what might be termed a kind of 'proto-reasoning' is the animal's marked ability to collate some of its sense experiences viewing them as temporally related. This makes possible a primitive learning process in the animal. It can profit from its mistakes. For example, it can learn from experience which kind of hunting procedures are successful and which are not. In this way, animals can learn the value of stealth for a successful hunt, and predators like the lion learn the importance of approaching their prey from down wind.

It can thus be seen that the nonhuman animal shares to a limited extent in the *'experimentum'* which is proper to the human. As noted earlier, the 'experimentum' results from the activity of the 'cogitative' sense in the human (which Aristotle sometimes referred to as the 'particular reason'). By the animal's comparing individual experiences, a kind of provisional template is fashioned through a recognition of similarities and dissimilarities in those things experienced. In the human this sense activity is performed against the backdrop of intellective knowledge where it is universals and particulars that are joined and contrasted. Through the merging of the activities of these two powers, the cogitative sense and the intellect, and with the resulting mosaic of awareness, the human can grasp the universal in the particular, and formulate singular judgments in which a predicate signifying universally is aligned with a singular subject. In the following chapter on language we will have occasion to consider how the animal's inability to formulate judgments of this type constitutes for it an insurmountable obstacle to the development of a true language.

Without the special intellective ability of the human to back up and uncover the meaning of singulars by an intellective comparison of them with universals, the nonhuman animal is able only to formulate tentative generalities based on the comparison of individual sensory experiences and thus to learn from them, accordingly as they are perceived to be similar or dissimilar. Of course, sense memory plays an important role here in permitting the animal to compare its experiences, for, if it were incapable of retaining and remembering what it had previously experienced, there would obviously be nothing specific with which it could compare a newly acquired experience. Through this kind of 'reasoning' process or sense comparison the animal is able to learn from its past ventures, to the extent that they were successful or unsuccessful, by a kind of trial and error method. This is why, as Darwin observed, animals that are older are much 'wiser' and more cunning than are their juniors. Accordingly, the younger animals are much easier to trap and to hunt successfully (*Descent of Man*, Pt. 1, ch. 3, p. 295a).

The nonhuman animal, then, shares in this limited way in the human *experimentum,* yet what it can accomplish is adequate to its needs. Its activity centers around two main objectives: self-preservation and the preservation of the species. Thus Aquinas states in his commentary on *The Metaphysics* of Aristotle:

> Because from many sensations and from memory animals become accustomed to seek or avoid something, hence it is that they seem to share in something of the *'experimentum'* even though but little. But men have, above the *'experimentum'*, which pertains to the discursive power, a universal reason, by which they live, as by that which is chief in them. (*In Meta. I,* lect. 1, no. 11)

Following Aristotle, Aquinas draws a parallel between the *experimentum* and art, indicating that the former arises out of the memory of singular events as the latter arises from an all-inclusive universal grasp of similar singular acts of knowing.

> The way in which art arises from *'experimentum'* is the same as the aforesaid in which the *'experimentum'* arises from memory. For, as from many memories there arises one 'experimental knowledge', so from many of these there arises an universal taking of all similar cases. And so art has this advantage over the *experimentum:* that the latter deals only with singulars, but art with universals. (*In Meta. I,* lect. 1, nos. 17–18)

HUME, DARWIN, AND ANIMAL BEHAVIOR

Without access to the world of the universal, the animal lacks the knowledge of the artist, who in creating a work of art applies universal concepts to the world of singular objects. Yet Aristotle and Aquinas instruct us that in a restricted way the animal does possess a kind of knowledge of the universal, for, though the senses are confined in their activity to knowing singulars, the singular they grasp is itself a particular instantiation of a universal.

The fact that the senses do in a limited way grasp the universal in knowing singular things affords us significant help in explaining how the nonhuman animal, who, though its highest cognoscitive act is a sensory one, is still capable of limited tool use. For such sensory awareness would be sufficient, it appears, for the animal to discern at some level a 'causal' relation between one thing and another. Yet it is most important to recognize that we employ the term cause in this instance in the same restricted sense as did Hume. Not, that is, as implying a necessary connection between one thing and another, but rather as simply recognizing their being spatially and temporally related, in that one event follows successively upon another. This is not of course, as Hume so clearly saw, a 'cause' in the ordinary or traditional sense; it is nothing more than a descriptive grasp or imaging of what has been sensibly experienced. Such an awareness remains a purely sensory act, for it does not truly transcend the singular. It does not support the causal claim that, e.g., since 'B' exists, 'A' necessarily exists, as Hume, too, consistently argued, but merely describes what is visually experienced, namely, that one event follows repeatedly upon another. The awareness of 'A' as 'cause' of 'B' is the result of repeated sensory experiences which are collated provisionally to form a particularized configuration, or *Gestalt*. Thus, in this instance there is for Hume no recognition that the sequencing and collocation of events must occur in the way they do; causal connections simply consist in the factual observation that there is a certain sequence between events. The attribution of the 'term' cause/effect to what is experienced is, consequently, purely a matter of custom, or of one's having experienced things in a similar sequenced pattern a number of times previously (cf. Hume's *A Treatise of Human Nature*, Bk. I, Pt. 4, p. 260).

The ability of the nonhuman animal to experience 'cause' in the Humean sense seems sufficient to explain in a satisfactory manner the tool-using phenomenon attested by Goodall, Darwin, Rengger, and others. Quite a few species make use of various objects to "serve their purposes." I have observed, for example, a German shepherd drop a stick directly in the path of a lawn mower, thereby requiring the gardener to pause to remove it, and then toss it aside. The satisfied animal would them chase after the stick, retrieve it, and, despite admonitions to the contrary, drop it again directly in the path of the mower. It is likely the animal had previously observed the gardener stop the mower in order to remove small objects such as sticks or stones which happened to lie in the mower's path, throwing them to one side.

Did the dog grasp a 'causal' relation between objects dropped in front of the mower and their being tossed away? Yes, in the Humean sense of 'cause'. This explanation fits with Aquinas's attributing to animals the ability to 'reason' in this weak sense of the term, whereby singular events are connected temporally and spatially. Thomas Aquinas states expressly: "The sense too is a kind of reason, as is *every cognitive power*" (*ST* I, q. 5, a. 4, ad 1). Such instances of animal behavior, however, do not display an intellectual awareness of the causal principle, which views the relation between events as more than one merely of repeated factual occurrence, but as necessary and universal, transcending space and time. Animals do pick up the 'causal relations' of contiguity

and succession, and so are able to make 'judgments' that effectively assist them to adjust to their environment and to exploit it to their advantage. This is true of all animals to some degree, even insects, for this is
for them part of life in their real, physical world. Such organisms are
not completely isolated units, but have a social dimension to them requiring that they interact with the world in which they live. The examples brought forward by Darwin, Jane Goodall, and others regarding
the chimpanzees and other primates are in substance no different from
countless instances of 'causal' activity on the part of members of many
other species of animals. But these activities give no indication that
they transcend the singular event. The organisms performing them
spend no time devising other better methods of doing what they do.
They do not penetrate into the significance of the causal chain of events
which they 'observe' concretely, and hence they are not concerned with
developing new strategies for performing old tasks. They do not operate on the level of reflective consciousness, and hence do not become
aware of a cause as a cause.

Though the animal is in an important sense as truly a cause as is the
human—for as an individual, independent organism, the animal is a
genuine cause of events, being physically responsible for their occurrence—animals are, in an equally significant sense, not as fully causes
of these events as is the human. The latter alone freely makes the decisions that direct it to do what it does. Somewhat restrictedly, then, one
can affirm that the human is an *uncaused cause* of the events it causes,
in the sense, namely, that nothing causes it to react to its environment,
both physical and intellectual, *the way* it does. The human alone controls that. The nonhuman animal, on the other hand, is a *caused cause,*
in that the judgments it makes and the decisions it arrives at are a spontaneous response to an automatic internal calculus determined by its
own nature. This accounts for the great dissimilarities noted in the activities of the various species of animals. Having different natures, they
possess different operational codes and matrices, for what is suitable
for one species of animal can easily prove to be most unsuitable for another.

If the foregoing analysis is seen to have merit, it enables one meaningfully to assess the performance of the human theory of knowing. For it allows one to see with singular clarity that the Humean analysis of knowledge is an accurate portrayal of knowledge on the animal level. In fact, Hume should perhaps be credited with providing philosophy with its most detailed and unified account of the nature and achievement of sensory knowledge. How sad, however, that Hume supposed, as have many of his subsequent disciples, that he was giving an adequate account of human intellective knowledge as well.

The Humean notion of cause—the mere factual observation that one thing frequently follows upon another, without any inplied necessary interconnection between them—seems to me to offer the most reasonable explanation of how the nonhuman primates and other animals are able to make use of various objects to perform simple tasks, most of which have as their sole purpose the obtaining of food. If one would wish to affirm, then, that animals are capable of reasoning in this sense, there should be no difficulty in agreeing, as long as one also recognizes that there is in this instance a marked difference in the meaning of 'reasoning' as applied to humans and to the nonhuman animal. It would in fact, appear to be the case that when Darwin and others, in support of their belief in a genetic continuity between all species of animals, particularly the primates, contend that animals are capable of reasoning in the same sense as are humans, it is in the Humean sense that the word 'reasoning' is employed. One need only eliminate the intellective dimension of one's conception of human knowledge in order to agree. This is seemingly a high price to pay in order to render credible the philosophical underpinnings of evolutionary theory.

Darwin asks whether animals possess a power of abstraction and the ability to form general conceptions. Disclaiming the requisite knowledge for addressing such a question, he nonetheless agrees with a certain Mr. Hookham, whom he quotes approvingly as having written to the *Birmingham News* in May, 1873: "It is pure assumption to assert that the mental act is not essentially of the same nature in the animal as in man" (*Descent of Man,* Pt. 1, ch. 3, p. 296b). Further, his own subsequent comment provides clear evidence that Darwin's underlying epistemology is decidedly Humean; that he mistakes the indefinite perception of a singular thing for an authentic immaterial, abstract conception:

> If one may judge from various articles which have been published lately, the greatest stress seems to be laid on the supposed entire absence in animals of the power of abstraction, or of forming general concepts. But when a dog sees another dog at a distance, it is often clear that he perceives that it is a dog in the abstract; for when he gets nearer his whole manner suddenly changes if the other dog be a friend. (Ibid.)

Darwin seems not to recognize that the dog, encountering another dog, experiences it as a particular dog, even though it may not at first recognize that this particular animal is the dog that is its friend. There is no need to invoke an abstract conception of dog in order to account for the event Darwin has described, although it obviously serves his purpose to attribute to the dog the power of abstract thinking. As we have seen, the dog does in some limited sense grasp the universal in the particular, as one sees Callias or Socrates not only as Callias or Socrates but as this particular *human*. Yet this entails not an intellective but merely a sensory act.

This is of course a very significant point, even one of watershed dimensions. What one understands by 'reasoning' controls the whole argument regarding coming to recognize an individual as human. One needs to be consistent in working out a philosophy of the human person which fully respects and embraces the presuppositions of one's definition of the reasoning process. If the human and the animal are viewed as basically similar in nature, there are some very grave consequences emanating from this view; especially those pertaining to questions of 'rights' and ethical theory, which must be faced squarely and honestly. Some of these questions we have yet to treat of. They will be taken up in the chapter on human and animal rights. Darwin, of course, by no means wishes to deny the value of moral conduct in the human, though he seems hard put to provide a creditable explanation in support of this conviction. Moreover, he certainly does not offer support for his implied claim that animals, too, have a conscience (*Descent*, Pt. I, ch. 3, p. 294b). Darwin, though, is persuaded that those writers who deny that "the higher animals possess a trace of reason" appear to explain away by "mere verbiage" all those facts to the contrary which he has presented (ibid.).

THE HUMAN TOOL MAKER

Because of the human's universal perspective, his use of a tool is not restricted to a single task, such as prying open a termite mound or cracking open nuts. The human can employ the same tool in many different ways, using it to perform a great variety of distinct tasks. Garden tools and cooking utensils provide excellent examples taken from everyday living. These can be used in many different ways even though they are singular, individual things in themselves, because the human intellect is able to visualize or grasp them as potentially pluralistic. The intelligibility of the tool, once it is removed from the restricting conditions of its materiality, can be joined mentally with other intelligibilities, which in turn derive from experiences likely of a quite different kind, thus finding new applications and uses. In addition, certain 'specialized' tools such as the monkey wrench or a carry-all van have a certain universality implanted in them, granting them an inborn flexibility, even though they themselves are singular objects. This allows them a greatly expanded usefulness. The number of tools, moreover, which the human has ingeniously fabricated and now has at his disposal, is almost beyond calculation, and there is clearly no limits to the kinds of tools or machines the human might succeed in creating or inventing in the future. A simple reflection on the inventory of tools and machines that have emerged during the past century alone serves to bring home to one the awesome creative power of the human mind. Arguably, no new tool has had a greater ability to effect global change than has the computer. It is supremely ironic that those who today seek to equate the mental powers of the nonhuman animal with those of humans do so at a time when the radical disparity between human and nonhuman achievements has never been more obvious.

The phenomenal burst of mental energy which the twentieth century especially experienced, and which continues unabated with our space programs and telecommunications breakthroughs; the new science of microbiology and genetic engineering, coupled with spectacular advances in chemistry, astronomy, nuclear physics, medicine and other fields too numerous to mention, has all been made possible by what Aristotle insightfully referred to as "the tool of tools," namely, the human intellect. With its unique and uncanny power to penetrate into the inwardness of things, the human mind is capable of 'seeing' both the myriad connecting and interlocking relations which exist *in* things as well as *between* them. By this same vision the human mind is, despite the recent uncovering of the unimaginable complexity of physical reality in all its forms, also at the same time able to lay hold of the oneness and unity of the universe. The proliferation of artifacts and tools emanating from the human mind, though mediated by the human body, has been made possible by, and is at the same time an infallible sign of, the mind's prior mastery of the causal manifold operative within the heart of reality either as presently experienced, or experienced in the past, or imaginatively projected and anticipated for the future.

The human's ability, then, to penetrate the experience initially offered it by the sensory powers, and to grasp that experience in a timeless, spaceless, manner, explains why the human is able to fashion myriad varieties of even the most complex of tools, and, preserving them, keep them ready to hand, so that they might again be made use of as occasion demands. These are tools of tools, and their *proliferation* has spawned the vast industrial complex of our modern world. While the human application of intelligence to the environing world goes on at a breathtakingly accelerating pace, especially with the advent of the computer—which allows the human to perform computations of unbelievable complexity within milliseconds—the animal world remains statically in an ineluctible state of inflexibility. Only knowledge that is universal could succeed in the self-expansion of knowledge we presently witness. Moving within a world of incalculable complexity, yet confidently soaring above its purely physical limitations of all planet earth's conscious inhabitants, the human alone commands a vision of the world as one, as a universe. Animal behavior reveals none of this explosive creativity which so signally characterizes the mature human regardless of economic levels, or social conditions.

DARWIN AND PRIMITIVE PEOPLES

There still are today, of course, some peoples whose level of culture and civilization has not appreciably advanced beyond the primitiveness of the stone age, who live a more or less nomadic existence. Darwin himself experienced one such people at first hand, on his around-the-world trip on the *Beagle*. As the ship cautiously made its way through the treacherous waters of the straits of Tierra del Fuego, there appeared along the shores a people who must be numbered among the most primitive of all humans. Their environment was almost certainly among the most hostile to human living. On the next to the last page of his last major work, *The Descent of Man*, written toward the end of his life, Darwin recollects the impressions made upon him by this primitive, forlorn tribe some forty-five years earlier:

> The astonishment which I felt on first seeing a party of Fuegians on a wild and broken shore will never be forgotten by me, for the reflection at once rushed into my mind—such were our ancestors. These men were absolutely naked and bedaubed with paint, their long hair was tangled, their mouths frothed with excitement, and their expression was wild, startled, and distrustful. They possessed hardly any arts, and like wild animals lived on what they could catch; they had no government, and were merciless to every one not of their own small tribe. (Pt. 3, ch. 21, p. 596b)

Darwin utilizes this moving experience to undergird what he states is "the main conclusion arrived at in this work, namely, that man is descended from some lowly organized form." He took the fact that there are humans living at such a bare subsistence level culturally as strongly supportive of his contention that the difference between the human and nonhuman animal is not nearly as great as generally supposed. The revulsion he experienced during his encounter with the Fuegians led him to remark that he would as soon be descended from certain animals as from humans such as they (*Descent*, Pt. 3, ch. 21, pp. 596b–597a).

> He who has seen a savage in his native land will not feel much shame, if forced to acknowledge that the blood of some more humble creature flows in his veins. For my own part I would as soon be descended from that heroic little monkey, who braved his dreaded enemy in order to save the life of his keeper . . . as from a savage who delights to torture his enemies, offers

up bloody sacrifices, practices infanticide without remorse, treats his wives like slaves, knows no decency, and is haunted by the grossest superstitions. (Ibid.)

One can but wonder why Darwin takes the worst possible human scenario and contrasts it with the best possible animal scenario, in order to establish some thread of continuity between the human and the nonhuman animal. But even more startling is his failure to recognize that Fuegians and other similarly deprived native peoples are quite capable of learning new ways and customs, of bettering their condition if provided with the opportunity.

Here it seems most apposite to observe that just such a turn around had been achieved in the most spectacular manner a full two hundred years before Darwin's trip on the *Beagle* and in an area fifteen hundred miles north of where the Fuegians lived. There, in the early part of the seventeenth century, in an area comprising parts of what is now Argentina, Paraguay, and Uruguay, Jesuit missionaries from Spain and Italy established '*Reductiones*' or settlements among nomadic tribes of Indians, some of whom had never before beheld a white man. These were tribes incessantly at war with one another, noted for their cruelty and lack of compassion for their enemies, whose wives were enslaved, who were nomadic and totally dependent on their prowess as hunters to avoid starving to death. Within a comparatively few years they had transformed themselves into peaceable, law-abiding communities. With the Jesuits as their mentors, the Indians learned how to cultivate the land, to grow their own crops, to lay out entire townships of two to three thousand people. They built elegant churches and other structures, some rivaling those of Europe. They became adept as well at choral singing, the playing of the musical instruments then known in Europe, including the organ, the harpsichord, and the harp. The last of these is today the national musical instrument of Paraguay. In a few decades, this very primitive people had been able to assimilate to a remarkable degree much of the best of European culture, because they were humans having an unrestricted capacity for learning, which needed only to be tapped and nurtured. This spectacular transformation occurred because they were intelligent beings to begin with, who were able through the medium of language to penetrate to the meaning of things, understanding reflexively the purpose of things and the desirability of what was proposed to them. It was their 'native' intelligence that made the difference, their intelligence that separated them from the wild animals they hunted.

Language, of course, played a most important role in this remarkable success story. It is to this perhaps most intriguing human phenomenon that we next turn our attention. Before doing so, however, one final comment on Darwin's definitive assessment of the human as it relates to the nonhuman animal. While Darwin cannot but admit the superiority of the human animal, he does so grudgingly, and needs to remind us that man's achievement was not truly of his own doing. Though man has attained "to the very summit of the organic scale" (*Descent*, Pt. 3, ch. 21, p. 596a), Darwin opines that this attainment is not owing to his own exertions, but simply to the fated interaction of genes through natural selection. The final words of his treatise are this baleful reminder to us of our lowly origin.

> Consequently, we must acknowledge, it seems to me, that man with all his noble qualities, with sympathy which feels for the most debased, with benevolence which extends not only to other men but to the humblest living creature, with his god-like intellect which has penetrated into the movements and constitution of the solar system-with all these exalted powers— *Man still bears in his bodily frame the indelible stamp of his lowly origin.* (Ibid., p. 597b; italics added)

The biological similarities the human shares with other animal species led Darwin to downplay the qualitative differences between them and to explain these latter as acquisitions having accrued to the human through the interplay of custom and environmental influences and the genetic changes brought about by sexual selection. Despite his intellective and moral qualities, the human is at bottom nothing more than an animal that has admirably succeeded in working its way to the top to become the CEOs of the evolutionary ladder. By an ironic twist of logic, precisely because man has so spectacularly arisen from such "lowly origins," Darwin envisages the likelihood of his achieving yet greater powers, for no limits can be placed on what natural selection

can accomplish through time. As a consequence, Darwin thinks m
evolutionary development from lower life forms to higher ones through
natural selection provide him with a greater hope for "still higher destiny" than could the view that he was "aboriginally placed" here by a
divine power. Clearly, Darwin is suggesting that if man's origin were
traceable to a divine being, his ability to develop in any significant way
would be, if not altogether precluded, at the least severely hampered.
As he states: "The fact of his [man's] having thus risen, instead of having been aboriginally placed there, may give him hope for a still higher
destiny in the distant future" (ibid.). It is indeed hard to see how the
hope Darwin here holds out for mankind could be of much value or
consolation to those of us still living, or to our progeny or our ancestors, to whatsoever species they may belong or have belonged. But this
is the price one must pay for an incoherent epistemology and a philosophy of nature that rests on the shifting, sandy foundations of evolutionary theory.

4 Human and Nonhuman Language

Perhaps the most intriguing of all the activities humans perform is the employment of language. The renowned linguist Benjamin Lee Whorf considered language "the greatest show man puts on." If one gives serious attention to what language is and what it enables the human to accomplish, it is difficult to fault this characterization. Certainly during the past century searching questions encompassing the human language phenomenon have occupied center stage; many of these questions were earlier raised by the two intellectual giants of the ancient world, Plato and Aristotle. They were also probed with great skill by St. Augustine, especially in his *De Magistro,* as well as by St. Thomas Aquinas, most notably in his commentary on Aristotle's treatise *On Interpretation.*

If we inquire into the reasons why language is such a fascinating and at the same time elusive phenomenon, we have not far to look. Language might be likened to a bridge linking two disparate worlds— the world of mind and the world of things. Indeed, language may be said to co-inhabit these two distinct but intimately related worlds. Though language is not, properly speaking, the same as idea, it does reflect and even in a sense contains idea. Though language is genuinely a sensible phenomenon possessing qualities shared by sensible objects, it is much more than an object of sense. Much as motion entails a fleeting fusing of potentiality and actuality, language is neither mind nor thing taken separately, but a rather 'unstable' union of the two.

In a mysterious and tantalizing way language accomplishes what appears as an unstable joining of two incompatible elements, for it brings together mind and matter in symbolical union. Though in part a physical, tangible reality, language is, at the same time, laced with thought. The sounds uttered when language is spoken, no longer remain merely sensible phenomena, for from these 'sounds' thought is 'magically' communicated from one person to another. This is what renders language a distinct phenomenon, for, as a reality immersed in the sensory world, it is capable of transmitting a reality not of itself sensible, much as a (visible) copper wire has the capability of transmitting electrical energy, not of itself visible.

Through the medium of language, then, intellective communication between persons becomes possible, and conversation, the process of linguistic interplay between individuals, comes to occupy center stage in the ongoing drama of human existence. Through conversation one gains access, though limited, into the hidden, conscious world of other humans. Through language the dark, opaque minds of others light up. The human is a notoriously language-dependent animal. Without it the human is condemned to live in a world interspersed with deep shadows. Having no other recourse we humans must employ language to express our thoughts to others, receiving from them in return some share in their own inner world.

Although one ordinarily first thinks of language in terms of the spoken word, the written word is also a powerful communicatory instrument. In addition to spoken and written language, gestures and actions of various kinds can also be regarded as a true form of language, especially if accompanied by a conscious intent to communicate something of one's inner world. For the present, however, we restrict our reflections on language to the spoken word, as the language form most universally employed by humans, embracing as it does all races, ages, and cultures, as well as both genders. Subsequently we will also briefly consider these other forms of language, for they provide the occasion for further nuancing our understanding and appreciation of the great richness and complexity of human language.

Spoken language, then, by providing us with the means of conversing with another, enables one to share something of his or her inner world of consciousness with another. This is accomplished simply through the utterance of patterned sound. Without some form of language each individual person remains cut off from social intercourse with others. One cannot know what another is thinking, feeling or wishing unless language somehow comes into play. The inner sanctum of the other remains unknown to us, since sensory perception alone is incapable of penetrating into the hidden recesses of another's mind or spirit. Unless the thoughts of the mind are expressed through sensible symbols, they lie beyond the pale of the most keen sensory perception. One person cannot *see* what another is thinking, nor *hear* their ideas.

The aim of our present investigation is twofold: first, to examine the nature of human language, and, second, to contrast it with the 'language' or 'languages' of the nonhuman animal. In recent years the animal language issue has attracted considerable interest. Numerous language experiments have been conducted with the larger primates, particularly with gorillas and chimpanzees. Yet animal language is not a question that has only recently attracted attention, for Darwin himself saw the issue of animal language as highly important for his evolutionary theory, and took considerable pains in attending to it. It is indeed owing to language that, in Darwin's view, the human has been able to outdistance the other primates both in his bodily development and in mental achievement. This same view is accepted today by a considerable number of evolutionists. But before an effective comparison can be made between the speech of the human and the nonhuman animal, a more indepth account must be given of the human as a speaker.

THE HUMAN AS SPEAKER

Despite the fact that speech is a daily phenomenon in the lives of all humans, excepting those suffering from serious physical or mental disability, it appears that many remain unaware of its true nature and of its great complexity. Though professional linguists of the past two centuries have given meticulous attention to this topic, the phenomenon of language still generates a large list of unanswered questions about its nature and its origin. Involving as it does the orchestration of the activities of the special senses, as well as the functioning of the vocal organs, together with the memory, imagination and their coordination with intellective concepts and ideas, human speech is a breathtakingly spectacular achievement which largely eludes any adequate explanation. This view we find reiterated by one of the most accomplished linguists of any age, Wilhelm von Humboldt, who arguably mastered more languages than any other single human. In his work *On Language: The Diversity of Human Language-Structure and Its Influence on the Mental Development of Mankind*, first published in 1836, von Humboldt, in discussing the complexities of human language and the difficulties one faces in attempting to describe it, writes:

> Apart from refinement of the ear and vocal organs and the impulse to give maximum diversity and most perfect elaboration to the sound itself, the merits of a language, with regard to its *sound-system*, depend quite especially on the latter's relation to *meaning*. To represent outer objects, that speak to all senses at once, and the inner motions of the mind, entirely by impressions on the ear, is an operation largely inexplicable in detail. That a connection exists between the *sound* and its *meaning*, seems certain; but the nature of this connection is seldom fully stateable, can often be divined merely, and far more often still is wholly beyond conjecture. (Peter Heath, translator, 1988, p. 72; italics added)

Such candidness on the part of one so accomplished in language study is indeed commendable. Von Humboldt has surely pinpointed the key as well as the most obscure element of the speech phenomenon: its bringing together, in one complex activity, sensory and intellective awareness. Precisely because speaking, sensing, and understanding are distinct activities, operating at very different levels, it is difficult or impossible to obtain a clear conception of their union. Sensory experience as such cannot transcend itself; it can never succeed in uncovering the intelligible in the spoken word.

The intellect, for its part, cannot fully penetrate the sensible component of human language. Since the sensory act deals with the singular as singular and the intellective act knows the singular as universal, they are and remain distinct though complementary acts. This accounts in good part, I would claim, for the difficulty we experience in attempting to analyze the nature of language. Yet, since language is perhaps the most typically human of all human activities, its obscurity can also account in good measure for problems we encounter in seeking to understand what it means to be human. Language provides us with a near-perfect mirror image of our human condition, reflecting back upon us who gaze upon it a reflection replete with enigma.

The essential building block employed by all languages is the word. Restricting our present consideration to the spoken word, one can define a word as an articulate sound or series of sounds signifying and communicating an idea. The meaning attached to a given word is conventionally assigned; that is, the connection between and the idea it expresses is not natural but conventional or assigned arbitrarily. From this it follows that there is no inherent reason that can be given why any particular word should convey the meaning it does. It is precisely this conventional dimension of the word that makes possible the proliferation of languages, allowing very different sounds or series of sounds to convey the same idea or meaning, thus making possible the inter-translatability phenomenon we find exhibited among all human languages. There is, therefore, nothing to prevent any particular language from being translated into any other language, no matter how diverse the vocabularies and syntaxes of the two languages may be. Nor is there anything to prevent the entire vocabulary of the English language, for example, undergoing a radical shift in meaning, so that none of its words any longer stands for the meaning currently and traditionally assigned. There is no special rule, that is, which requires that the word 'car' should continue to mean 'vehicle', or that 'dog' should designate a particular species of quadruped. The sound 'rac' could just as well signify 'car' and 'gan' could just as well signify 'dog'. Instead of saying: "My dog is very friendly." One could as well say, "My gan is very friendly," and one would still know what the statement was affirming.

THE MECHANICS OF SPEECH

Speaking is physically and biologically extraordinarily complex, a fact often overlooked or underestimated. When one speaks, one is emitting a stream of air from the mouth while carefully controlling and modulating its flow. The lungs provide the immediate source of the air flow. Without their supportive function, speech would be impossible.

When out of breath, one cannot speak. The vocal cords serve to control the air's rate of flow. By the contraction and expansion of vocal cords the airflow is widened or constricted, and it is the task of the lungs to provide the supply of air that passes through the voice box at any given moment. The mouth functions to mold and constrict the sounds coming from the larynx, where the vocal cords are situated. To accomplish this, the tongue, teeth, and lips are employed—the precise sound to be made determining the degree of participation of each of these, as well as its specific contribution. Thus the lips play a quite different role when the letter 'p' is sounded than when the phoneme 'g' is expressed. Also, the teeth figure more prominently in the expansion of sounds of the phoneme 'd' than with the letter 'x'. Those phonemes depending almost entirely for their expression on the larynx alone are termed vowels. That vowels hardly require the use of the mouth for their formation one can easily demonstrate to oneself by holding one's tongue in a completely immobile position and noting that one experiences no difficulty in the pronouncing any of the vowel sounds in use in the English language. Upon further experimentation one quickly discovers that any of the variant forms of these vowels can be pronounced in the same fashion. For example, the three different 'i' sounds as found in the words 'pol*i*ce', '*i*ce', and '*i*t' can be accurately pronounced without the slightest movement of the tongue.

All the remaining phonetic sounds made by the human, other than vowel sounds, are formed in the mouth and are called consonants. The word 'consonant' itself derives from the Latin word *consonare*, which means literally 'to sound with'. Consonant sounds are made possible by a breath of air passing through the mouth. Yet the consonant phoneme cannot be uttered without some employment of the tongue, teeth, and lips, as again anyone can easily confirm by trying to pronounce the letter 'q' without repositioning the lips in a very noticeable way. The result is similar if one attempts to pronounce 'r' or 'h' or 'j', or any of the other non-vowel phonemes, without movement of the tongue or lips and, in the case of dentals such as 'd' or 't', without a pronounced thrusting of the tongue against the teeth.

It should be noted, however, that in the employment of speech, vowels and consonants do not truly appear as isolated units, but rather unite to form one sound which we term a syllable. Deriving from the Greek, the word 'syllable' simply means 'sounds spoken together'. It is the result of two or more sounds elided or fused to form but one sound, as, for example, the sound 'syl' in the word 'syllable'. Von Humboldt points this out when, in discussing the nature of the articulated sound, he states:

> Through this simultaneous pair of sound-procedures the *syllable* is formed. But in this there are no longer two or more sounds, as our mode of writing might seem to suggest, but really only one sound expelled in a particular manner. The division of the simple syllable into a consonant and vowel, insofar as it is sought to think of them as independent, is merely an artificial one. In nature *the consonant and vowel mutually determine each other* in such a way as to constitute, for the ear, a quite inseparable unity. (*On Language*, p. 67; italics added)

But a syllable which is articulated as a single sound is not necessarily a word, though some words are monosyllabic. A word always conveys meaning, whereas a syllable need not, although there are languages in which many words are monosyllabic. All words made up of more than one syllable are of course composed of more than one sound, although these sounds are articulated and pronounced in such a way that a special quality of unity is evidenced.

A polysyllabic word is in effect, therefore, a harmony of sounds which, taken as a whole, constitute a particular meaning. Many words in the English language are polysyllabic, that is, they are constellations of different sounds which meld together to form one 'lengthened' sound. Thus the word 'constitution' comprises four syllables, i.e., con-sti-tu-tion, none of which have meaning in isolation from their positioning in the one complex unity of sounds which make up the polysyllabic word 'constitution'. To indicate the special union these four phonemes or sounds form with one another in making one word in ordinary speech we pronounce them almost as one sound, uttering them with much greater rapidity than one would when articulating them as individual sounds.

Now this segmentation of sounds is essential to a clear articulation of our thoughts. If a speaker fails to provide the clue that these particular sounds are meant to form one word, we say that person tends to run his or her words together. This makes it difficult to understand what they are saying, for the hearer is left to his own devices to bind these sounds together into word bundles and to construe their intended meaning. This is the precise situation one finds oneself in when conversing in a second language which one has not yet mastered. If one's interlocutor is speaking at a normal pace for a native speaker, one receives the impression that that person is speaking very rapidly, with the result that one is unable to keep up with the sound flow and to grasp clearly what is being said. When this is the case, one must often guess at the meaning of the sounds being uttered, separating and uniting the sounds heard into words as best one can, and perhaps requesting that the individual repeat what was said or speak more slowly.

But of course speech is much more complex than the comparatively simple task of word-formation, for words are only a halting first step toward the realization of geniune speech. It is only through the collecting of words into bundles or sentences that the intended meaning of the single words or phrases becomes clear. Just as syllables are formed by harmonizing diverse sounds, and words appear from the uniting of syllables, so, similarly, phrases and sentences result from the uniting of several, or even many, words into distinct bundles of meaning. Words do convey meaning, but the meaning of words taken singly is very incomplete. Single words suffice in their isolated state neither to communicate a complex human thought nor to express subtle nuances. There are, admittedly, limited instances when the utterance of a single word in a particular context may convey a very simple but complete thought, but this occurs only because the context makes clear to the interlocutor what words essential to the speaker's full meaning have been omitted. This will often occur in the speech of very young children who are still in the process of learning the fundamentals of linguistic speech and for whom the formation of a complete sentence is still a formidable task. Thus a child may point to an object while exclaiming "Dog!" "Cat!" which we readily interpret as meaning "There is a dog" or "There is a cat."

A grouping of words that signifies a meaning, but a meaning that is obviously merely a partial expression of a thought, is either a phrase or a clause. "In the morning" is an example of the former. In itself the thought expressed is altogether incomplete unless from the context the fuller meaning is apparent. Thus, if "In the morning" is a response to the question "When are you leaving for Hawaii?" it is a perfectly intelligible statement because it is clear from the context that it is but part of a sentence, the rest of which is left unexpressed, i.e., "We are leaving for Hawaii in the morning." The phrase "in the morning" does nothing more than indicate the time of the act; it does not indicate of itself what the action in question is. There can be no "in the morning" unless there is an action of some kind to which it refers. Its meaning, and this is similarly true of all phrases, is purely referential. If fully divorced from that to which it refers, it lacks meaning.

A clause differs from a phrase in that it signifies an action or a state of being. It contains, therefore, a subject and a predicate. Now clauses can be of two kinds, dependent and independent. The independent clause is a group of words containing a subject and predicate which expresses of itself a complete meaning; that is, would express a complete meaning were it not actually conditioned by a dependent clause. The dependent, like the independent clause, contains a subject and predicate, but is unable to stand of itself, for it must be complemented by an independent clause upon which it depends for its meaning.

An example will serve to render the foregoing easily intelligible. Let us take the following sentence to illustrate: "We will be leaving for Hawaii next Sunday, if we are able to book a direct flight." The sentence is composed of two clauses, the first of which is independent, while the latter cannot stand of itself. "If we are able to book a direct flight" expresses a conditioned but not a complete thought, for it does not include that which it conditions. As to the independent clause, "We will be leaving for Hawaii next Sunday," this does express a complete thought in itself, if taken absolutely, even though when joined with the conditional clause in the example cited above, it actually loses its independent status in terms of its meaning. Hence the clause "We will be leaving for Hawaii next Sunday," does not convey the true meaning of these words in context, for the one uttering them does not mean to say without qualification that they will be leaving for Hawaii next Sunday, but only that they will do so *if* they are able to book a direct flight. Hence, in the example given, two clauses join together to form but one sentence; the meanings of the separate clauses fuse to constitute but one full meaning.

A sentence, then, is a collection of words that join together to give expression to a complete, self-contained thought. From a strictly physical standpoint, however, a spoken sentence is nothing more than a succession of different sounds. As such, there would be no basis whatever for distinguishing a sentence from a clause or phrase or even from a single word, if the latter should happen to be polysyllabic. A word such as 'uncooperatively' or 'extraordinarily', contains more syllables than many simple sentences, yet it is not a sentence or even a phrase.

MIND, THE CATALYST OF LANGUAGE

What is it that creates a sentence, uniting many diverse sounds and joining them together as so many pearls on a string, aligning them in a specific order? Nothing else than meaning itself. Without the addition of the supreme catalyst of meaning, there simply is no language. Meaning is the golden thread, invisible to the outer eye, that shepherds sounds into a fold of specific configuration, magically transforming them into the expression of a complete thought. It can thus be seen that language is, at bottom, the embodiment of thought or meaning in sensible symbol. This coincides with the express view of von Humboldt, who states: "The immediate aim of language is the reproduction of thought" (*On Language*, p. 88). He further comments:

> Originally, in the invisible motions of the spirit, we should in no way think separately of that which relates to sound, and that which is demanded *by the inner aim of language—the power of designating,* and the power that produces the designandum. The general capacity for language unites and embraces them both. (Ibid.; italics added)

And though language involves a marriage between mind and sensible reality, it is unambiguously mind that plays the principal directive role in its creation. It is mind that enlivens the sounds, transforming them into words, giving these latter meaning by elevating them to the symbolic level, simultaneously conferring upon them the place and order they are to assume within the complex sentence structure unique to each particular language. The power and versatility of the intellect confer upon language the transcending, universal ingredient by which it is able to give expression through lowly, sensible word-signs to whatever the human mind might conceive. Von Humboldt emphasizes well this intellective dimension of language and its irreplaceable role.

> All the merits of sound-forms, whatever their artistry and sonority, and even when coupled with the most active sense of articulation, remain, however, incapable of bringing forth languages worthily fitted to the mind, if the radiant clarity of the ideas relating to language does not suffuse them with its light and warmth. *This wholly internal and purely intellectual part of language is what really constitutes its nature;* it is the use for which language-making employs the sound-form, and this is why language is able, as ideas continue to take shape, to lend expression to everything that the greatest minds of the latest generations strive to entrust to it. (P. 81; italics added)

LANGUAGE AND SYNTAX

But there is more. Language does not occur through a haphazard conjoining of words. There are rules of expression or laws of symbolizing which must be adhered to, if the world of inner consciousness of the speaker is to emerge through spoken words. In every language there are laws or grammatical rules specific to that language. They govern the manner in which words may be ordered to form a meaningful sentence—that is, a collection of words so arranged as to convey a specific concept or idea. The laws governing the construction of correct sentences constitute, when taken as a whole, what the grammarian terms *syntax.*

Syntax, as a subset of grammar, treats of the ordering of the parts of a sentence. Deriving from the Greek, the word 'syntax' is composed of the prefix *syn* meaning 'together with' and *taxis* meaning 'order'. Hence etymologically it simply refers to the manner in which words are to be ordered and aligned within a sentence. Each language has its own syntax or set of word-ordering rules. For linguists it is syntax that most differentiates one language from another, not vocabulary. In determining to which family of languages a particular language belongs, the linguist looks primarily to the manner in which its sentences are constructed.

Why any given language has the syntactical rules it does, remains for the linguist an elusive question, and it may well remain an unanswered one. But that languages do differ, often markedly, is beyond question. The syntactical rules of English are almost at opposite poles from the those of Latin or classical Greek. Among contemporary languages other than English, those of Russian and Japanese differ almost as night and day. Whereas in the Latin and the Japanese languages, e.g., the verb is regularly placed at the end of the clause or sentence, while in English and Russian it follows immediately after the subject. In Latin, Greek and Russian, nouns are highly inflected, so that, depending on the function they perform within the sentence, a suffix will be attached to the noun stem indicating precisely what that function is.

Nouns are also classified according to gender, which in many instances will require a special suffix-ending to indicate this difference. Any adjectival or participial modifiers must also reflect both the case and the gender of the noun they modify. In Japanese, however, there is, except for some personal pronouns, no change whatever in the sound of nouns regardless of their special function in the sentence or whether the object to which they refer is masculine or feminine.

In English also, of course, no change in the noun is made for syntactical reasons, so that the noun-form remains constant whether it serves as the subject, or the direct or indirect object of the sentence. In this respect the English language is closer to the Japanese language than it is to Latin. In Japanese, nouns remain invariant, regardless of the manner in which they signify within a sentence, whereas in Latin they are variously modified according to their syntactical function. Contrarily, however, if one looks to the placement of the verb in Japanese and Latin, it regularly occurs at the end of the sentence, rather than immediately after the subject, as in English.

The gender of nouns is yet another point of great syntactical importance for most Indo-European languages. Whether the noun is masculine or feminine, as for example, in French and Italian, or masculine, feminine or neuter, as in the Greek, German or Latin languages, rigorously controls the word endings of the adjectives and participles which must agree in both gender and number with the noun they respectively modify. Thus in German, one says, *"Ein guter Mann"* (A good husband), but *"Eine gute Frau"* (A good wife). The adjective 'good' must change from 'guter' which is masculine, to 'gute', which is feminine. Similarly in Latin, one says *'Caesar locutus est'* (Caesar has spoken), where *locutus,* must agree with the noun to which it refers which is masculine. But in the sentence, *Roma locuta est* (Rome has spoken), the same participle changes to *locuta* because it must agree with the subject to which it refers, which is feminine. In Japanese, on the other hand, nouns are totally gender neutral, so that their modifiers remain completely unaffected by the nouns they refer to.

What is truly most remarkable about languages, however, is that, although the rules governing the word formation and sentence structure of various languages, e.g., Finnish, French, Arabic, Turkish, Bantu, Chinese, can differ in the extreme, these languages are all, nonetheless, intertranslatable. Each of the languages named above is a representative of one of the six major families of living languages listed by the noted Danish linguist Louis Hjelmslev (*Language: An Introduction,* translated by Francis J. Whitfield, 1970, pp. 68–91). Finnish is a member of the *Uralic* family, French of the *Indo-European,* Arabic of the *Hamito-Semitic,* Turkish of the *Altaic,* Bantu of the *Bantu* family comprising a large number of native languages in the southern half of Africa, and, finally, Chinese of the *Sino-Austric* family. In addition to these, there are numerous other languages, which, as far as modern linguistic research has been able to determine, share so little syntactically with the six major families of languages, that they are considered as single-language families. Among these are numbered Japanese, Korean, Dravidian (in southern India), Basque, Sudanese, Eskimo and American Indian languages, which latter are still classified into as many as a hundred or more different families (Hjelmslev, *Language,* 1970, pp. 80–81).

COMMONALITY OF LANGUAGES

In speaking of human languages we are referring to a very sizable number of substantially different languages, for most of the six families of languages listed above contain many family sub-sets. Thus the total number of distinct human languages is conservatively estimated to be approximately 4000–5000, depending, of course, on the criteria actually employed to distinguish a language proper from its various dialects.

If we recall what was said earlier at the beginning of this chapter, the phenomenon of the inter-translatability of languages should not, perhaps, strike one as surprising. If language is indeed a sensible, symbolic expression of thought, this would have to be true of any language, no matter how different the vocabulary and the grammatical mode of expression might be. And if this be the case, language, though in part a physical reality, and hence definitely limited, must, nonetheless, take on something of the unlimited, universal qualities of the very thought to which it gives expression. Every language, in order to be a language, must transcend the special modality of expression it employs. It is this transcending quality which renders all individual languages intertranslatable.

Now it seems reasonable to conclude that some languages, because of the breadth of their vocabulary and the notably refined manner in which their grammars may have been developed, provide a more highly sensitized instrument and thus allow for greater precision in the expression of thought. It seems reasonable to assume, then, that the more primitive societies, lacking a written language as well as an extensive vocabulary, including abstract terms, are incapable of formulating many concepts expressible in the more highly developed languages.

This would not, however, preclude the further development of the more 'primitive languages' to the point where they could match the accomplishments of those more advanced. Indeed, historically, this seems to have occurred as regards the development of the Germanic tongues into some of the modern world's most sophisticated languages. We referred earlier to the dramatic advancement made by the previously illiterate Guaraní Indians of Paraguay who, in a matter of five or six months were taught to read. This was possible only because human language itself opens out onto an unlimited horizon.

There is nothing that the human can think of that cannot find symbolic expression in language. This is not to say, as it is often interpreted to mean, that language is able to exhaust completely the intelligibility of that which it symbolically portrays. Thought can indeed never be fully domesticated in this way, for linguistic expression as a sensible reality must always fall short of the richness and breadth of human thought. But these limitations should not be taken to mean that language tells us nothing about the inner world of thought. Language, as seen, is the bridge linking two quite different environments. Were it totally immersed in the one or the other, language could not share in both, and thus it could neither introduce thought into the world nor bring the physical world to the threshold of thought. As a physical reality language cannot be thought itself, and as thought or immaterial reality it can not also be fully identified with the sensible or physical. But language is a hybrid of both, and the miracle of intercommunication through human speech provides irrefutable testimony that this is indeed true.

It is, then, ultimately owing to language's sharing in the world of thought that all languages possess a common core of meaning. Consequently, the content of one language can be assimilated and expressed by another. The symbols employed are indeed very different, and their modes of alignment differ, sometimes exceedingly, but thought as transcending the physical world, can comfortably accommodate itself to any mode of physical representation or existence. The number of possible human languages is thus theoretically unlimited. Indeed, any human individual could relatively easily create a new artifical language of his or her own simply by developing an original vocabulary and fashioning a new set of syntactical rules by which the vocabulary is aligned so as to give sensible expression to the immaterial reality of thought. Anyone would thus be able through study and proper instruction to master the newly formed language, even though that language might use an idiosyncratic alphabet, or even an original set of characters or pictograms, and follow a wholly new syntactical code.

WORD ORDER AND MEANING

In the preceding discussion we have emphasized the syntactical dimension of language; the fact that all languages proceed according to certain rules which control the manner in which words are aligned. To exemplify this, let us take a quite ordinary sentence—"When they noticed that the truck was apparently out of control and heading directly for their car, they quickly veered to the right barely avoiding a collision"—and turn it into a jumble of meaningless words by ignoring syntactical rules. Using most of the words found in this sentence, let us construct a new 'sentence' disregarding the rules of English word order and syntax. Thus one might say: "Avoiding quickly of out noticed for directly truck apparently when right to collision noticed the that and control heading car for they the veered a barely."

The result is linguistic chaos; one could derive no meaning from this rearranged 'sentence'. Whatever else it might be, it is assuredly not language. The reason we obtain no meaning from the "reconstructed sentence" *is not because the sounds expressed are meaningless.* They are all of them bonafide words of the English language, words to which a particular meaning (or meanings) attaches. We know the meaning each of these words has when isolated from the context of a grammatical sentence structure. But when we juxtaposed them in the above manner without regard for the syntactic rules governing English word order, we find that not only does the meaning of the original sentence disappear from view, but the meaning of the individual words seems for the moment to have been lost, owing to a false and misleading contextualization which renders each word ineffectual. As such, a word has no independent reality of its own. To illustrate this quite effectively we might take two words from the sentence employed earlier, viz., 'veered' and 'collision'. There is, obviously, no such thing as veering taken by itself. There must always be a subject which acts in this way. Only someone or something can 'veer'. Further, a thing can veer only if it bears some spacial relation to some other object. To veer also implies that something which once was moving in a certain direction now changes course, altering the direction in which it is going. Hence the subject that veers must always be understood as being in motion.

Returning now to the last portion of the 'reconstructed sentence' where the word 'veered' occurs, i.e. "for they the veered a barely" we look in vain for its contextualization. No intelligible relation to other words in the 'sentence' is noted, and our habitual notion of what it means to 'veer' is contradicted and frustrated by this lack of context. The result is, then, that the word 'veer' has lost its true meaning, as indeed do all of the other words employed in this randomly constructed 'sentence'. This brief experiment, carried out in English, could be analogously replicated in other languages predictably with the same or similar result. Unless the syntactic rules specific to a particular language are adhered to, no re-grouping of words at random will succeed in producing the meaning of the original sentence. Of course, for such highly inflected languages as Latin, ancient Greek, or Russian, the word order will play a much less demanding role than it does in English. But even in such highly inflected languages, failure to respect the syntactic rules will result in an almost equally garbled and meaningless collection of sounds.

THE COMPLEXITY OF HUMAN LANGUAGE

The meaning of a word is always contextualized by its place and role within the sentence of which it is but a part. As seen, the individual word is not an entity unto itself, for only through its links with other component parts of the sentence is it a word, and this in turn is traceable to the very nature of language itself. Though at first it might appear that words precede sentences, since the latter would not exist without them, the converse is true. The sentence takes precedence over words, for it is the sentence and not the words taken singly that reflect the thought expressed. Since language as a sensible symbolization of thought, which latter is immaterial, and hence both atemporal and aspacial, it can find expression only by employing multiple sensible symbols. The inevitable consequence is that many words are required to express a single inner act of knowing.

Thus the modality of thought expressed in language is quite different from the modality of the thought itself. The latter is a simple unity, while the linguistic expression of that same unity is multiplex. Consequently, one does not know what exactly is being expressed until the sentence has been completed, since the sentence alone provides a full expression of the thought to be communicated. Each word thus takes its precise meaning from its special role in the sentence structure of which it is no more than a part. Thus words are always understood within the holistic confines of sentence structure. In themselves they are mere abstractions divorced from the structural milieu in and from which originally they acquired their meaning. For this reason dictionaries often provide model sentences exemplifying how individual words fit into the language landscape. In this way one can come to understand the meaning or meanings a particular sensory symbol might have, for most word sounds have the capability of conveying more than one meaning. In addition, all words can be employed figuratively, including ironically. Which of the meanings in a given instance is being assigned to them can be ascertained only by encountering them within the palisades of sentence structure.

Despite the many real differences among any of the human languages and the high level of syntactical disparity found between Chinese and Hindi or between Japanese and Hebrew, or one of the African languages, each of these languages nonetheless contains a common base. All express meaning, and in the most fundamental sense the meaning languages express is identical. The fact that one language can be translated into another, no matter how different their syntactical grammars and vocabularies might be, bears witness to this. Moreover, any individual can experience this through the serious study of a second language, particularly if one succeeds in conversing in the newly acquired language at even a very elementary level.

This phenomenon of intertranslatability among languages is further witnessed to by the commonplace occurrence experienced worldwide viz., that the human child is capable of learning as its native tongue any one of the languages spoken anywhere on the face of the globe. This depends solely on what language its parents speak, and the language the child repeatedly hears during its early childhood years. Indeed, recent studies of how small children learn to speak point out that the child's first 'language lessons' actually occur while it is yet in the womb, as it accustoms itself to specific voice patterns and speech cadences, primarily those of its mother.

Then, after birth, before it even learns to pronounce a word, the infant struggles to communicate through vocal sound. This pre-linguistic exercise, which hitherto has usually been referred to as mere babbling, is also now recognized to be an inchoate, albeit very imperfect, form of speech. By babbling, the child is acquiring the control—over its facial and mouth muscles, voice box and throat—needed for future oral expression. The babbling stage prepares the child to utter sounds of a particular language with controlled precision, and in accord with that language's panoply of syntactical rules.

These several reflections on the diversity and complexity of the language phenomenon have been intended to make clear the stupendous achievement language truly is and the all-important role intelligence plays in its performance. A selection of words essential to the expression of thought must first be made from a vast vocabulary storehouse. It must then be assembled in a manner compatible with the grammatical rules proper to the particular language employed. The actual number of intellectual operations needed to compose a grammatically correct sentence of fifteen or twenty words is itself staggering, and, if one includes as well the selections which are made at every turn from a whole menu of options that could have been used to express the same thought, one is likely stunned to the point of silence.

All of which helps us to grasp why the nonhuman animal is incapable of language as humans practice it, and why, further, the various sounds uttered by animals are species-dependent, invariant, and extremely limited. As we shall soon observe, the animal lacks not only the mental endowment but the requisite physical capabilities as well which are needed to meet the awesome demands of human speech.

THE ROLE OF GRAMMAR IN LANGUAGE

We have considered the crucial role syntax plays in the formation of speech, and that vocabulary, though essential, makes up only a small part of what constitutes language. Few it any linguists would disagree, and this for reasons that are not hard to find. Unanimity among linguists, however, is notably lacking when it comes to discerning the *nature* and *origin* of the syntactic rules, because these latter are intimately connected with knowledge theory. As there is considerable controversy regarding knowledge theory, it is inevitable that one's basic view of the nature and origin of language should be affected.

Though it is grammar that primarily differentiates one language from another, serious questions arise as to the origin of grammatical rules. These questions singularly occupied linguists' attention during the past century, as the well-known philosopher of language Noam Chomsky points out ("The Current Scene in Linguistics: Present Directions," in *Selected Writings*, 1971, p. 4). There was, as he indicates, an earlier tradition of linguistics that spoke of a universal grammar common to all languages, but this view has all but been replaced in the twentieth century by a nearly exclusive concern with a comparative study and description of the structures underlying the various human languages. Structuralism, as this dominant school of linguistics came to be called, embodied the view that through a comparative study and cataloging of the various elements functioning in the sentence structure of a language, as well as their rate of incidence, a full grammar of the language could be obtained (p. 5).

Chomsky has vigorously opposed this structuralist view, contending that language cannot be explained merely on the basis of the individual grammars of each language, but that one needs to pry more deeply into the phenomenon of language itself, raising questions about the constraints all individual grammars exhibit and their very genesis as grammars. His major claim is that the task of the linguist cannot be considered complete without including the more fundamental questions of the origin of language as a human phenomenon, and without explaining the similarities among individual languages as well as demarcating their differences. Chomsky insists that the underlying elements common to all languages can be rendered intelligible only if one postulates a non-specific or universal grammar which somehow governs and gives direction to the grammars of all the individual languages.

How such a grammar arose, however, Chomsky is unprepared to say. He acknowledges that much more work needs to be done by conscientiously pursuing the question of precisely how these two grammars, i.e, the general and the particular, interrelate and intersect. He does seem to recognize, however, that the questions he is raising amount to a serious quest for nothing less than the ontogenesis of human language, and rest on the basic premise that human language transcends the world of a purely descriptive or structuralist linguistics. The universal grammar of which he speaks treats of what he has chosen to call the 'deep structures' of language. He employs this term in contradistinction to 'surface structures', which are treated by the individual grammars and which concern word formation and the specific rules of syntax which, taken as a whole, are proper to one individual language only.

CHOMSKY AND UNIVERSAL GRAMMAR

With the exception of the universalist grammarians of the eighteenth and nineteenth centuries, Chomsky is one of the few linguists, especially among the American school, who has vigorously opposed a strictly empiricist interpretation of language conforming to the model of a Lockean theory of knowledge. The more radical wing of the American linguistic school, under the aegis of Leonard Bloomfield (1887–1949), focussed its entire attention on a study of the comparative characteristics of languages, which are empirically observable and hence, in their view, are alone subject to scientific scrutiny. A descriptive study of language from this perspective, according to which one concentrates almost exclusively on the superficial differences between languages, is certainly not without its merits in Chomsky's view, but it is incomplete and hence inadequate. Where it fails, for Chomsky, is that it leaves untouched the essential component of language, namely, meaning (*Language and Mind*, 1972, p. 164).

The deep structures, which Chomsky postulates as necessary to make sense out of the underlying similarities of all languages and to render them intertranslatable, lie beyond the reach of sensory perception. These unperceived and unperceivable structures are, however, the ektypes or basic matrices of the syntactic rules through which, in all languages, the meaning is expressed. Chomsky refers to them as 'deep structures' since they always remain below the surface of sensory experience itself. He says that through these hidden structures similarities in languages appear. As John Lyons points out in his excellent study outlining the linguistic theory of Chomsky, not only are the similarities between languages as significant as the differences, but recent studies even support the conclusion that they are more significant:

> But there can be little doubt the "Bloomfieldians," as well as many other schools of linguists, in their anxiety to avoid the bias of traditional grammar, have tended to exaggerate these differences and have given undue emphasis to the principle that every language is a law unto itself. The grammatical similarities that exist between widely separated and historically unrelated languages are at least as striking as their differences. Moreover, recent work in the syntactic analysis of a number of languages lends support to the view that the similarities are deeper and the differences more superficial. (*Noam Chomsky,* 1970, p. 115)

Chomsky himself willingly grants that modern comparative linguistics have contributed in a significantly positive way to the advancement of language study ("Current Scene in Linguistics," in *Selected Readings,* ed. J. P. B. Allen and Paul Van Buren, p. 5). He acknowledges that without these advances, the exploration of a universal grammar could not be reintroduced into the linguist's agenda. "Given this advance in precision and objectivity, it becomes possible to return, with new hope for success, to the problem of constructing the theory of a particular language—its grammar—and to the still more ambitious study of the general theory of language" (ibid.). But with his penchant for speaking his mind openly, Chomsky concludes: "It seems to me that the substantive contributions to the theory of language structure are few, and that, to a large extent, the concepts of modern linguistics constitute a retrogression as compared with universal grammar" (ibid.).

Chomsky is convinced that linguistics alone cannot penetrate the inner *sanctum* of the language phenomena. Rather enigmatically, though, he holds out some hope that, with the proper methods and approach, some progress along these lines might be achieved, but he warns that "one should not expect too much from interchanges of this kind" (*Selected Readings,* p. 164). It should be noted that the interchanges he clearly has in mind here are those between the Analytic school of modern comparative linguistics and traditional grammarians. In criticizing Locke's rejection of innate ideas and Nelson Goodman's rejection of his own suggestions regarding a universal grammar, Chomsky makes an appeal for a balanced critique of the rationalist school of philosophy and a careful rethinking of its universalist claims (*Language and Mind,* pp. 172ff.).

The 'universals' Chomsky sees as operative in all languages can be distinguished, he believes, into two classes. One he terms 'substantive'; the other, 'formal'. The substantive universal finds application in the areas of phonology, semantics, and syntax, or in a combinatory relationship between them (*Aspects of the Theory of Syntax,* 1965, p. 28). It comprises a general class of items from which a particular instantiation must be drawn and applied to one of the three areas listed above; and this would be true of all languages. In other words, the individual rules found exemplified in any language are particularized forms of universal rules. As for the formal universal, it seems to differ from the substantive universal simply by the fact that it is more abstract and hence more general in nature. The formal universal would be found operative in the grammars of *all* languages, although, apparently, it would not control any of the specific characteristics of a language by which that language might be distinguished from any other (p. 29).

As might readily be suspected, these claims of Chomsky regarding a universal or generative grammar, as he also frequently refers to it (a term he himself traces to Wilhelm von Humboldt, who used the German word *erzeugen,* 'to beget', to express a similar notion), has proven to be extremely controversial. The strongest opposition comes from the radical empiricists, who not only maintain that Chomsky has failed to prove his claim, but argue further that the claim itself is unprovable. Chomsky commits a cardinal sin against the empirical method by intermingling a sizable portion of nonempirical philosophy with linguistics, which latter they view as empirical in the narrowest sense. In his evaluation of Chomsky's contribution to linguistic studies, Lyons concludes that on this matter the jury is still out. He grants that Chomsky's views have generated a great amount of interest and have been responsible for instigating renewed research into the nature of syntax. At the same time, he also feels that "the results that have been obtained so far must be regarded as very tentative; and this fact should be borne in mind when linguistic evidence is being used in philosophical arguments" (*Noam Chomsky*, 1970, pp. 127–28).

EVALUATION OF CHOMSKY'S 'DEEP STRUCTURES'

Chomsky's efforts in the domain of linguistics honestly to embrace all of the phenomenal characteristics of language and to seek to provide an explanatory model of language applicable to all of the human languages is, I believe, admirable. The questions he raises about the similarities between languages and about the genesis of language within the consciousness of individual speakers and listeners, are eminently good ones, and it is remarkable how much he has achieved during the past several decades in raising the awareness level among linguists and philosophers alike regarding the nature and origins of human language.

I believe, furthermore, that he is on firm empirical ground—despite the sometimes virulent opposition he encounters on the part of behaviorists and radical empiricists—in postulating a need for deep structures of some kind to account for the existence of human language. There is, clearly, much more going on in language performance than meets the eye (or the ear). To deny this is to propound a paradox. Denial involves much more than a mere string of sounds. Thus the very act of denial itself implies an affirmation of the thing denied, viz., 'meaning'. As concluded earlier, 'language without meaning' does not qualify as authentic language.

Now this is not to say that one cannot study language from the standpoint of its sensible aspects and qualities. Phonetics, for example, is assuredly a valid subject of study, and it can provide the student of language with important data that permit a deeper penetration of the linguistic phenomenon, but phonetics is not, withal, the whole of language and cannot be considered a wholly autonomous science. The same applies proportionally to the syntactical study of language. Both of these areas of language study are absolutely essential to the serious investigation of what language is and how it functions, but they lose their *raison d'être* if they are totally isolated from language's semantical

A.I. auditory research (handwritten margin note)

component. Apart from meaning, phonology becomes merely a branch of the science of physics, while the study of syntax itself becomes a subset of phonology. Apart from meaning, there can be no words at all but only sounds, related to one another only in terms of their comparative loudness or softness, tone frequency, or the rhythm with which they are spoken.

One final comment will be permitted on Chomsky's theory. One can be completely sympathetic to his contention that meaning is the enlivening dimension of language, or in von Humboldt's terms, its spirit, as well as his claim that the meaning of language cannot be coherently viewed as simply emanating either from the sound of the spoken word or from the manner in which these sounds are joined together. This is not to contend, however, that the meaning of language exists apart from either sound or syntax, but simply that meaning, expressed through spoken words with their unique syntactical arrangement, does not *primordially* emanate from them. Words are the messengers or bearers of meaning but not its authors.

This I take unqualifiedly to be Chomsky's position with regard to language, and I feel that in this he can and must be sustained by anyone who attempts to come to grips with the grounds of this wonderful human experience we call language. Yet, with some of Chomsky's further claims by which he seeks to 'characterize meaning' as 'deep structures', 'generative grammar', and 'transformational grammar', and to divide 'meaning' into substantive and formal universals, I surely have serious reservations. These may or may not involve substantive points, and may indeed prove to be mainly questions of a mere terminological nature. Of importance here is not the terminology employed but rather how, precisely, it is to be understood and interpreted. What causes one some uneasiness is Chomsky's reference to 'deep structures' and the grammatical transformations they undergo when employed in 'surface' language where they are 'actually' employed in speech. This seems to suggest that the difference between the 'two grammars' might be thought of as one more of degree than of kind. The retention of the word 'grammar' suggests this, even though it is always preceded by a modifier such as 'generative', 'universal', or 'transformational'. On the other hand, it is quite understandable that one who views himself as primarily a student of linguistics and secondarily of philosophy, should choose to speak in an idiom more congenial to his linguistic colleagues.

The crucial issue here, however, is the manner in which Chomsky understands the 'universal grammar' to be innate. His rejection of the naive manner in which innate ideas are handled by John Locke and his criticism of Nelson Goodman on this score (*Language and Mind,* 1971, pp. 172–73), certainly deserve commendation. Yet he does not state unequivocally just how determinate these universal rules of *first grammar* are before they find specific application in a particular language. From his critique of Locke's understanding of what Descartes understood by innate ideas, one is inclined to conclude that these universal 'rules' do not take on any recognizable specificity until the moment they are incorporated into the sensible structures of sound, formulated by the speaker. On the other hand, Chomsky's suggestion— that any future progress in better understanding the relationship between the two levels of grammar (the universal and the particular) might best be achieved through the application of mathematical insights to the language phenomenon—creates the impression that he is conceiving the universal grammar more as a mathematical entity than as an epistemological one.

Whether this suggestion accurately summarizes Chomsky's view I am not prepared to say, but, if it does, it would appear to skirt dangerously close to Descartes's vision of mathematics as the primordial, universal paradigm of all certain knowledge. If the universal grammar in question is mathematically analyzable, it would seem logically to follow that the grammar itself is a mathematical entity of some sort. And if this be the case, it is hard not to conclude further that mind and the dynamics of understanding are not qualitatively distinct from the act of speaking itself, in which the 'surface grammar' is, according to Chomsky, operative. In referring to the use of mathematical models to analyze grammatical structures, Lyons, in his book on Chomsky, offers this comment: "The mathematical investigation of phrase structure grammars, and more particularly of context-free phrase structure grammar, is now well advanced; and various degrees of equivalence have been proved between phrase structure grammar and other kinds of grammar" (*Noam Chomsky,* p. 70). Shortly thereafter, however, in commenting on the application of mathematics to the deeper level of grammar Lyons adds: "So far the mathematical investigation of transformational grammar, which was initiated by Chomsky, has made relatively little progress" (ibid.).

CHOMSKY'S CRITIQUE OF LINGUISTIC EMPIRICISM

However Chomsky may view the relation between mathematics and philosophy, there are certainly passages wherein he would seem to consider philosophy as a discipline distinct from, albeit no doubt intimately related to, mathematics. There can be no question but that philosophical investigations are seen by Chomsky as distinct from the study of linguistics (*Language and Mind,* p. 161). Nonetheless, the latter is important for philosophy's natural and full development (pp. 171–74). Indeed, much of the final chapter of *Language and Mind,* entitled "Linguistics and Philosophy," is an indirect justification for his claim that the linguist should not shirk basic questions concerning the nature of language, nor the manner in which it is used and understood, nor the conditions for its acquisition. At the same time, Chomsky feels that linguistics can provide data that will be of help in the development of the philosophy of mind. His last comment clearly confirms this:

> It seems to me that the study of language can clarify and in part substantiate certain conclusions about human knowledge that relate directly to classical issues in the philosophy of mind. It is in this domain, I suspect, that one can look forward to a really fruitful collaboration between linguistics and philosophy in coming years. (P. 194)

In fact, Chomsky expressly says that the uncovering of the nature of mind is one of the important reasons for pursuing the study of language (p. 182). Chomsky, then, has rendered linguistic studies and philosophy a signal favor in strongly emphasizing the dual nature of language. While he by no means underestimates its empirical side, he has forcefully drawn attention to its mental, transcendental dimension, since that side has been so woefully ignored by the overwhelming majority of professional linguists of the past century, and by a great many philosophers as well. Chomsky clearly recognized that those other aspects of language, which extend beyond the restricted horizon of scientific enquiry, needed emphasizing.

It is worth noting here that Chomsky defends his proposal of a universal grammar precisely on the grounds that such a conclusion is mandated by the very empirical performance of language itself. Responding to Nelson Goodman's claim that language acquisition does not require any amount of 'innate schematism', he states forthrightly that it does, affirming that the basic task of language is to explain:

> ... how a rich and highly specific grammar is developed on the basis of limited data that is consistent with a vast number of other conflicting grammars. An innate schematism is proposed, correctly or incorrectly, as an empirical hypothesis to explain the uniformity, specificity, and richness of detail and structure of the grammars that are, in fact, constructed and used by the person who has mastered the language. (*Language and Mind*, 1972, p. 174)

Seemingly irritated by what he perceives to be the out-of-hand rejection of his proposal as to how language is first individually acquired, Chomsky openly accuses his empiricist colleagues of dogmatism:

> Surely it is not reasonable to be so bound to a tradition as to refuse to examine conflicting views about acquisition of knowledge on their merits. But this is to treat empiricist doctrine (immune to doubt or challenge) as articles of religious faith. (P. 179)

LANGUAGE AND THE A PRIORI

If one grants, as I believe the facts of language acquisition make it reasonable to do, that there is something at least akin to an a priori dimension to language, how is this best construed? Is it necessary to appeal to a universal form of grammar which gives shape and direction to the surface grammar (to employ Chomsky's own term) in order to explain the emergence of language from a continuous flow of sound? I believe these to be *the* most fundamental questions that can be raised regarding the philosophy of language, and hence perhaps the most important as well.

The general direction of response to this question was given in the previous chapter when the nature of human intellection was explored. The consideration most germane to the query concerning the a priori dimension of language points to the special role of the 'particular reason', also previously referred to as the cogitative or discursive power. It is this power—which, it will be recalled, is sensory and not intellective—that for Aquinas provides the crucial cognative link between the intellect and the senses. Since, as our previous discussion has emphasized, the language phenomenon involves a combination of the intellective and the sensible, the cogitative sense must play a key role in the genesis and development of language.

The intellective act as such we have characterized as an immaterial or non-physical act. That is, by nature it is directly concerned with the universal and not the singular. On the other hand, the act of the cogitative power is essentially sensory, and hence is restricted in its knowing to what is singular or particular. Although these two powers differ specifically, they are united in their acts, inasmuch as each knows the same thing, but from a different perspective. The cogitative power knows a particular experience precisely as particular, while the intellect knows the same experience not directly as particular but rather as universal. An analogy of sorts can be found in the acts of seeing and hearing where we have two distinct powers, each of which grasps a different aspect of the sensed object.

We return, then, to the original question: is there need to postulate as a priori a universal grammar to account for the phenomenon of language? Or, in Chomsky's words, is an innate schematism needed "to explain the uniformity, specificity, and richness of detail and structure of the grammars that are, in fact, constructed and used by the person who has mastered the language?" (*Language and Mind*, p. 174). The quick answer to this question I would take to be, 'no', if the question is to be taken literally. But there is still some hesitation in so responding, since Chomsky seems reluctant to attempt to characterize in further detail the form or nature of this 'innate schematism', appearing to be content with recommending that this should be the focus of future linguistic studies. But to explain why I replied to the above question negatively, and why I believe that more studies of the type seemingly envisaged by

Chomsky will avail little toward answering the question of the a priori in language, I offer the following consideration.

If one assumes that the intellect is an immaterial power, then there can be no question of its being constrained by any schema of universal grammatical forms or of any kind of specific categories as understood, for example, along Kantian lines. Such a priori forms would place limitations on the intellect, constricting its activity much as instinctual knowledge constricts the activity of the nonhuman animal. That is to say, the a priori grammatical schema would serve noticeably to restrict the intellect's flexibility. Intellect would no longer be fully open to the totality of experience, i.e., of being itself, which is precisely the quality specifically characterizing it as intellectual and distinguishing it from all other cognoscitive powers. In short, the intellect's total indeterminacy, or non-specificity as a knowing power vis à vis its object, is what opens it up to everything that is, to all of being, making an unrestricted scope of inquiry possible.

This same indeterminacy, however, is not antithetical to the first principles of being, frequently referred to as the first indemonstrable principles, for, precisely because these are principles of being in the unrestricted sense, they impose no limitation on the activity of intellect. Indeed, 'being' enters into the very definition of intellect, for what distinguishes it from all other knowing powers is precisely that its object is being without limit or restriction. Consequently, there is no inconsistency in allowing that these first indemonstrable principles, which are of the most general nature possible (since coextensive with being), are somehow present to the intellect prior to all experience.

To avoid repeating the error of Locke, who badly misunderstood Aquinas's position regarding the a priori nature of the first principles of knowing and of being, it needs to be emphasized that these principles are not held to be *actually known* in their pure, a priori state. Though they are habitually present from the beginning, the first principles remain, primordially, unknown. They can become known only when the intellect is itself actually informed and hence specified by a known object. Yet, any determinate intellective act is then capable of providing the basis for such an awareness, since the intellect can understand whatever it understands only through the active presence of these self-same first principles.

It was Locke's misfortune to have failed to recognize the distinction between an ontological and a psychological intellective presence of these first principles. This led him to misconstrue the status of their a priority, and consequently, to deny their transcendent status as principles (cf. Locke's *Essay Concerning Human Understanding*, 1689, ch. 1, pp. 12–13). Locke's attempt, then, to prove the reality of these principles through a posteriori reasoning alone was doomed from the start, as the subsequent history of Western philosophy bears witness. As we shall have occasion to remark later, Merleau-Ponty also fell victim to this same misunderstanding in his analysis of the nature of language.

How, then, does what has just been claimed with regard to the nature of the first principles, relate to the issue of universal grammar? In just this way: in all intellective acts the first principles must be operative, for they are, as remarked above, nothing less than an integral part of the intellective act itself, and hence inseparable from it. But all other principles of necessity must in some way be restricted, in that they find application only within a limited area of the scope of human knowing, which of itself is unrestricted.

Though the reader may well wonder about the relevance of this excursus into the area of metalinguistic theory and the metaphysics of the first principles of speculative reason, such a consideration does relate intimately to the question Chomsky raises regarding language and universal grammar. All sciences have a first principle upon which every inquiry and conclusion within that science depends, a first principle being nothing more than an initial starting point. But no first principle of any particular science can be a first principle absolutely. It itself is dependent upon other principles more basic or more general. But such regression can not go on indefinitely, and that is why there must be one principle which is absolutely first, and which, therefore, is indemonstrable.

All of which brings us back to the question of language and its origin within the individual. It is clear that linguistics or the science of language must have a first principle which is proper to itself and hence directly related to its definition of itself. If language is acceptably defined as the expression of thought through sensible symbol, then it follows that the first principle governing the study of language must be that language is not *what is thought,* but rather *the means by which what is thought* is communicated. It follows that the semantic element of language represents its most important and basic dimension; the other properties that language possesses are subordinate and subsidiary to it.

Language is also syntactical, but the syntactical rules of language which govern the manner in which words are united only exist to enable the phonemes employed to convey meaning. Without the meaning communicated through symbol, syntax is deprived of its sole source of intelligibility. The alignment, whatever form it may take in specific instances, of sounds into words, of words into phrases, and phrases into sentences, is the work of what we have seen von Humboldt refer to as the inner form of language, namely, its meaning. It is this inner form which breathes organization into the various constellations of sounds, providing them with the power to manifest, however imperfectly and inadequately, the thought that otherwise will continue to lie hidden within the mind of the knower.

What we find to be the case with syntactic rules applies also, proportionately, to phonetics. The sounds that are expressed by the action of the throat and mouth are not sounds simply emitted at random, but are clustered according to definite though still flexible patterns unique to each language. These sound patterns, of course, are reflections of the conceptualizations of the mind, to which the speaker strives to give expression through the employment of language. Phonetics and syntax, then, as mediators of meaning, are an integral part of language, although they do not constitute its most formal aspect which, as seen, is the meaning symbolically conveyed through the syntactical ordering of sounds. The priority of function which we must attribute to the semantical aspect of language is thus a priority of nature but not of time. There is, that is, no such thing as spoken language without the sensible symbol of sound, nor without the word-sounds being ordered and organized in some way by the underlying meaning and occurring simultaneously with it.

It would appear, then, that there can be no such thing as a *universal grammar* in any proper sense of the word—that is, a set of rules, however generic, that govern the incorporation of thought in matter, i.e. sensible symbols, and that are, properly speaking, innate or a priori. What is innate, I would submit, are the first indemonstrable principles, but these are of such a general nature as to transcend all categories, since, lacking specific content, they could not in any appropriate way be considered as constitutive of a grammar or even as general 'grammatical' rules, since this would be to constrict them. They are universal rules of being which can find their application in any existing thing or action, including, of course language.

The unrestrictive universal rules, then, are indeed found to be operative in language and to make possible the intelligible ordering of sounds and words into meaningful sentences. But, as incorporated in language, these universal rules have become particularized, and in this sense as particularized, cannot be considered to be innate or a priori. The specific grammatical rules that the linguist is able to uncover through a careful analysis of individual languages reflect the intelligibility of the first principles as operative in language, so that all languages, no matter how disparate in form and structure, are equally, as languages, sensible instruments for the expression of thought.

In this respect all languages are *one* in that they share in a common root, which is the intelligibility of thought, and it is this shared common ground which alone explains the intertranslatability of any and every single language into any other. The vocabulary and syntactical rules of languages differ, but all languages possess a lexicon of words, and all function uniformly according to rules of intelligibility uniquely embodied in each individual language.

From this it follows that a search for Chomsky's universal or 'generative grammar' is foredoomed to be fruitless, unless one is using the term 'grammar' here in a quite equivocal sense, meaning by it merely the first intelligible principles of being, and hence of meaning, which necessarily find application in every linguistic form and expression.

The ultimate key to language must lie in its disclosing meaning, for meaning cannot be disclosed save in an ordered, uniform way, which explains why all languages must employ a vocabulary and an ordered structure to achieve their goal of communication. But having said this, one must grant that the transformation of thought into language, which is evidenced by the fact that we do communicate what is within our minds through the medium of orchestrated sound, still remains for us, and will doubtless remain, an area of the greatest obscurity. The confluence of these two diverse streams—the mental and the sensible, each of considerably different density and quality, constitutively present in any linguistic utterance—is an event that can, it seems, never be fully encompassed by the human mind.

There is something about the union of mind and thing which, then, occurs in all knowing, as well as in the act of speaking, which lies beyond the scope of articulate human comprehension. This can be considered a necessary corollary stemming from the abstractive manner in which the human intellect knows. Owing to this mode of knowing, whereby the singular, material thing the intellect grasps is always known immaterially and universally (cf. St. Thomas Aquinas, *ST* I, q. 85, a.1), the knowledge obtained through the abstractive process cannot be communicated without resorting to the use of language to serve as the intermediary between mind and sense. Since all human knowledge is drawn from the sensible world, its truth lies formally in the acknowledgment of that fact (cf. Aquinas, *De Veritate,* q. 1, a. 9), and language is the instrument through which the concordance of mind and thing is recognizably acknowledged. If there is to be a communication of thought, it needs, as it were, to be cloaked with the mantle of sensibility, that it might achieve reentry into the sensory world from which its content originated.

The function of language, then, is always transformational and hermeneutical. It translates meaning from one mode of existence to another; from the actual to the potential, if viewed from the side of the speaker; from the potential to the actual, if viewed from the side of the listener. Language is the only forum in which two human minds can meet, and only minds are able to meet in the forum of language.

THE PROBLEM OF INTERCOMMUNICATION

Though all humans make use of language in one or more of its various forms and, unless impeded by a physical or mental defect, this even at a very tender age, it remains a performance of all but unimaginable complexity. For the moment we shall focus on the speaker, considering some of the elements entailed in the performance of a relatively simple act of speech. The following sentence will serve to illustrate the point: "Because I am not feeling well today, I will not be able to keep my appointment this afternoon." The meaning of this sentence is clear for anyone having a basic command of the English language. It is one sentence, because it expresses one complete thought. Yet this sentence is composed of two clauses which, if taken separately, express two complete thoughts, viz. "I am not feeling well today," and "I will not be able to keep my appointment this afternoon." Because, however, the meaning expressed by the first clause is subordinately related to that of the second, providing the explanation for its truth, the two clauses join together to form one sentence. In order to formulate and understand this sentence, one has to have grasped the causal relation of the dependence of one clause upon the other. In addition, there are eighteen words which compose this sentence. Each of these words plays its part in helping to express the thought contained in the sentence. These words have in turn been drawn from a minimal pool of several thousand words. In speaking, one strives to draw out from this sizable pool those which will appropriately fit the meaning intended.

Moreover, one must complete the selection of words within a time frame of seven or eight seconds, for this is the approximate length of time it will take the normal adult speaker to express orally a sentence of this length. Further, each word involves at least one sound or syllable. The articulation of these sounds entails a most complex series of coordinated physical actions involving the lungs, larynx, vocal cords, lips, tongue, teeth, and mouth. This means that in selecting the correct words to convey the thought, one must simultaneously concentrate upon the expression of the sounds in order properly to enunciate the words selected.

Nor is this all, for we have yet to include what is likely the most difficult aspect of articulated speech, namely, the syntactical organization of the words. In many languages, and most especially in English and Chinese, both highly uninflected languages, the correct ordering of the words is crucial to the meaning of the sentence. For in English it is primarily through the ordering of words that the relation of one word or phrase to another is indicated. Only if this relationship is clearly provided can one, upon hearing the words spoken, correctly surmise what was intended, or indeed make any sense at all out of what was said.

It should be borne in mind, then, that all of these elements of human speech—the words, the sounds used to express them, and the syntactical ordering peculiar to each language—are essential to language's single important task of communicating thought. Other accompanying qualities make a significant even though secondary contribution by providing further nuance and precision to the expressed thought. These secondary characteristics would include the rhythm with which the words are spoken, the tone of voice employed, the emphasis given to certain words or phrases, the pitch used in uttering them, the stressing of a word or syllable and, finally, the clarity with which the words themselves are enunciated. Yet, all of these secondary characteristics perform the same function of contributing to the communication of thought, though they do not possess the same importance in all languages. In Chinese, for example, and some other related languages, such as Vietnamese, the pitch or tone employed in uttering a word can alone serve to differentiate one word from another. This is because in these languages the same sound, but uttered in a different tone, is frequently used to express three, or four quite different meanings, and in some instances as many as eight or nine tones are employed as in the Cantonese Chinese dialect.

There are, in addition, still other secondary characteristics of the spoken word which assist the speaker, along with his or her preferred choice of words, to nuance the thought in a particular way, as well as to reveal subtle emotional overtones. To this same purpose the rhythm or cadence of one's language can also be effectively employed to add special nuances of meaning to one's speech.

All of these facets of spoken language contribute in some way to communicate the speaker's accompanying emotions along with the meaning intended. This is particularly true of word choice, even though such choice must respect the limitations imposed by the syntactical rules of each language. And while there are grammatical rules that must be followed, these still permit a wide range of flexibility for the creative construction of sentences. The degree of such flexibility is idiosyncratic both to each speaker and to the particular language employed. The ensemble of such speech qualities characteristic of each speaker we commonly refer to as that individual's style of speaking or writing.

These reflections on the complexity of human speech emphatically underscore that language is indeed, as Benjamin Whorf remarked, "the best show man puts on." The array of operations, both mental and physical, carried out simultaneously by one uttering a single complete sentence is truly imposing. Had we humans been aware of all these requirements before having developed the power of speech, we would likely have been dismayed by the enormity of the task and dismissed it out of hand as an unachievable goal. When Michelangelo, awed by the magnitude of his own work, tapped his statue of Moses on the forehead with his mallet and exclaimed "Speak!" little did he realize the nature of his demand!

To anyone who has lived in a country where a language distinct from one's own is spoken, it can be a constant source of both amazement and amusement to observe very young children speaking with ease in their own tongue. Without hesitation and with consummate fluency they chatter away. Regardless of race or language, all children, once they reach the age when children normally begin to speak, rapidly improve their language skills both in terms of increased vocabulary and of being able to speak with ever increasing precision and rapidity. With time the child gradually masters the more basic grammatical rules proper to its language. This clearly shows that all humans have a natural aptitude for language, and despite the enormous complexities of human language, children from their earliest years show themselves to be singularly adept as linguists. As we shall shortly examine, this is in stark contrast to what we observe among the many various species of nonhuman animals.

What was earlier noted regarding the act of speaking and what its normal performance entails, is also true, though inversely, of the one who listens. Because listening has every appearance of being a much more passive activity than does speaking, one might easily conclude that to listen is a considerably less complex activity, and hence much less demanding than the act of speaking. But this would be a superficial conclusion, drawn from an uncritical acceptance of first impressions, which, in this case, can easily mislead. Deriving meaning from the spoken 'word-sounds', which is the purpose of listening in the first place, demands many of the same highly refined language skills recently enumerated, all of which must be deployed successfully within an extremely short time frame.

THE RAPIDITY OF HUMAN SPEECH SOUNDS

Philip Lieberman, professor of Cognitive and Linguistic Studies at Brown University, has recently pointed out that human speech sounds occur at an "extremely rapid rate," anywhere in the range of fifteen to twenty-five sounds per second. He also states that the best we can do in decoding "non-speech sounds" is seven to nine items per second (*Uniquely Human,* 1991, pp. 37, 38). Such a rapid rate of transmission in speech is essential to its intelligibility, for, as Lieberman indicates, if lengthy sentences are spoken slowly, our memory is unable to hold all of the sounds in place long enough to follow what has been expressed. A fairly lengthy sentence may contain well over one hundred different speech sounds. Two sentences appearing in a recent paragraph of this text, for example, contained three hundred forty-four different sounds. As Lieberman also notes, one adept at the use of Morse code can transmit as many as fifty words per minute, which approximates five sounds per second. But at this rate the code expert is so absorbed in the transmission of the sounds that he is unable at the same time to comprehend the meaning of what he is transmitting (p. 38).

For one whose only authentic experience of the spoken word is with his or her native tongue, it may be very difficult to appreciate the staggering intricacy of the 'simple' process of listening. One who has patiently struggled to learn a second language has experienced at first hand the challenge the task of listening offers. On numerous occasions such a one is tempted to despair of ever being able to 'catch on to' what is being said. This will be especially true if those whom one is listening to are speaking at a rate normal for native speakers. Even though one may have already acquired a rather sizable passive reading vocabulary, these same words, when correctly spoken by a native speaker, will often appear to the beginner to be spoken with such rapidity that they all seem to run together, making it impossible for the listener to separate out the individual words so as to identify them.

Yet, unless the singular words can be identified, the novice listener has no prospect whatever of fitting them together to form a sentence, and of thus providing a secondary, contextualized meaning to them as they lie submerged in the overarching word unity we refer to as the 'sentence'. For the fledgling learner of a foreign language the unhappy result is that too often even the simplest sentence appears initially to be nothing more than a string of unconnected sounds signifying absolutely nothing. Only after considerable exposure to a spoken language does one gradually arrive at last at the point of being able to separate out the individual sounds, of grouping them together into distinct words, and finally of fitting them together as a distinct unity so that they reveal a meaning fully transcending the meaning of any of the words taken separately.

At the same time it needs to be stressed that, although the individual words sacrifice their own identity in uniting to form a sentence, each makes a unique and significant contribution to the meaning of the complete sentence. Thus grasping the unity of all the sounds expressed within the parameters of the sentence, the listener lays bare the thought expressed by the speaker. And even then one's grasp of the meaning of the sentence will often remain provisional, for it is usually further conditioned by the context in which it is uttered as well as by what preceded it and what is likely to follow.

When one considers the intricacy of the listening process, and the few microseconds one has to identify the flowing stream of multiple sounds, joining these together as 'one meaning sound' and, finally, forming these into a single complex expression of meaning, it is not surprising that the challenge of learning another language requires considerable time and persevering effort.

ALL HUMANS LEARN TO SPEAK

Yet, despite these formidable obstacles, all humans learn to speak and understand a language even from their earliest years. The underlying explanation for the phenomenal linguistic ability the human possesses, i.e., that everyone is by nature potentially a polyglot, can only be traced back to the roots of language itself. As the sensible symbolization of thought, all languages convey meaning regardless of the sounds that are used or the manner in which the language is syntactically structured. Through familiarization with the vocabulary and structures unique to each language, one can penetrate to the meaning they symbolically contain. This is why the native language of each individual is ordinarily the language of their parents or social millieu, and it explains as well why this same individual is capable of learning still other languages over time.

Languages, of course, are deeply immersed in the culture of a nation and a people, so that in learning a second language one is introduced at the same time into a new culture; in acquiring an idiomatic grasp of the language one unconsciously imbibes much of that culture. But the cultural differences attached to languages merely modify their mode of expression and do not radically affect the meaning conveyed by the language itself. The fact is that all languages witness to some manifestation of being; they can become intelligible to all humans; and they are intertranslatable.

With eloquence and great acumen Wilhelm von Humboldt describes what the act of human speech involves. His description of language—a fitting summary to this section focusing on the nature and complexity of human speech—serves as a springboard to our next inquiry into the language capabilities of the nonhuman animal. Von Humboldt writes:

> We have spoken at length of the compounding of the *inner thought-form* with the *sound*, and perceived in it a *synthesis* which—as is possible only through a truly creative act of the mind—engenders from the two elements to be combined a third, in which the particular nature of each vanishes. It is this *synthesis* whose *strength* is at issue here. . . .
> Since the synthesis we are speaking of is not a state, nor even properly a

deed, but itself a real action, always passing with the moment, *there can be no special sign for it in the words, and the endeavour to find such a sign would already in itself bear witness to a lack of true strength in the act,* in that its nature was misunderstood. The real presence of the synthesis *must reveal itself immaterially,* as it were, in the language; we must become aware that, like a flicker of lightening, it is illuminating the latter, and like a fireflash from regions unknown, has fused the elements that needed combination. . . .

We may call this act in general—as I have done here in this case—the act of spontaneous positing by bringing-together (synthesis). It recurs everywhere in language. We see it most clearly and manifestly in sentence-formation, then in words derived by inflection or by affixes, and finally, in general, in all couplings of the concept to the sound! (*On Language,* pp. 183–84; italics added)

As this account given by a true linguistic genius attests, the unifying and synthesizing process carried out through the reality which is language can be explained only by an activity that can not itself be a word or sign, but, as von Humboldt remarks, as an immaterial act, *a fireflash,* that brings together the elements of language into a cohesive whole, *leaving no sensible trace.* What the words express cannot themselves be seen or heard but only understood. The inner world of language transcends the outer world of time and space, the world of the sensible.

LANGUAGE AND THE NONHUMAN ANIMAL

Having closely examined the phenomenon of human speech and language, we now turn our attention to the nonhuman animal, and particularly to the nonhuman primate. Through a reflection on animal communication we wish to ascertain the extent to which animals might be said to possess the faculty of speech. This question has received considerable attention and publicity during the past several decades, perhaps more than its fair share. There can be no question of the importance that Cassirer attaches to the issue of animal intelligence as well as to the concomitant issue of the animal language, for he characterizes the former as one of the greatest problems confronting the philosophical anthropologist (*An Essay on Man,* p. 32). Cassirer further contends that a disciplined analysis of human speech will lead one to conclude that it contains an element not found in the animal world. Interestingly enough, he hastens to add that such a view is not in any way opposed to a general theory of evolution, although he is careful to qualify this position by asserting that modern biology no longer speaks of evolution in the same terms as did the earlier Darwinism, nor does it offer the same explanation of its causes. Though Cassirer hints at the position we have earlier discussed and identified as the theory of "punctuated equilibria," he does not employ that specific term, but he does speak of evolution as involving sudden mutations and as emerging. He *Visions* provides nothing beyond this to explain why there is no phenomenon parallel to human speech to be found anywhere in the animal world (cf. ibid., p. 31).

In recent years there have been ongoing language experiments with primates, notably with gorillas and chimpanzees. These are now restricted to seeking to teach primates to converse through sign language, since earlier attempts to teach them to convert sound into speech proved to be fruitless. The results of the sign-language experiments are tentative at best, and many linguists are sharply critical of the claims made by those conducting these experiments. But before we examine these efforts to instruct nonhuman primates in the rudiments of sign language, some preliminary remarks are in order which touch on the overall question of animal language and its relation to the evolutionary hypothesis.

From the very beginning of modern evolutionist theory the question of animal language loomed large. Those who argued for strict continuity between all living things would need to account for the significant difference between human language performance and the 'language' of even the highest nonhuman primates. Why should the latter, so physically similar to humans, be completely unable to utter sounds other than primitive call signals? The high level of linguistic development in the human, who in some ways differs little physically from the higher primates, when contrasted with the latter's almost complete inability to utter sounds other than primitive call signals, was sure to raise serious and persistant questions as to the reasons why.

Darwin himself clearly saw the importance that language played in his whole scheme of human evolution, and he expended considerable effort in trying to show why the nonhuman animal had made so little progress toward matching human linguistic accomplishment. The strategy he adopts is rather peculiar, and it appears to be internally inconsistent. First, maximizing the animal's ability to know, Darwin claims that animals do understand articulate sounds, in support of which he cites the comportment of some dogs. He concludes from this that the difference in linguistic accomplishment is attributable not to differing knowledge capacities but to differing physical structures (*Descent of Man*, Pt. 1, ch. 3, p. 297b). Earlier we saw where Richard Leakey picked up on this argument.

Darwin argues that the ability of talking parrots to make 'intelligible' sounds means that the ability to utter such sounds cannot be what distinguishes the human from the nonhuman animal. So he concludes that "the lower animals differ from man solely in his almost infinitely larger power of associating together the most diversified sounds and ideas" (p. 298a).

He argues further that bird song provides the closest analogy to human speech to be found in the animal world (298b). In his view, the power of associating sounds and ideas, i.e. the power of language, is what truly demarcates the human from the nonhuman animal. It is not alone the ability of the human to think and understand that is the high point in human evolution, but the human's superior ability in bringing together ideas and sounds. Such a conclusion must strike even the most ardent supporter of Darwin's teaching as strained. What is equally puzzling is the allegedly logical sequel Darwin draws that the human's linguistic ability depends directly upon "the high development of his mental powers" (ibid.). In Darwin's view, then, the human differs from the nonhuman animal not precisely in that the human is intelligent while the nonhuman is not, but rather in that the human has attained to a *yet higher level* of intelligence than his nonhuman counterpart, since the human is capable of a higher level of speech. To quote Darwin: "The fact of the higher apes not using their vocal organs for speech, no doubt depends on their intelligence not having been sufficiently advanced" (300a).

—

Within the animal world Darwin finds among birds the closest analogy to human speech. Yet he also grants that the cries of birds, as well as their power to make song, are instinctive, even though he affirms that the song they sing and the call notes they use have been learned from their parents. "The sounds uttered by birds offer in several respects the nearest analogy to language, for all the members of the same species utter the same instinctive cries expressive of their emotions; and all the kinds which sing, exert their power instinctively; but the actual song, and even the call-notes, are learnt from their parents or foster-parents" (ibid.). Yet, what Darwin assumed to have been learned, i.e., the actual songs sung by birds, have in recent years been shown definitively to be instinctive. Contemporary ornithologists have shown that the melody of the songs sung by various species of birds is independent of any instructional role on the part of either their parents or foster parents. The respected biologist W. H. Thorpe, whose specialty is birdsong, relates an experiment made with a species of finches found in South Africa. A number of eggs of these birds, which, as Thorpe remarks, possess a song altogether unique to their species, were removed from their nest and flown to Australia, where no such species of bird is found. The eggs were then placed in a foster nest and carefully monitored by tape recorders set up nearby. When the hatched finches reached the age at which finches commence to exercise the power of song, it was found that the young finches produced a melody identical to that of birds of their own species. As the tapes were later replayed there was found to be no similarity whatever between the song of the displaced finches and that employed by their foster parents. One must conclude, then, that the birdsong itself, and not merely the power of song, is instinctive (cf. *Animal Nature and Human Nature*, 1974, pp. 112ff.).

Now there were at Darwin's disposal strong indications that 'the facts' did support his interpretation. He was aware that the first attempts of the nestlings to sing were highly rudimentary, and that it was only after some trial and error that they finally hit upon the melody which, once recognized, would remain with them for the rest of their lifespan. "Their first essays," Darwin states, "show hardly a rudiment of the future song; but as they grow older we can perceive what they are aiming at; and at last they are said 'to sing their song round'" (*Descent of Man*, Pt. 1, ch. 3, p. 298b).

In his day, of course, Darwin did not have the technology for carrying out experiments such as the one Thorpe describes. But the experiment with finches shows that the nestlings, in their "first essays," are testing out various combinations of sounds against an inborn template. The sounds that do not fit the pattern of song at which they are instinctively aiming are simply rejected. Gradually the young birds adopt those sounds and the particular alignment that produce the instinctually sought-for melody. Once the melodious sounds they make coincide with the inborn template, they become locked in. The testing is over. The melody these nestlings will sing from then on is never to be modified or exchanged for another.

The contrast between this manner of 'learning' and the human child's ability to learn any one of over 5000 different 'foster languages' merely through continued exposure to them is stark; there is no possibility of confounding one with the other. And the fact that the repertoire of songs of any particular species of bird is extremely limited itself bears unmistakable witness to the instinctive nature of the songs birds sing. From the fact that the child is not so restricted regarding the language or languages it is capable of learning, it becomes apparent that the use of language in the case of the human is not at all instinctually determined, but is rather a matter of free acquisition. Thus Darwin is pressing and very much wide of the mark when he gratuitously states that "the songs of allied, though distinct species may be compared with the languages of distinct races of man" (p. 298b).

Accounting for language in the human seems to have presented Darwin with the most formidable challenge to his effort to provide consistency to his theory of the origin and development of the human species. He acknowledges that the superior human brain accounts for human acquisition of language. He theorizes that articulate language "owes its origin to the imitation and modification of various natural sounds, the voices of other animals, and man's own instinctive cries, aided by signs and gestures" (p. 298b).

Yet Darwin grants that even before primitive speech forms could have arisen in the human, there must have developed in him a higher form of mental powers. He writes: "The mental powers in some early progenitor of the human must have been more highly developed than in any existing ape, before even the most imperfect form of speech could have come into use" (p. 299b). If one asks why it is that the early progenitors had acquired superior mental powers relative to other primates, Darwin acknowledges that nothing more than general causes can be given, and that it would, moreover, be unreasonable to expect more. What these general causes might possibly be, however, he does not hazard to speculate. His own words are revealing:

> If it be asked why apes have not had their intellects developed to the same degree as that of man, general causes only can be assigned in answer, and it is unreasonable to expect anything more definite, considering our ignorance with respect to the successive stages of development through which each creature has passed. (P. 300a and b) *different species of apes.*

The ignorance we possess with respect to the 'successive stages' of development' through which the *human* has passed as well as the acquisition of articulate speech by him, do not, however, present for Darwin, "any insuperable objection to the belief that man has developed from some lower form" (p. 301b). Earlier Darwin had stated with an air of undoubting assurance and, but for the seriousness of the matter, almost amusingly logical ingenuousness that "the lower animals differ from man solely in his almost infinitely larger power of associating together the most diversified sounds and ideas; and this obviously depends on the high development of his mental powers" (p. 298a).

It is clear, therefore, that Darwin attributes speech in the human to the fact that they have attained to a higher level of intelligence. At the same time, however, he believes that the human has acquired this superior level of intelligence by his assiduous use of language. Here it is singularly important to note the considerable difference between saying that through language one has attained to intellective powers, and saying that through language one enhances that power by acquiring further knowledge. Darwin's argument trades on an ambiguous employment of the term 'intelligence', for he fails to acknowledge a firm distinction between intelligence as a permanent natural power or capacity for knowing universally and abstractly, and intelligence as a habit or ability acquired through the use of that same power. In short, the distinction between intellect as power and acquired intellective knowledge is clearly either denied by Darwin or simply ignored.

Because of this ambivalent use of the term 'intelligent', Darwin momentarily avoids the fallacy of circularity in affirming that the human gradually became more intelligent through the employment of language on the one hand, and that humans are able to employ language because they have attained to a higher level of intelligence on the other. What, then, has apparently blinded him to the underlying circularity of this manner of reasoning is his firm belief that there is no definite line distinguishing sensory knowing from intellective knowing. Relying on an alleged parity in kind, though not of degree, between sensory and intellective powers, Darwin claims to see no reason why it should not be possible for animals that are, as he sees it, also intelligent, to become ever more so, attaining eventually to the intellective level of the human.

The crucial question, however, which goes unasked, but whose answer is essential for grounding the belief that animals and humans both share in intelligence, though not equally, is how one then explains the marked difference in their achievement. Darwin acknowledges the huge disparity in this regard, but offers no plausible explanation for it other than that the human has become more intelligent through the use of language. Why, then, for a similar reason, the nonhuman animal has not so advanced is acknowledged by Darwin to be something of a mystery. He can only restate the fact that they indeed did not develop linguistically as did the human, and this even though some animals at least have the physical ability to imitate some human speech sounds. Yet the very fact that no animal species has even begun to rival either the human's linguistic or intellective achievement, surely should at least instill seeds of doubt regarding the nonhumans' alleged status as 'intelligent' beings. One cannot be a little bit intelligent; and if one is *possessed of the power of intelligence*, one can *acquire the art of speech*. Yet no one becomes intelligent because they speak; rather, because one is intelligent, one can learn to speak. Language and intellective ability are conversely related. We need to be very aware here of the marked difference between *recognizing* that someone is intelligent and one's *be-*

ing intelligent. One clearly has it backwards if one claims that the power of intelligence itself is acquired through the use of language. Intelligence is so *manifested?* Yes. *Acquired?* No. Rather, language is irremediably dependent on intelligence. Its singularly primordial function is not to provide intelligence but rather, through the medium of sensible symbol, to manifest and communicate it, as discussed earlier in this chapter. Without intellective awareness there would be no thoughts to communicate, and hence no need of human language. A very strong clue, of course, as to why animals have not learned to speak.

A similar argument is employed by some Darwinists, who claim that the human developed a larger brain than did other primates because, upon leaving the jungle for the savanna, the human became bipedal, which in turn freed his hands for more 'intellectual' tasks. What seems to be overlooked in arguing in this manner (and it is an oversight often afflicting Darwinist reasoning) is that, while the entire theory of evolution rests on random selection which is inherently directionless, purposefulness and direction are continually introduced to *explain* an organism's need to adjust to its changing environment in order to survive. Thus bipedalism was allegedly developed by the prehuman primates in order to accommodate to their changed savannah lifestyle, which in turn led to a change in the structure of their hands, and eventually the development of language, for which changes in the mouth, tongue, and lips were required. Finally, the use of language led to the development of larger brains and this to greater intelligence. There is hardly need to remark that this line of argumentation is tightly interlaced with teleological threads clearly alien to the orthodox Darwinian contention that all life forms randomly emerged from lower or less complex life forms.

Latter-day Darwinists simply sustain this same incongruity in availing themselves of the classical Darwinian sleight of hand that actually employs a teleological form of argumentation in seeking to justify a non-teleological theory. "We may confidently believe," Darwin writes, that the continued use and advancement of this power [of speech] would have reacted on the mind itself, by enabling and encouraging it to carry on long trains of thought. A complex train of thought can no more be carried on without the aid of words, whether spoken or silent, than a long calculation without the use of figures of algebra" (p. 299a). But why might we "confidently believe this"—namely, that the use of language would lead to an advancement in mental powers—unless there were, from the very beginning, an intimate relation between speech and thought? Unless mind already existed as subsoil from which speech could emerge? Yet such assumptions, though commonplace, are meaningless outside a teleological context according to which order and design are holistically embedded in the very heart of nature itself. We are faced here with a mental sleight of hand having the most serious consequences as concerns any coherent accounting for the most ordinary data of experience.

DIFFERENCES IN ANIMAL AND HUMAN VOCAL TRACTS

Darwin considered it beyond question—mistakenly as it turns out—that the vocal organs of the higher animals are basically similar to ours. This raised for him the puzzling question as to why it was that these organs do not perform in any way qualitatively comparable to the speech organs of the human. Why would animals have developed highly particularized speech organs which they, lacking intellective development, were incapable of using? The question seems, at least indirectly, to have troubled him, for he has this to say:

> As all the higher mammals possess vocal organs, constructed on the same general plan as ours, and used as a means of communication, it was obviously probable that these same organs would be still further developed if the power of communication had to be improved; and this has been effected by the aid of adjoining and well adapted parts, namely the tongue and lips. (P. 300a)

Philip Lieberman has pointed out (*Uniquely Human,* pp. 51ff.) that in all mammals save the human, the larynx is positioned high in the throat, while in humans, the larynx is located lower in the throat. For each, the positioning of the larynx, when combined with certain other anatomical features peculiar to each, accomplishes its purpose.

The larynx of the nonhuman mammal rises up at the back of the mouth like the air tube of a scuba diver, making it most unlikely that foreign objects will be able to block the windpipe. The tongue is long and thin and lies wholly in the mouth. This physical arrangement allows such animals to breathe and drink at the same time, a feat of some urgency to them, since it reduces the time spent in a highly vulnerable drinking position. Mouth, tongue, and larynx are all designed to facilitate the intake of food and transmit it quickly into the stomach chamber (p. 54). What Darwin did not realize and, what it seems no one of his era even suspicioned, was that the vocal tracts in the human and the mammal are only superficially similar. As was just noted, in all mammals save the human, the larynx is positioned high in the throat. Its principle function is to protect against objects falling into the larynx, thus blocking the passage of air to the lungs. This effectively results in the nonhuman animal having two separate entry passages, one which leads directly to the lungs and the other to the stomach. This permits the animal to swallow food and breathe at the same time. The positioning of the human larynx is lower in the throat. Both the breathing tract leading to the lungs and the food intake tract leading to the stomach are in the same throat area, and they remain together until the latter branches off at the larynx. This explains why humans cough when something falls into the larynx, for this is the only effective way we have of clearing the air passage.

Because of this shared use of the esophagus, there not infrequently occur fatalities in the human owing to a blockage of the windpipe. Lieberman puts the number of such fatalities at tens of thousands each year (ibid.). In the nonhuman animal such a fate is very unlikely, since its larynx is located at the back of the mouth; its positioning prevents the passage of food into the air passage and permits air to flow where it should without being blocked by food or other objects. It also allows the nonhuman animal to breathe and drink at the same time. This arrangement requires, however, that the tongue of the nonhuman animal lie wholly in the mouth and that it be long and thin. The mouth, the tongue, and the position of the larynx of the nonhuman animal are all designed, as it were, to facilitate the intake of food and to transmit it as quickly as possible into the stomach chamber (ibid.).

It is thanks to the position of the human larynx in the middle throat rather than at the back of the mouth—as well as the rounded thick tongue extending back into the throat and the peculiar rounded shape of the mouth—that human speech is possible. This is true not only as regards the formation of the many sounds the human is capable of making, but also with regard to the rapidity with which these sounds can be uttered. Animal sounds are throated and restricted to vowel sounds, principally a, e, and u. The vowels 'i' and 'o', which nonhumans do not utter, require a facial musculature to move the lips and shape the mouth that they lack. Consonants, uniquely human sounds, require an active movement of the lips and of the tongue. Thus the nonhuman primate *cannot* formulate most of the sounds the human can, because it lacks the physical equipment to do so. This explains in part the failure of all efforts to teach primates to express the sounds of human speech, which even animal "speech therapists" now generally acknowledge. A detailed account of the differences and similarities of the various vocalizations of the human and nonhuman primate is provided by Lieberman in *Uniquely Human*. One significant conclusion he draws from his study is that "monkeys and apes, lacking functional neocortical vocal control, likewise cannot produce the muscular maneuvers that are necessary to produce human speech" (p. 85).

The speech organs of the human are admirably designed to provide a rich panoply of tones and sounds to enable the human to express symbolically in sound what has been conceived in the mind. Indeed the entire human facial physiognomy is custom-made to enable the human to utter the sounds essential to human speech. In order for speech to be possible, the mouth *cannot* be elongated, as is the case for most animals; the tongue *must* be round, thick, comparatively short, and extend down into the throat; the larynx *must* be located in the middle of the throat rather than at the back of the mouth; high cheek bones and forehead are needed to permit the musculature essential to a highly flexible jaw and lips. Clearly, all of these characteristics of the human physiognomy are ordered to human speech.

Failing to see thereby an inherent inconsistency, Darwin attributes language to the human and not to the animal on the sole basis of the former's intellective superiority, all the while apparently accepting as true the factually gratuitous assumption that there is no appreciable disparity between their powers of vocalization. At the same time, as seen, Darwin supposes that it is through his use of speech that man's mental powers have developed and been so noticeably enhanced.

Since an understanding of the differences between human and non-human vocalization became generally appreciated only as the result of the efforts of phonologists in the last century, it is not surprising that Darwin attributed the apes' inability to vocalize as humans do exclusively to their lack of intellective development. If one were to assume, then, that the intelligence level of the nonhuman primates had reached a point comparable to that of the human, there would in Darwin's view be no reason to doubt that the primate animal would also enjoy the power of human speech. This we may conclude from Darwin's own surmise:

> The fact of the higher apes not using their vocal organs for speech, no doubt depends on their intelligence not having been sufficiently advanced. The possession by them of organs, which with long-continued practice might have been used for speech, although not used, is paralleled by the case of many birds which possess organs fitted for singing, though they never sing. (*Descent of Man,* Pt. 1, ch. 3, p. 300a)

Yet this circular dilemma seems at bottom endemic to Darwin's evolutionary theory, for the dilemma can hardly be avoided without introducing a view of development that is unabashedly teleological. The felicitous alignment of the human mind with the physical powers of speech, which man obviously possesses, can hardly be a matter of mere fortuitous occurrence, i.e. natural selection. The human is, after all, the only truly intellective animal, and the language he employs is, as Bickerton observes, like no other form of animal communication. "If," he asks "so many species were at a stage of potential readiness for some primitive version of language, how was it that out of all those species only our ancestors attained the goal?" (*Language and Species,* 1990, p. 129). Somewhat ironically, Bickerton, who describes himself as an anti-continuist, offers no direct response to his own question. Nonetheless he fully recognizes the uniqueness of human language, emphasizing its limitless versatility:

> Though in themselves the sounds of a language are meaningless, they can be combined in different ways to yield thousands of words, each distinct in meaning . . . a finite stock of words . . . can be combined to produce an infinite number of sentences. Nothing remotely like this is found in animal communication. (Pp. 15–16)

In addition, Bickerton notes a stark contrast between the limited capabilities of animal communication and the vast communicative potential of human language, the former having merely limited aims which directly benefit animals as individuals or their own troop, herd, or flock, while the latter is open to the unlimited experiences the human has of his outer environing world and of his inner world of consciousness.

> All other creatures can communicate only about things that have evolutionary significance for them, but human beings can communicate about *anything*. In other words, what is adaptive for other species is a *particular set* of highly specific referential capacities. What is adaptive for our species is the *system* of reference *as a whole,* the fact that *any* manifestation of the physical world can (potentially at least) be matched with some form of expression. (P. 15)

Because, then, of the wide gulf separating human language from animal forms of communication, Bickerton is critical of what he refers to as the 'naive continuism' of the conservative Darwinians. These latter, he contends, mistakenly seek a direct link between these two kinds of communication, and, in seeking to build their case, have all but left out of consideration the representational dimension of human language which is its true differentiating characteristic. He states:

> In large part, failure to resolve the Continuity Paradox has resulted precisely from what one might call the 'naive continuism' of the antiformalists, who have tried in a variety of ways to establish a direct line of development from animal communication to human language. Although all their efforts have signally failed to produce a convincing 'origins' story, their rejections of more formal approaches has left them without any viable alternative. (P. 9)

The antiformalist view, as Bickerton refers to it, would all but eliminate any importance for the syntactical aspect of language, placing the major emphasis on its lexicological dimension. He laments the fact the linguists have surrendered the study of primate communication as well as discussions on the nature and origin of language, to the nonlinguists (pp. 105, 108). It is not his contention that the study of syntax alone is sufficient to respond adequately to these kinds of questions, but rather that it does have an important and irreplaceable role to play. On this point he leaves little doubt as to his views: "It seems reasonable to stand the antiformalist position on its head and say that it is quite senseless to study the origins and functions of language without at the same time studying the formal structures that underlie those functions" (p. 9).

Though critical of the naive continuists and antiformalists who underemphasize the place of meaning and syntax in order to narrow the gap between animal communication and human language, Bickerton is himself a 'punctuated equilibria' evolutionist. His view, while continuist in a still fundamental sense, rejects a naive interpretation of evolutionary development, for he sees as undeniable a qualitative difference between human language and animal forms of communication. He speaks glowingly of the power of human language and what it has enabled the human to accomplish:

> Only language could have refined the primitive categories of other creatures and built them into complex systems that could describe and even seem to explain the world. Only language could have given us the power to manipulate those systems through the power of constructional learning, designing futures different from our past and then seeking to make those imagined futures real. (P. 256)

Yet, in all but the same breath, Bickerton still attributes the emergence of this wondrous power of human language to "evolutionary chance":

> Only evolutionary chance triggered, in our ancestors, the emergence of the first stumbling attempts at language. And even these were no more than a haphazard stringing together of meaningful elements, effective in helping our forefathers to exploit the environment more efficiently. (Ibid.)

BICKERTON AND THE EMERGENCE OF LANGUAGE

Bickerton differs from mainstream evolutionary theorists in including a certain measure of purposefulness or teleology. He accepts as a "plausible assumption . . . that wherever life evolves, it will sooner or later produce creatures that think, and communicate—even if they do not look like us—more or less as we do" (p. 253). He thus attributes a particular goal to evolution.

He is critical of the traditional evolutionary view for having focussed mainly on the physical aspects of living things rather on the cognitional, or what he terms "representational systems" (ibid.). By favoring a representational system Bickerton is allying himself with Locke's epistemological theory, whereby what we know directly are the impressions of things and not the things themselves. It appears that he interprets this to mean that human knowledge, by being representational, and thus at one remove from the world represented, allows the human to systematize and arrange its knowledge according to general epistemologial categories. He takes this form of knowing as a marked advance over the grossly pragmatic knowing of the nonhuman animal. He does not indicate how such a higher form of knowledge evolved, other than to cite the varying levels of representational systems at work in today's world as proof of the 'fact' that such systems have evolved and continue to evolve. As to how such systems originally got their start, overcoming the nonrepresentational and purely communicative systems he acknowledges to exist among animals, no hint is given, although he does postulate a primitive kind of protolanguage as the forerunner of present day human language.

In the final analysis, Bickerton's position does not seem to differ appreciably from the original Darwinian theory, if one accepts the view that Darwin did indeed see evolutionary development as inherently progressive, a point, we have seen, some orthodox Darwinians (e.g., Mary Midgley, Stephen J. Gould) will not concede. Though he holds the development of man and of human language to be somewhat teleological, Bickerton does not seek to explain why this is so, though this is an issue which can hardly be held to be inconsequential, if one is seeking, as he purports to be doing, the origin of human language.

Bickerton assumes that, whatever life forms evolve will eventually develop into creatures who think and communicate in a way similar to the way humans do, even though in appearance they may not look like us at all. This is curious because, as seen, even if the nonhuman animal possessed human intelligence, it would still be unable orally to give expression to its knowledge because it would be incapable of uttering the sounds essential to human speech. Thus any creature possessing the power of communicating in a way similar to our way—which could, in short, speak a 'human language'—must also have physical features similar to ours. This is simply a requirement of phonetics. We might well reiterate here that human language, the expression of thought through sensible symbol, requires for its formation a close intertie between mind and body. It is not a matter of pure coincidence that the talking animal, man, possesses the physical features he does.

BIPEDALISM AND HUMAN SPEECH

Some have claimed that the human's bipedalism is a development which occurred simply because primitive man traded an arboreal existence for a terrestrial one. Today this view has been largely abandoned. Bipedalism is, to begin with, 'necessitated' by the human's comparatively large head, which could never successfully pass through the birth canal if the pelvic orifice in the female were not considerably larger than that of any other primate. It is precisely the upright posture which bipedalism allows which in turn makes possible the enlarged pelvic bone of the human female. In short, the human must stand erect and walk on two feet as a matter of course precisely because his cranium is larger, and this is so because, as an intelligent being, the human requires a more sophisticated brain.

The human physiognomy is specifically designed to meet the needs of an intelligent animal. Clearly, it would be altogether disadvantageous to man's development and survival not to have been gifted with the power of reason. The structure of the human ankle, which is the most complex joint not only of the human body but of all known skeletal organisms, is specially designed to withstand the enormous pressure placed upon it by the human body, by reason of the human's bipedal gait and posture. The erect posture of the human results in the full weight of the human body's being placed squarely upon the two feet. The striding gait as the human walks necessitates an extraordinary flexibility in the ankle joint in order that it might quickly and smoothly adjust to the rapidly changing pressures and movements bipedal walking and running entail.

The comparatively broad pelvic bone found not only in the human female but in the male as well, is not an historical accident or freak of nature. It is directly related to the human's bipedalism, for the pelvic area serves as fulcrum and anchor for a massive complex of muscles that are needed for erect posture. The nonhuman primate is unable to stand or walk erect for any length of time because it lacks the strong musculature in the pelvic area needed to hold the upper body in an upright position. Because standing erect is not fully natural to the nonhuman primate, most primates lead an arboreal existence. That is why Jane Goodall, Dian Fossey and others concerned with the survival of the chimpanzee and the gorilla in the wild, urge that the forests providing these primates their natural habitat be left intact. They realize only too well that, without its arboreal habitat, the nonhuman primate could not long survive, since they are not by nature bipedal. On the savannah they would quickly succumb to the attacks of strong, agile vertebrates such as the lion, leopard, and hyena, and they would be deprived of their irreplaceable food source of fruits and nuts.

The human hand, as well, is unique among all living things for its structure. With the wondrous flexibility the compatible thumb provides, the human hand can perform a wide variety of tasks, and what it cannot manage in its natural unaided state, it can manage through the use of tools and machines. Aristotle remarked on this fact many years ago as he referred to the human hand as the tool of tools, because of its unique capacity to adapt itself to whatever object one might seek to grasp or manipulate (cf. *On the Soul,* Bk. III, 8, 432a).

Furthermore, the intricate tasks the human hand can accomplish— for example, carving, doing needlework, painting, the playing of musical instruments, etc.—make it plain how wonderfully well the hand compliments the human mind, which alone possesses the creative power to conjure up the limitless tasks the human's hands are called upon to perform. Yet the versatile hand of the human would not only not be advantageous to the nonhuman animal, but would prove to be a severe, if not fatal, handicap, if the human lacked intelligence. To be of use, the human hand must be at the disposal of an intellect and open to multiple-task performance, which in turn renders its grasp far less strong than the hands of other primates, or the paws of most large vertebrates. Further, a quadruped equipped with the tender hands of a human would be grossly disadvantaged.

MIND AND THE HUMAN BODY

We have previously discussed the unique structure of the human mouth, face and throat, all of which are exquisitely ordered and fashioned for the transmission of thought through vocalized sound. The head of the human, moreover, is larger than that of any other primate, because it must house a significantly larger brain. What, then, the physiological features of the human bring home to us most strikingly is the uniqueness and complexity of the human body when compared to the anatomies of other animals.

The human is not merely a primate with a larger brain; it is not a nonhuman primate plus something else. The human is unique in every respect, and because of the unparalleled intricateness and interdependency of its various parts, all of these had to come into existence at one time. Man could not have evolved bit by bit, since all parts must be in place in order for the human to function as it is designed to function. To say that the human's larger brain developed as a result of his efforts to communicate, as does Darwin, is like saying that man became bipedal because he preferred walking to crawling. This kind of reasoning would be rejected out of hand by the scientist as woefully unscientific and unrealistic, if employed in any other area of scientific enquiry save that of the human, or if proposed by anyone other than scientists themselves. That this manner of reasoning goes largely unchallenged in scientific circles surely poses one of the great academic paradoxes of the modern era. It is refreshing to hear a respected linguist of Derek Bickerton's stature forthrightly assert the failure of Darwinian theory to explain the mental evolution of man, even though he considers himself a Darwinian evolutionist in the mould of Stephen Jay Gould, Niles Eldredge, and Steven Stanley. Bickerton writes: "A century and a quarter after Darwin expounded the mechanisms of physical evolution, the mechanisms of mental evolution are still without a history and without a convincing explanation" (*Language and Species,* p. 4).

Yet, as must be apparent from the foregoing discussion, the distinction Bickerton makes between the mental and physical evolution of the human is not, on the basis of the available data, either justified or satisfactory. The hidden suppositions of such a dichotomization are much too redolent of a Cartesian anthropology to present an acceptable synthesis of all relevant human phenomena. Mind and body cannot be properly viewed as two separate semi-independent entities, but need to be seen as joined, constituting one reality, one nature, one human being.

As Bickerton acknowledges, there are but two ways (according to evolutionary theory) that enhanced life-forms may emerge; i.e., either through the recombination of genes that occurs through sexual selection, or through a mutation directly affecting a gene (p. 7). And when a mutation takes place "the instructions for producing part of a particular type of creature are altered. Instructions for producing a new part cannot simply be added to the old recipe. There must already exist specific instructions that are capable of being altered, to a greater or lesser extent" (ibid.). Applying this line of reasoning to language, Bickerton then concludes that "language cannot be without antecedents of some kind" (ibid.). To maintain the continuity and cohesiveness of evolutionary theory, the antecedent must consist in some form of protolanguage, as previously indicated (cf. p. 128). However, if this be true with regard to language, it must also be true with regard to other aspects of evolutionary development, and certainly to the physiological changes that had to have occurred in order for language to have emerged as a unique human phenomenon. But Bickerton does not extend his critique of evolutionary theory to encompass the physical realm; he accepts the emergence of more sophisticated life forms without their having been prepared for by "antecedents of some kind." This is the paradox that relentlessly pursues the phantom of evolutionary orthodoxy, which pins all its hopes on a fortuitous recombination of genes through sexual selection or a direct mutation of genes effected by an unknown agent. Yet in all such cases there must clearly be genes to be recombined or mutated; the real possibility of change and development must first be there, which is a point, as remarked earlier, expressly made by W. H. Thorpe (*Human Nature, Animal Nature,* p. 57). Nothing can result through the recombination of genes that is not already potentially present in the genes themselves. Whatever is going to emerge from their alliance will not be emerging from 'nothing'. It is quite clear that Chomsky is of like mind, for he terms as fallacious "the belief that the principles of evolutionary biology commit us to some doctrine of 'continuity' of a sort that has never been made at all clear but is commonly invoked" ("Human Language and Other Semiotic Systems," 1978, p. 438).

THE HYPOTHESIS OF A NONHUMAN PROTOLANGUAGE

We have already noted that those seeking to demonstrate a strict continuity between the nonhuman and the human animal by establishing a link between human language and animal forms of communication have met with total disappointment on the level of spoken language. The recognition that animals cannot be taught to speak, however, has not meant a cessation of efforts to show that animals are, nonetheless, capable of language in a not-insignificant way. The focus in recent years has been shifted from oral to non-oral communication.

There has, then, been a concerted effort in recent decades to impart to the higher primates, e.g. the chimpanzee and gorilla, an ability to communicate their 'thoughts' through the silent world of unspoken signs. This form of experimental research was undertaken in the conviction that the nonhuman primate truly inhabited a thought world but was hindered from giving it outward expression by its inability to produce the full range of human sound.

Bickerton is willing to entertain as a real possibility the claim that apes possess a form of primitive protolanguage. Such a protolanguage, he believes, undeveloped as it might be, nonetheless provides a basis for finding some common ground between human and nonhuman language (*Language and Species,* p. 190). He strives to establish as a working hypothesis that human communication originally operated on the level of a similar protolanguage (pp. 191, 156ff.). It slowly graduated from there to the full-blown syntactical language of modern man. His assumption leads to circularity: Bickerton had previously argued that the human mind itself is a development of language usage, but here he maintains that the use of language was rendered feasible by the development of a larger brain.

In the end such circular reasoning can be avoided only at the cost of minimizing the differences between the human and nonhuman mind, to the point where the human's enhanced mental powers owe their origin to the human's use of language in a more focused and amplified manner than that found in the nonhuman animal (p. 257). But this concession, made to salvage a continuity flow between species, so blurs the status of both as to render unintelligible the reasons behind the marked differences between the human's and the animal's language ability, behavior, creative skills, and physical attributes.

THE AMBIVALENCE OF EVOLUTIONIST REASONING

Instances of ambivalent reasoning by those seeking to confer respectability on the evolutionary claim that the evolution of the human from the nonhuman animal was the inevitable result of the working out of sexual selection, are not as uncommon as one might suppose. Mary Midgley attributes the failure of the nonhuman animal to reach the same level of language proficiency as the human to a "voluntary" deficiency. The nonhuman animals allegedly lacked the willpower to advance, "have *not wanted to enough* and so have never developed their powers beyond a certain rudimentary point" (*Beast and Man,* p. 226; italics added). Richard Leakey confidently informs us that "our ability to speak is just one aspect of the evolutionary drive to create a more accurate world in our heads" (*Origins,* p. 204a) and concludes, "This is the heart of the cognitive advancement that created the human mind." Eugene Linden tells us that the language experiments with apes "are changing our notions about the nature of thought and about the difference between animals and humans" (p. 109 "Talk to the Animals", *Omni,* Jan. 1980). Such comments sow confusion, whose source is the lack of a common, stable understanding of what is meant either by language or by mind. As Chomsky sagely observes, "To determine whether music, or mathematics, or the communication of bees, or the system of ape calls, is a *language,* we must first be told what is to count as a *language*" ("Human Language and Other Semiotic Systems," p. 430).

Because such terms as 'language' and 'mind' have no clear identifiable meaning for many, it becomes acceptable to reason even in the most fanciful manner about them while still retaining a semblance of scientific respectability. Yet the failure to clarify terms often results in conclusions that are either ambiguous or trivial or bizarre.

Imprecision in the use of such terms perhaps appears in its more pronounced form among those engaged in instructing the nonhuman primate in the use of sign language. One can admire the tenacious dedication to what must be a most trying and tedious task, but the progress made by those seeking to teach sign language to chimpanzees and gorillas has been, by any standard, painfully slow and significantly limited. Though some of the primates receiving instruction have allegedly managed to 'learn' in the neighborhood of two hundred hand signs, it seems generally agreed by linguists that it would be an imprecise use of terms to call what they learned 'language'. What it appears that these primates have learned is simply signs whose direct referent is individual objects such as 'chair' or 'banana' or some such sensory object.

Bickerton believes that it is logical to assume that the ape has a 'concept' of a 'banana' which contains all members of a class (*Language and Species,* p. 107), but this is to confound sensory image with an immaterial idea. The higher mammals, certainly, have the power of retaining images of things they experience, but these ought not be equated with concepts. Were they possessed of the latter, they would be found inquiring into the nature of the banana, its origin, etc., and would be able to employ words (signs) that refer to the signs themselves and not merely to their immediate, sensible referents.

When Sue Savage-Rumbaugh lays claim that the chimpanzees, with whom she and her associates have conducted numerous experiments over a great many years, are capable of employing terms such as 'tool' or 'foodstuff' even when the referents are not present, this, even if accurate, does not establish that they are thinking abstractly. Though the physical referent may not be actually present, there is nonetheless, I would submit, an imaginative referent present in the animal's consciousness which is evoked by the sign, which perhaps refers in an indeterminate way to its original physical referent. And though this internal 'image' referent can be associated with several or many physical objects, it is still merely a sensory image and not an immaterial concept. One might refer to it as a material universal which, as will be recalled, is the way in which it is labeled by Kant. It is not a strict universal, because its meaning extends only to those few things that have actually been experienced. Thus, when one ape signs to the other that it wants something to eat (foodstuffs), it is clearly 'thinking' of one of the food items that it has had in the past. The referent is there, though in a vague sort of way.

APES AND THE HUMAN CHILD

One cannot be a little bit intelligent. If the apes with whom the above experiments were being conducted were truly capable of conceptual thought or universal ideas in the unrestricted sense, there would be no inherent limitations upon what they could be taught. Rather, their minds would be open to the unlimited horizon of human knowledge. Yet, according to Lieberman's estimate of the linguistic ability of some twenty or thirty apes to whom ASL (American Sign Language) has been 'taught', their ability seems about equal to that of a two-and-a-half-year-old human child (*Uniquely Human,* p. 155). The fact that the ape's progress is fully arrested at this comparatively very early developmental stage for humans is of appreciable significance. At that age the human child has not yet developed its full phonetic powers, and hence would expectedly be unable to make many of the human language sounds correctly. In addition, its brain is still developing and will continue to develop for many more years. And, though the human child has not yet developed all of the physical and mental powers which make full human speech possible, it has, nonetheless, already begun its linguistic odyssey.

Indeed, recent studies of the behavior of human infants strongly suggest that, even as babies, and long before they are capable of formal speech of any kind, the human child is already hard at work learning the language basics of the tongue spoken by its parents. This is the conclusion drawn by Patricia Kuhl, a psychologist at the University of Washington, who has been conducting studies of very young human infants, focusing on their first steps at language acquisition. According to a recent account of her findings appearing in *Life* magazine (July 1993), she has reached what, for many, will be a startling conclusion. According to Kuhl, infants during the first four months of their lives are already in the process of learning words. They are able to sort through the jumble of words they are hearing and pick out those sounds which for them have meaning. Kuhl further claims that from birth to four months "babies are 'universal linguists' capable of distinguishing each of the 150 sounds that make up all human speech."

During the succeeding two months, again according to Kuhl, the infant begins the process of shifting its attention from the full array of sound that can be uttered by humans to those actually employed in their own native tongue. Hence, at the tender age of six months the human infant has already begun to assemble a primitive working vocabulary which it can employ in a passive manner. Nonetheless, these early beginnings of language instruction are not only important but strictly essential to the infant's later being able at the age of twelve to fourteen months to begin utilizing these word-sounds in an active manner by speaking. Of course, this 'protospeech' is bereft of any formal grammar, which can come into play only when the child begins to fashion the words into sentences. These findings of Kuhl are certainly significant, and they confirm what we have hitherto remarked about the nature of language and the physical conditions that must be fulfilled if human speech is to occur.

The difference of vocabulary between the human child and the apes who were instructed in sign language is thus altogether striking. No mature ape has acquired a 'vocabulary' of more than a few hundred words, while the vocabulary of the small child far exceeds that number. Further, during these early years the human child has received no formal linguistic training or education, whereas there is no evidence of chimpanzees or other apes in the wild making use of signs as language. The apes that have learned to communicate in the most primitive way through the use of signs have been taught to do so by humans. Moreover, the child's linguistic skills increase dramatically from year to year, without benefit of formal teaching.

Even more significant than the sizeable vocabulary employed by the child is its ability to speak in sentences. Although during its early years the child's sentences are, by comparison with a human adult's, quite simple, they nonetheless reflect the fundamental syntactic exigencies of its particular language. Though these grammatical rules have not yet formally been taught it, the child spontaneously assimilates them by the simple experience of listening. Only later, in school, will the child be taught to identify the abstract syntactical rules that all along it has unconsciously been observing in forming meaningful sentences.

What must also be emphasized with regard to the child's learning practices is that, as all parents and teachers commonly experience, the child ceaselessly questions why this or that occurred, or why something should be done in this way or that. The child is always eager to learn more, and early on exploits the power of language to obtain further information about what it has experienced. The small child soon becomes a 'talking machine' during its waking hours.

None of this is found to occur with regard to nonhuman primates, even after they are well beyond childhood. For the nonhuman animal, language learning is not a spontaneous undertaking, nor are apes autodidacts; they must be taught by humans to sign. The effort expended by their human teachers is such as to try the patience of Job. Even to impart a very restricted vocabulary is an extremely laborious task, and one demanding much forbearance and ingenuity on the part of the instructors. Nor can the primates truthfully be said to express themselves in sentences, for there is no evidence that the few signs they have learned are ever put together in a manner which could be described as displaying a syntactical order. The apes simply fail to grasp such an interrelationship between signs. Furthermore, the object of the interest exhibited by these primates is uniformly directed toward what might satisfy their immediate needs or wants, and/or their emotional state; it is egocentric in the extreme—which is why such a large portion of the 'vocabulary' they do learn relates to food and such like items, in which they have a consuming interest.

These and other language limitations displayed by those primates who have been the recipients of labored human instruction lead Chomsky to conclude that human language "is outside of the capacities of other species, in its most rudimentary properties" ("Human Language and Other Semiotic Systems," p. 439). Indeed, it seems to be the case that these student primates are learning a set of signals, and not a language at all, unless one wishes to define language in the broadest sense as the mere linking of one sensation with another. The hand signs the apes have been taught to identify and employ differ little in nature from the sound signals given to dogs and other similarly intelligent mammals, which are also capable of learning a fairly large repertoire of voice commands. And in the case of the seeing-eye dog, there is a question not only of their learning sound signals but tactile signs as well.

Furthermore, what indeed the apes are actually accomplishing is difficult to determine, for it is not an easy task to sort out and separate what the researchers working with them actually observe from their interpretation of what is observed. This is in no way to call into question the integrity of the instructors, but simply to point out how very difficult it is under the circumstances for them to maintain the optimum psychological distance from their work which is required to safeguard an objective standard of interpretation. Animals can be trained to respond to signals in most extraordinary ways without their having, withal, acquired any authentic linguistic skill. It is, indeed, the researchers themselves who are the final arbiters of the progress the animal primates are making in acquiring language skills. Nor can there be much doubt but that their instructors are subtly drawn to evaluate their pupils in the most favorable way, for the unspoken personal goal of their work is undeniably to establish 'scientifically' the linguistic skill of the nonhuman primate.

Further, as Chomsky notes, "The system taught to apes and other species differs from human language at the most primitive and elementary level. As far as has been reported, they are strictly finite systems in principle. . . . with no significant notion of phrase and no recursive rules of embedding or structure-dependent operations" (p. 435). John Limber, professor of psychology at the University of New Hampshire, is no less critical of the results of language teaching experiments with the nonhuman primate. He observes, "One curious aspect of the literature dealing with teaching human language to infrahumans is a lack of concern for the actual structures of human language" ("Language in Child and Chimp," 1980, p. 218). He also points out that any normal three-year-old child "has far surpassed even the most precocious ape in language structure" (p. 204), and that the "naming paradigm," which is excessively concerned with the word rather than the sentence, is "widely recognized as a very unsatisfactory model of human language" (p. 200).

Because of the highly complex nature of language, one cannot rely upon the mere external trappings of language alone to determine whether what appears to be language actually is. If we ask a four-year-old a question, to which it quickly replies in a certain way, and from its reply we strongly suspect that the child did not really understand our question, we would normally repeat the question, perhaps rephrasing it. The point being that there are ways by which we can readily ascertain whether or not the child has truly understood our question, i.e, whether the verbal response the child gave really made any sense at all to the child itself. The child could have pretended to understand our question and simply blurted out the first thing that came to its mind. But further questioning of the child will allow us to determine its actual mental state, allowing us to correctly interpret 'what' the child did or did not say, and indeed whether what it said it truly understood. As Cassirer has remarked, "A genuine human symbol is characterized not by it's uniformity but by its versatility" (*An Essay on Man*, p. 31).

The foregoing example illustrates the close tie-in between language and understanding. One who displays no capability of translating the language into deed or action cannot be said to be using language, for the same intelligence that is at work in employing language functions in the performance of a task. In each instance it is insight and understanding that is communicated to the outer sensory world through an ordering of physical components. In the case of language, those components are the sounds or signs made in accordance with grammatical and syntactical rules; in doing or making, it is a series of multiple physical actions coordinated hierarchically, forming a dynamic unity directed toward the achieving of a particular goal. Since intelligence is common to both language and action, interaction between them is possible. Thus one is able to explain in language what it is that one has done or intends to do, and, to accomplish through action what one has proposed.

This explain why it is that all humans are able to speak, and why it is that they are likewise able to learn to perform intricate tasks. And it is precisely here that we see a sharp contrast between the human and the nonhuman primate. Any successful claim that animals are capable of employing language in the human sense must rest on a demonstration of the nonhuman animal exhibiting the same reciprocity of language and action that is observed in the mature human.

Yet animals cannot be induced to perform even the most 'simple' useful tasks that a three- or four-year-old child can perform. Such tasks might include setting the dinner table, drying dishes, hanging up one's clothes, straightening one's room, folding up laundry, helping weed a flower bed, etc. By invoking the 'law of reciprocity' between language and action, we show that any claim that apes or other mammals are capable of language in the human sense can mount to a reasonable level of credibility only when accompanied by evidence that these animals are also capable of performing actions oriented toward language-directed tasks. If, as alleged, animals are able to 'speak' or make visual word signs, then these same animals must demonstrate that they can perform tasks comparable to the simple household chores mentioned above and toward which the appropriate word signs direct then. Only by demonstrating such ability can the animal effectively remove all grounds for suspicion of any involuntary bias on the part of their instructors in evaluating their students' progress.

Duane M. Rumbaugh and Sue Savage-Rumbaugh, well known for their efforts to instruct chimpanzees in the rudiments of sign language, report some success in teaching two apes to perform menial tasks, such as scrubbing their cages, helping to hose them down, and preparing dishes for washing. Very few details are given, however, and the fact that they "help" to hose down their cages suggests that they would not perform this duty if left entirely to their own devices ("Apes and Language Research," 1983, pp. 207–17, esp. 209). It is noteworthy that the Rumbaughs clearly accept the importance of the interconnectedness between linguistic ability and task performance. They recognize that there is little use touting the 'linguistic accomplishments' of the nonhuman primates unless these are accompanied by goal-directed activities of some complexity that are truly learned and not merely attributable to instinct.

Similarly, the assertion that the chimpanzees engage in "preparing dishes for washing" is not particularly informative. The use of the word "prepare" would suggest that this is all they could be coaxed into doing, and is hardly comparable to actually washing the dishes and putting them away afterwards, which a small child would readily do if the watching of a favorite television program afterwards depended upon the successful completion of the task.

The detailed contrast we have been drawing between the comparative linguistic and task-performance record of small children and even the most mature of the higher primates is fully supported by Philip Lieberman, even though he does not reject the Darwinian explanation of these differences. Lieberman states very clearly: "Although they [chimpanzees and other apes] can acquire a limited vocabulary by using sign language, they are incapable of grasping the rules of syntax that even three-year-old human children master. Nor do chimpanzees or other animals ever create works of art or complex devices, or convey 'creative' thought. Nor do they, in their natural state, adhere to the most basic aspects of higher human moral sense" (*Uniquely Human,* p. 1).

It is to be lamented that so many who heroically work with the higher primates, and seek to instruct them in the ways of the human, fail to accept the full weight of their own empirical findings, which so clearly indicate a marked qualitative difference between the world of the human and the world of the animal kingdom.

Clearly accepting the important connection between linguistic ability and task performance, the Rumbaughs recognize that only empirically tested performance allows us to determine the quality of knowledge that lies behind the language signals animals are thought to have mastered. Here again, of course, the only action-forms that are relevant are those that have been learned.

Whatever may have been the hopes and aspirations of those who have conducted studies of the linguistic skills of nonhuman primates, these studies have not narrowed but rather have widened the gap between these apes and the human. Indeed, perhaps unintentionally these studies have served forcefully to draw our attention to the great intricacy of the language phenomenon. That the adult human is able to call up from a vast reservoir of thousands of word symbols those few that are apt and requisite for expressing an inner mental experience and arrange them in the precise order essential to communicating that thought with clarity, is an unbelievably wonderful achievement. No one has ever succeeded in fully explaining or understanding it. And the rapidity with which the human accomplishes this, is, as seen, altogether beyond our powers adequately to express. Nor is the linguistic ability of children of nine or ten years of age any less remarkable, even though their vocabulary be more limited and less recondite than that of the adult human. Even at that early age the extent of their passive vocabulary is singularly impressive, since they are able to follow conversations on a broad range of topics as well as understand commands (though not necessarily accept them) with notable ease.

Finally, in assessing the linguistic competency of youngsters, one should not overlook the fact that they are perfectly capable, where the opportunity presents itself, of learning other languages than the one they learned to speak as a very young child. Among the upper class families of Europe, for example, it was, and doubtless still is, not uncommon for families to engage foreign nannies in order that their young children might learn a different language even before reaching school age. With this arrangement bright youngsters can learn one or even two additional languages while still at a very tender age. There are, of course, various regions in the world where children grow up in multicultural societies and where, as a matter of course, they simultaneously learn to speak two, three, even four widely differing languages.

When we inquire into why it is that children are capable of learning more than one language and, why any human has, given the opportunity, the ability to learn any one among some four or five thousand different languages, we are led back to the single conclusion that all languages, despite differences of syntax, vocabulary and phonetics, share something in common, namely, meaning. Meaning is the common thread which unites all human languages, thus rendering them inter-translatable. Words, however phonetically sounded or syntactically arranged, are alone bonded together through the common intelligibility they integrally express. It is owing to this mental capability—lacking, as experience teaches us, in all but humans—that humans alone are able to converse.

As seen, Richard Leakey has stated that it is social life which plays the important role "in the evolution of human intelligence" (*Origins*, p. 189b). Better had he said that it is intelligence that plays the important role in the evolution of societal life, for it is through language that the seeds of human civilization are first planted and then cultivated. Language is, indeed, not thought itself, but the sensible, symbolic expression of it. Leakey would agree, for with candor he grants, "The externalization of names and concepts, either in coded sounds or in specific gestures, is of course what turns thoughts into language" (p. 204a). Unfortunately, in the next sentence, he compromises his own definition by flatly denying any distinction between the 'thought' of the human and the 'thought' of the other inhabitants of the animal kingdom. "And biologically," he continues, "it is the communication of thought, rather than the thought itself, that separates humans from the rest of the animal kingdom" (ibid.). This is all but a literal restatement of Darwin's view given at the conclusion of his final major work, *The Descent of Man*. For both Leakey and Darwin the linguistic impotency of the nonhuman primate does not in the least seem to gainsay its being basically as intelligent as the human.

THE ORIGIN OF HUMAN LANGUAGE

Earlier we alluded to the question of the origin of language. We conclude our discussion of human language with a brief consideration of this most difficult problem. Relying on scientific and historic data alone, we will perhaps never be able definitively to answer the question. However, for those committed to a Darwinian perspective, the question of the origin of language presents an exceptionally sticky wicket, as already indicated at the beginning of this chapter.

There is a common tendency among Darwinians to see language as the inevitable result of an enlarged cranial capacity of the human or his ancestors on the one hand, and as the product of growing social and environmental constraints placed upon the human and his prehuman ancestors on the other. These, it is alleged, both allowed and necessitated a higher, more sophisticated level of communication. Derek Bickerton, though excepting his own theory of the evolutionary origin of language from this view, sees the phenomenon of language as being "an alarming and embarrassing accident, as much from its apparent lack of evolutionary antecedents as from its unexpected power and complexity" (*Language and Species*, p. 102). As noted earlier, Bickerton puts forward a protolinguistic hypothesis which, as he acknowledges, grew out of his studies of various pidgin and creole languages, which he theorizes to have been the universal antecedent of the 'modern', fully developed syntactical languages (pp. 130ff.).

How this might have occurred remains shrouded in mystery. The postulated move upward from the primitive protolanguage, which is all but free of syntactical rules and only faintly resembles the status of any pidgin language, to the level of a mature human language, seems to rest on the premise that refinement in the human art of communication became either necessary or highly advantageous to the early human. But, in the end, this view seems to differ more in detail than in substance from the original view of Darwin, who rather ambivalently saw the emergence of modern human language as both owing to and the cause of the human's larger brain and more complex nervous system.

Further, what Bickerton seems to have overlooked in drawing the analogy between the more primitive state of human language (protolanguage) and pidgin is that the creation of any pidgin is dependent upon the prior existence of at least one fully developed human language. Bickerton provides an interesting description of how a pidgin language was formed in Hawaii out of the need of Japanese, Chinese,

American, and Polynesian immigrants for a *lingua franca,* since, for the most part, none spoke the others' languages (*Roots of Language,* 1981, chaps. 1 and 2).

But of course, all of the pidgin interlocutors here in question were already fluent in at least one other language, and the pidgin they developed among themselves was a kind of patchwork quilt of words and expressions from two or more of the four languages represented. Pidgins, which begin as nonsyntactical are the product of (and hence dependent upon) the existence of syntactical languages; a pidgin moves irresistibly and irreversibly to the status of a full-blown language having its own unique syntax and grammar. Creole thus becomes the ultimate destiny of all pidgin languages. And this is because the latter were initially pointed in that direction by the syntactical languages of which they were already an imperfect imitation. It is difficult to see how a protolanguage, if it is genuinely, i.e. fully, 'proto', has much chance of emerging out of a melange of raw jungle calls, which, for those making them, appear already to have proved quite adequate to their purpose. Perhaps we should resignedly rest content with the modest assessment of two leading primatologists regarding language's origin, Sarel Eimerl and Irven DeVore, who tell us that the question of how language came to be remains "a total mystery" (*The Primates,* 1965, p. 183).

5 The Human, the Nonhuman Animal, and the Right to Life

In the preceding chapters we have discussed basic questions concerning the human and the nonhuman animal—their origin, their knowledge, their level of freedom, and their ability to communicate. These considerations were intended to serve as a preparation for the further questions to be examined in this chapter. These include the nature of life, the various levels of life, the question of the right to life, and how the lives of the human and the nonhuman animal relate to each other and intertwine. Our immediate intent, then, is to examine the moral dimension of humans, the manner in which their moral status involves rights and duties, and the extent to which moral considerations may find application in the human's dealings with the nonhuman animal.

That an indepth reflection on the moral implications of the interaction between the human and nonhuman animal is timely and warranted is made clear by the vigorous discussion that has ensued during the past few decades regarding the role and moral status of the nonhuman animal in our world. Responding to the not-infrequent instances where animals have been and perhaps still are being mistreated and brutalized, those designating themselves often as 'animal activists' have insistently sought through various means, including forceful intervention, to draw the general public's attention to such mistreatment, and seek definitively to remedy the abuse by corrective legislation.

Within recent years there has arisen a new field of enquiry termed 'ecology' or 'environmental studies'. To say that it is a new field of enquiry is of course not altogether accurate, since ecology is really nothing more than the elevation of the original study of economics to the level of global concerns. Ecology might be said to differ from the study of economics, which traditionally has been viewed more as an objective science with slight emphasis on the human dimension, in that ecologists seem intent on including by design the moral implications of humans' interaction with their environment.

This is evidenced by the fact that those involved in the study of ecological and assorted environmental questions frequently seek to exert moral pressure on the populace to alter its way of thinking about and interacting with the environment, whether in urban, rural, or totally undeveloped areas. This is particularly true when the questions touch on the life and well-being of the nonhuman animal.

THE ANIMAL LIBERATION MOVEMENT

As just suggested, animal liberationists have become very vocal of late in calling for a reevaluation of the place of the animal in the human's world. While some have put forward balanced and moderate proposals, others have made radical demands that would, if adopted, entail almost kaleidoscopic changes in the behavioral patterns of most humans, particularly, but not exclusively, those residing in the first world or industrialized areas of our planet. Perhaps the most far reaching and uncompromising of these is the demand that the nonhuman animal be regarded as a 'person' and accorded most or all of the basic rights we presently and routinely grant to humans. These include, above all, the right to life. Animals are to be considered, it is claimed, *persons* because they are sentient beings, conscious and capable of experiencing pain.

This view is advanced by, among others, Peter Singer, a philosopher originally from Australia but presently on the graduate faculty at Princeton University, who with Tom Regan is among the most widely known and admired of the animal rights apologists (cf. "Animals and the Value of Life," appearing in *Matters of Life and Death: New Intro-*

ductory Essays in Moral Philosophy, ed. Tom Regan, 1980, p. 240). Since the chimpanzee, and perhaps other primates as well, should, according to Singer, be accorded the status of 'persons', they possess the right (plus an accompanying responsibility?) of being counted, along with humans, as moral beings.

To restrict personhood to the human while denying it to the nonhuman animal is, in Singer's view, a form of elitism which we can well do without. Singer has coined the word 'speciesism' to describe this bias in favor of one's own species, which he views as a fault analogous to racism, and, one may presume, to male chauvinism (p. 234). The alternative to species-bias, he concludes, is "to abandon the belief that human life has unique value" (ibid.). He takes the position, then, that "there is no ethical basis for elevating membership of one particular species into a morally crucial characteristic. From an ethical point of view, we all stand on an equal footing, whether we stand on two feet, or four or none at all" (*In Defense of Animals,* 1985, p. 6).

The view espoused by Singer, however radical it might appear, is not without its ardent supporters. This is dramatically evidenced to by a recent issue of the Italian journal *Etica & Animali* devoted exclusively to the theme of *Nonhuman Personhood* (1998, Paola Cavalieri, ed.). In the opening paragraph of the lead article, entitled "Speciesism and Basic Moral Principles," Michael Tooley spells out his thesis: "that all basic moral principles should be formulated in a 'species-free' way: they should not involve any reference to any particular species, nor should the general concept of a species enter into basic moral principles in any way" (p. 5). Tooley ultimately concludes that it would be wrong to confer rights on humans while denying them to animals on the grounds that the latter are not persons. Personhood, he affirms, ought not be restricted to the human alone (p. 35).

Writing in the same issue of that journal, John R. Searle asserts that it would be "breathtakingly irresponsible," based on the knowledge we presently have about the brains of the higher mammals, to conclude that human brains "can sustain intentionality and thinking, and animal brains cannot" ("Animal Minds," pp. 39–40). Earlier he had stated that many species of animals "have consciousness, intentionality and thought processes" (pp. 37–38). Searle expresses bewilderment over the fact that many have denied that numerous species of animals possess these qualities. Not surprisingly, he includes his own dog, Ludwig, within the circle of mindful beings, rejecting all epistemological distinctions between different levels of consciousness (pp. 49–50).

Juan Carlos Gómez, in an article in the same review entitled "Are Apes Persons? The Case for Primate Intersubjectivity," concludes that Apes are indeed persons because they manifest a form of second-person intersubjectivity, which suffices to demonstrate that they do possess intentionality (pp. 58–60). Though he grants that second-person intersubjectivity is not fully equivalent to human intersubjective interaction, since the latter can also communicate in first- and third-person terms, Gómez concludes—ingenuously, it would appear—with the claim: "I am not a person in so far that *I think* I am a person; I am not a person in so far as *another thinks of me* as a person. I am a person in so far as I and another *perceive and treat each other* as persons" (p. 61; italics added). Thus following the lead of Singer on personhood, whether consciously or no, all of the above recognize that in order to justify the granting of rights and prerogatives heretofore attributed only to humans, one must alter the grounds upon which such special consideration is predicated. This they proceed to do by broadening the definition of *person* so as to include at least some species of nonhuman animals. In short, they conclude that *there are persons who are nonhumans.*

Other articles included in this special issue of *Etica & Animali* focus on yet other species of nonhuman animals, variously calling for a recognition of these as persons on the basis of their specifically diverse communication skills. Thus one article entitled "Dolphins and the Question of Personhood," by Denise L. Herzing and Thomas I. White, would argue that Dolphins, too, should be granted the title of persons. Another article titled "An Exploration of a Commonality between Ourselves and Elephants," contributed by Joyce H. Poole, would extend personhood to the elephant. The final article, contributed by William O. Stephens and entitled "Masks, Androids, and Primates: The Evolution of the Concept 'Person'," argues that the historical, core meaning of the term is "neither intensively nor extensively coincident with the concept of 'human being' but is rather species neutral" (p. III). Hence Stephens sees no reason why personhood cannot be attributed to at least some nonhuan animals (pp. 122–23).

The foregoing should make clear why, with regard to the issue of rights, the notion of personhood and its presuppositions are very much in need of a closer analysis. If personhood can be extended to some species of nonhuman animals, the pandora's box seems to have been opened, and it becomes difficult to imagine circumstances which would prevent its being extended not only to primates, but to many or all species of quadrupeds and mammals, and even beyond. In the end, the term 'person' would come to lose its meaning, the distinction between species become impossible to retain, and the entire ethical project of rights be devastatingly undermined. In addressing this very important issue of personhood and rights, it will be found most helpful, I believe, to turn our attention to the case made on behalf of the nonhuman animal as subject of rights by Tom Regan. No one, I believe, has offered a more sustained and challenging argument against the prohibition of extending rights and personhood to the nonhuman animal than he.

Though differing in significant ways from Singer's reasons for recognizing animals as moral beings, Tom Regan's final assessment of the political and 'philosophical' status of animals is in practice very much the same. He concludes that "like us, animals have certain basic moral rights, including in particular the fundamental right to be treated with the respect that, as possessors of inherent value, they are due as a matter of strict justice" (*The Case for Animal Rights*, 1983, p. 329). Since Regan is claiming that animals have certain moral rights owed them in 'strict justice', he is clearly making the further claim, along with Singer, that animals, at least some of them, like humans are persons. Arguing in this same vein, Harriet Schleifer has expressed her opposition to the view that the human species is inherently superior: "The animal liberation ethic," she affirms, "demands a basic shift in moral consciousness, a repudiation of human superiority over other species through force" ("Images of Death and Life: Vegetarianism," in *In Defense of Animals,* ed. Peter Singer, p. 67).

Those advocating 'animal liberation', and there are many, often strive to build their case by playing down the differences between the human and other animal species, thereby claiming to find a common ground in the humans' and animals' sentience, and in their mutual ability to experience pain as well as emotion. From this some conclude that the nonhuman as well as the human animal is rightly viewed as possessed of moral rights, even though the former may not be a moral agent (Regan, *The Case for Animal Rights*, p. 331). As a consequence, it is argued, humans cannot consider themselves to be the masters and, above all, ought not, under ordinary circumstances at least, treat the animal as their inferior, and certainly not as a source of their own sustinence.

For this reason, philosophical vegetarianism becomes, for many animal liberationists, at the very least an ideal to be striven for and, maximally, a duty to be acknowledged and effectively observed by all. The extent to which the higher animals can be said to possess a 'right' to life is a question that has occasioned much give and take among animal liberationists themselves. Though there seems to be no consensus among them as to whether animals possess such a 'right', there does seem to be agreement regarding the obligation the human has to avoid using them as a food source, save under circumstances of dire necessity to sustain life (cf. Regan, *The Case for Animal Rights*, pp. 330–53, and Daniel A. Dombrowski, *The Philosophy of Vegetarianism*, 1984, pp. 121–33). In certain life and death scenarios, exceptions to the vegetarian canon are admissible—even one as committed to philosophical vegetarianism as Tom Regan, will allow them—but such exceptions are held to be extremely rare (p. 351). The issue of vegetarianism is deceptively complex, and hence we shall return to a further consideration of it in the concluding portion of this chapter.

The general thrust, however, of the animal liberation movement aims at revolutionizing the manner in which we humans view the animal world. To accept that we are basically kindred spirits with animals and that our behavior toward them should fundamentally parallel the kind of behavior we recognize as proper and befitting, though not always realized, toward our own fellow humans, is perhaps the fundamental tenet of animal liberationism. Before examining the detailed claims made by those supporting this view, we direct our attention for the moment to the underlying assumptions of this movement, and attempt to ascertain some of the causes of its genesis.

There are, it seems, three phenomena of our contemporary age that have provided an effective stimulus for the active promotion of the animal liberation movement. One of these is the role played by the media. Although the manner in which the animal has been treated during recent decades is arguably not much worse than in times past and more than likely less inhumane, the concerted effort of the movie and television industries vividly to portray and record animal life in its natural habitat, thus bringing it to a massive public audience, surely has had much to do with the dramatic shift in attitude in the general populace regarding the animal world in general. The presentation of animal life in such a natural and direct way, often depicting the hardships inflicted on the animal by the constant incursions of human civilization into its hitherto undisturbed habitat, and the focus on humans' harsh treatment of animals in urbanized society, could not but have stimulated feelings of profound sympathy for and even outrage at the often sad and pathetic plight of many animals. This would prove to be true especially among city dwellers, who on the whole have had little opportunity to experience animal life in the wild, and who likely have even had but scant contact with domesticated animals. The sudden exposure to this new world through the visual media, and particularly to the human's often cruel exploitation of animals, whether domesticated or not, could not but effect a profound sense of revulsion in many. Undeniably, there has been no lack of instances in which animals have been poorly treated and abused by humans, a first-hand viewing of which was bound to arouse justified indignation among those experiencing them for perhaps the first time on the screen.

Some of these aroused viewers would subsequently translate their indignation into concerted action, by joining animal activist support groups and by employing the written word in ardent defense of animals. This is altogether understandable and deserving of respect, even commendation. But the manner and rationale of this defense has not always been without serious defects and shortcomings, for it is often at the human's expense that a case has been built for acquiring justice for the animal. We have noted Peter Singer's contention (and in this he is not alone) that the animal merits our humane concern precisely because 'it', too, is a person, even though not a 'human' person.

The second phenomenon that seemingly contributed singularly to the fueling of the animal rights movement in its present form was the 'continuist' view regarding the human and the nonhuman animal, which constitutes the keystone in Darwinist evolutionary theory. Most animal liberationist philosophers seem enthusiastically to embrace Darwin's continuist theory as related to the differentiation of species. One who also shares Darwin's view regarding the origin of the human is *prima facie* disposed as a matter of course to look upon the other primates as his distant cousins entitled to a respect not unlike that traditionally accorded humans.

The third factor contributing to the recent upward surge of support for the philosophic view assumed by animal liberationists is almost certainly the state in which contemporary philosophy finds itself. Modern philosophy, having been slowly but surely eroded in its underpinnings during the past three centuries, has, in its post-modern phase, reached what many consider to be a crisis status; philosophic certainties are few and far between, and the intellective qualities of the human have become markedly blurred. As a result, the very existence of nature, and hence of human nature itself, is something contemporary philosophers often find they no longer believe in or accept. The state of contemporary philosophy, then, provides a fertile opportunity for the revolutionary upheaval and overturning of many ideas so central to the traditional views regarding mankind and human civilization. At the same time, the continuist theory created a climate favorable to the growth and maturation of other ideals of the sort commonly promulgated by the animal liberationist movement, especially those providing its philosophical underpinnings. Again, however, this does not gainsay everything the liberationist movement has either supported or achieved on behalf of the nonhuman animal. Rather, it merely points out the weakened state of the philosophical framework within which animal liberationists have sought to achieve their aims, many of which any informed, fair-minded person would readily support.

ARE ANIMALS PERSONS?

We return now to the liberationists' contention, brought forward by Singer and others, that the nonhuman animal deserves to be considered a person, and hence is entitled to the same or similar respect we ordinarily accord to humans. The major basis upon which this claim is argued is that the animal is a sentient being possessing a nervous system and thus subject to pain as is the human. Hence both do indeed share the same fundamental nature.

As for the intellective ability of the human, this, it is contended, does not, as seen, differ qualitatively from the sensory powers that the nonhuman animal shares with the human. Singer asserts, for example, that among the higher animals "the situation is the same as it is with humans" ("Animals and the Value of Life," in *Matters of Life and Death*, ed. Regan, p. 224) and Mary Midgley has remarked that "intelligence is a matter of degree" ("Persons and Non-Persons," in *In Defense of Animals*, ed. Singer, 1985, p. 56).

Further, Stephen R. L. Clark is fearful that, if we attribute special reasoning powers to the human which we do not accord to the higher animals, we will be alienating ourselves from other living things. He concludes, "It seems very much more likely that our minds as well as our bodies resemble those of other animals" ("Good Dogs and Other Animals," in *In Defense of Animals*, ed. Singer, p. 46). What the above intend, therefore, when they designate some living thing a 'person' is that it is an individual capable of conscious activity. Since, they argue, animals obviously exhibit through their actions that they are conscious and aware of what they are doing, they, as much as we humans, are entitled to be considered persons. It will be by equating the intellective and the sensory act that the 'alleged' gap between the human and the higher nonhuman animal is all but closed. The human is, at best, a more advanced form of animal. (Midgley, of course, rejects all 'high-low' nomenclature. Differences between things are merely 'differences', and they cannot be separated off into hierarchically related categories.)

But Darwinian evolutionists in practice generally consider the animal to be a higher life-form than the plant; the primates a higher life-form than insects, and the human a higher life-form than the nonhuman primate, even though for them 'higher' designates merely a quantitative and not a qualitative difference. This, as considered earlier, was the view of Darwin himself, who clearly looked upon the evolutionary spiral as mounting ever upward, but did not view the human as the definitive endline of evolutionary development.

All of this talk, then, by the leading proponents of the animal liberation movement and by philosophical ecologists, of 'equalizing' the human and the nonhuman animal is not incidental merely to their ultimate interests, for it is calculated to serve as the quasi-metaphysical platform upon which the edifice of a philosophy of animal 'rights' theory is to be erected. In their view, the restriction of 'rights talk' to the human alone is a flagrant form of 'specieism' or 'anthropocentrism'. The latter term is often employed by animal liberationists to describe the traditional view that only humans can be the subject of rights; that is, possess a life that has an inherent and not merely an instrumental value. It is just this view that is stigmatized as "human chauvinism" by William T. Blackstone in his article "The Search for an Environmental Ethic" (in *Matters of Life and Death*, ed. Regan, p. 303). According to many contemporary thinkers, the continuist view of living things, which constitutes the backbone of classical Darwinist theory, renders both feasible and necessary the reformulation of ethical theory to include the nonhuman animal within the larger circle of beings having natural rights. Indeed, some (cf. Feinberg, "What Sorts of Beings Can Have Rights," *Southern Journal of Philosophy*, 1976) would even argue that rights ought also be extended to plant life.

Yet this view proposing the extension of rights theory to include all living things is not widely held by those promoting animal rights. It does appear, though, to be a view taken by a considerable number of ecologists and environmentalists. The best known proponent of an ethics embracing all animal life is undoubtedly Albert Schweizer, according to whom there should be "no sharp distinction between higher and lower, more precious and less precious lives" (as quoted by Blackstone in "The Search for an Environmental Ethic", in *Matters of Life and Death*, ed. Regan, p. 306).

The call for this expansion of the moral rights theory asks not a minor readjustment of the classical rights theory, but rather a reshaping of its very foundations. Indeed, at stake is the very notion of what it means to be human. This is so because ethics itself is not, *pace* Kant, an autonomous science, but a corollary of a metaphysics of the human person.

Ethics and morality are corollary to the human nature underlying human behavior, for it is always this particular human who lives, who becomes angry, who consoles. What one understands the human being to be has everything to do with one's assessment of whether or not what this individual does is good or bad, praiseworthy or deserving of criticism. It also has everything to do with one's metaethical theory as well as with one's conception of exactly what a right is. Manifestly, the typical animal liberationist's view of rights lies deeply imbedded in the Darwinist account of the human. Daniel Dombrowski openly acknowledges that Darwin "explicitly holds, as anyone must hold who understands the theory of evolution, that 'there is no fundamental difference between man and the higher mammals'" (*The Philosophy of Vegetarianism*, p. 17). Dombrowski adds, with evident assent, that Darwin "demolished the intellectual framework of speciesism," and "signals a revolution in our perception of animals: we are animals!" (ibid.).

One can surely agree with this assessment of Darwinism as regards its devastating implications for "the framework of specieism" even *if* one not accept Darwin's theory as providing a viable synthesis of the human condition. The conclusion, however, that "we are animals" needs to be understood within the context in which Aristotle himself understood it, namely, that we are animals in the sense that the term 'animal' is taken as signifying indeterminately the essence or nature of the human. 'Rationality', however, as Aristotle understood it, does not lie *outside* the essence of human, but is a constitutive characteristic of humanity and differentiates it from the generic term 'animal.' Hence by the statement, "The human is a rational animal," Aristotle is not suggesting that 'rational' relates to 'animal' in the same way that 'two-footed' relates to it. In the proposition 'The human is a two-footed animal', 'two-footed', being a descriptive characteristic of 'animal', does not enter into its essential definition.

But it is precisely the Aristotelian view of the human, later concurred with by St. Thomas Aquinas, that Darwin must deny in order to level out the hills and fill in the valleys separating the human from the nonhuman animal, standing as it does as a formidable obstacle to his continuist view concerning all life-forms. For Darwin, the human is merely another animal. In ascribing to him powers of reasoning, Darwin intends nothing more than a refined and more fully developed kind of sentience. Contrarily, for Aristotle and Aquinas the human is not a being 'composed' of 'rationality' and 'animality'. These are, as such, logical constructs. Rather, the human is a being whose very animality *is* rational. The human's rationality totally penetrates and is suffused throughout his animality; it is not a distinct 'quality' added to it. This union of rationality and animality clearly differentiates the human from all other sentient beings whose animality is not a rational animality.

Understood thus, the statement "the human is an animal" is not acceptable as true *unless* 'animal' is understood as referring incompletely to the full human's essence. Similarly as in the case of water, one cannot correctly say that water (H_2O) is oxygen, unless one understands thereby that, though water contains oxygen, it is not constituted by it alone (cf. St. Thomas Aquinas, *On Being and Essence*, chaps. 2 and 3). In short the human and the nonhuman are not univocally animals, but analogically.

DARWINISM, ETHICS, AND METAPHYSICS

If one accepts Darwin's view of the human, as apparently Dombrowski and most or all of the animal liberationists do, then the intellectual framework of 'speciesism' has indeed been demolished, there remaining no essential difference between the human and the nonhuman animal. But much more than 'speciesism' has then been demolished. Darwinism, as applied to the human, sounds the death knell for philosophy itself, reducing metaphysics to mythical flights of imagination, while restricting ethical theory to an inexorable calculus of utility. *Ethical relativism is the definitive, unavoidable result.*

One ecologist-environmentalist, William T. Blackstone, who, although an animal rights advocate, clearly saw the dangers lurking behind the façade of a utilitarian-based ethics, suggests that perhaps the best place to look for a firm basis for the grounding of an environmental ethic is the metaphysics of Aristotle, Spinoza, or St. Thomas Aquinas ("The Search for an Environmental Ethic," in *Matters of Life and Death,* ed. Regan, p. 312). Blackstone indicates that this is the direction also suggested by Tom Regan and John Rawls, who believe that a reinvestigation of the metaphysics of Aristotle and St. Thomas Aquinas might prove especially rewarding (pp. 324–25). But it is already clear to Blackstone that in order to develop a consistent theory of our environment capable of answering the environmentalists' questions about "value, right conduct, and rights we require some *theory of the scheme of things,* an enlightened metaphysics" (p. 325).

Given the contemporary philosophical climate, these are audacious words and all the more welcome for their seeming rarity among philosophical environmentalists. Not often does one encounter in contemporary discussions of moral or ethical questions any reference to the relevance of metaphysics. The Kantian critique of metaphysics still remains very much a part of the contemporary scene. But one cannot successfully build an ethical theory on the foundation of autonomy alone, as Kant sought to do, nor can ethical judgments be grounded on utility alone, for ethics is not economics. Yet, more often than not, it is precisely utility that has been the chosen path of those supportive of 'rights' for the nonhuman animal.

If we wish to speak of indigenous rights, whether of the human or of the nonhuman animal, we absolutely cannot escape employing a metaphysics of some kind, which necessarily grounds the assumptions from which our argument proceeds, whether or not overtly expressed. To affirm that there is no significant difference between the human and the nonhuman animal is an assumption-laden statement, involving both a theory of knowing, as well as a metaphysics of being and becoming. In his impressive 1939 work, *Language and Reality,* W. M. Urban expressly refers to Darwin's underlying philosophy as one of "naturalism" which has resulted in what Urban terms the complete "naturalization of the intelligence." From this has followed "the naturalization of language with which intelligence is bound up" (pp. 30–31). He adds: "The naturalistic and ultimately behaviouristic view of language which has developed of necessity from Darwinian premises, has brought with it a scepticism of the word, a distrust of language more fundamental than any hitherto experienced." And this in turn has made of language "merely a method of adaptation to and control of environment, and denies to it *ab initio* all fitness for apprehending and expressing anything but the physical, all of those functions which have belonged to it by virtue of its traditional association with reason and with *Geist*" (ibid.). But the very attempt to deny metaphysics ultimately entails its affirmation, for only through an appeal to principles that are themselves metaphysical could the alleged case against metaphysics be carried forward. Like the legendary phoenix, metaphysics continues to reemerge from its ashes. As the eminent Thomist philosopher Étienne Gilson expressed it so aptly, "Philosophy always buries its undertakers" (*Unity of Philosophical Experience*, 1937, p. 306). In a somewhat lighter vein, the trenchant reposte of Mark Twain to an obituary notice he read announcing his own demise—"The report of my death is somewhat premature"—echoes the centuries-old reply of metaphysics to its would be undertakers.

THE BASIS OF ETHICAL THEORY: THE MORAL ACT

Before taking up the question of moral rights, it is necessary first to examine the nature of moral theory and its relation to the human person who performs moral acts, for rights cannot be meaningfully discussed as entities isolated from moral theory. What, then, is a moral act, and what formally distinguishes it from other human activities? How explain the fact that, although all moral acts proceed from a human subject, not everything a human does is a moral act, i.e., characterizable as either good or bad?

We consider an act as moral, first of all, only when one is aware of what one is doing. If a person sneezes or stumbles while walking, no one views these as moral acts, since they are not intended. Though we commend a person for telling the truth and reprimand him or her for telling a falsehood, we would neither praise nor blame anyone who sneezed or accidentally stumbled. The reason is obvious. To perform an act that is in a fundamental sense moral, one must first be aware of what one is doing, and one must likewise be doing what one does freely. Conscious awareness and freedom are indispensable components of any moral act. Whether the moral act, whatever that act may be, is judged to be good or evil is quite another matter.

This much seems to be agreed to by all those affiliated with the animal liberationist movement. None, however, contend that animals are ever blameworthy for any of their actions. That is, they openly recognize that the animal is *incapable of performing a true moral act,* for the animal lacks that degree of knowledge necessary to act morally; to be able to judge right from wrong, before acting.

I hasten to add, somewhat parenthetically, that not all who are supportive of animal liberation actually contend that animals have rights, though they do affirm that the human is bound by certain obligations toward them, yet for some even this is a controversial issue. This, indeed, is the position espoused by Peter Singer and R. G. Frey. For purposes of our present discussion, however, this difference of viewpoint has no significant bearing on the question of the nature of the moral act. Hence these two differing views will be treated as one, inasmuch as both are supportive of the general claim that humans should treat animals with respect. It should be born in mind, nonetheless, that Singer does not accept the notion of natural rights ("Animals and the Value of Life," p. 238) and that R. G. Frey regards rights as "superfluous" ("On Why We Would Do Better to Jettison Moral Rights," in *Ethics and Animals,* ed. Harlan B. Miller and William H. Williams, 1983, p. 289), and as "mere appendages to moral principles" (ibid.).

THE NATURE AND ORIGIN OF RIGHTS

We come, then, to the key issue of 'rights'. What are they? How does one come by them? and who has them? These are the questions we will attend to before confronting the question of whether or not the nonhuman animal can be viewed as the subject of rights. Our plan here is first to lay the groundwork for the discussion by presenting the classical position developed by St. Thomas Aquinas who, of course, relied heavily on the ethical theory of Aristotle in developing his position. This view will then be contrasted with the contemporary animal rights position.

Etymologically, the word 'right' derives from the Latin word 'rectus' which means straight, as in 'straight line'. The noun 'rectitudo', straightness, is the abstract term taken from the adjective and, as applied to a line, refers to the property a line possesses by which it is designated as straight. As this term is applied to a free human act, a 'right' action is one that matches or harmonizes with the nature (acting subject) from which it flows. This explains why, in everyday English usage, we refer to a 'right action' as one that is morally 'correct', i.e., *correctus.*

Etymologically, then, the term 'correct' designates the existence of a harmonious relationship between the subject acting and the activity performed. Put another way, a 'correct' mode of acting is simply an action that 'affirms' or reinforces the subject performing it, since the two are harmoniously aligned, forming as it were, a straight line. Thus an action which is morally correct reaffirms the subject performing it, precisely inasmuch as it is an imperfect but authentic representation of the acting subject. The subject is rightfully, then, considered parent to the action, while the actions themselves are the progeny of the subject. When an act is said to be properly aligned with the subject from which it proceeds, it is considered to be a *right* or *correct* action. This is, then, what is meant by referring to an action as 'morally good'. A morally good action is simply one that harmonizes with the nature of the subject performing it; a morally good action perfects the subject of the action, making the subject be more fully what, fundamentally, it already is.

It should be further noted that a 'right action' is closely related to a just action, though not actually identified with it. All *just actions* are necessarily *right actions,* while it is not true that all right actions are necessarily just ones. A just action is restricted to an act that "accords to others what is rightfully theirs" (St. Thomas Aquinas, *ST* II–II, q. 58, a. 1). Here a just action is taken in its stricter or more proper sense. Thus it is clear why not all right or correct actions are 'just', but all just actions are right, i.e., correct. Thus, for example, to recount something about an other that is true is a "right" action, but necessarily "just." It might harm that person's reputation.

Justice, then, according to Aquinas, differs from the other moral virtues in that it alone is overtly social in nature, ordering one's actions not only to the subject that performs them but to other individual subjects as well (*ST* II–II, q. 58, a. 2). By nature, then, justice entails the notion of equality, and equality can be had only when two or more things are somehow related and compared one to another. Obviously, the same thing cannot be related to itself. Other virtues, such as prudence or temperance, are more subject-directed, in the sense that they aim directly at perfecting the individual exercising them; justice, though, extends outward to the needs of others. Thus, if there were but one individual in the world, there would be no such thing as a just or unjust act, since there would be lacking all basis for needing to treat others fairly. To perform a just act, therefore, there needs to be a certain proportion between the act and the way in which it might affect others. If there is a proper proportion between one's act and others, the act is said to be just, for then the very aim or intent of the act performed is to render to another that to which they have a right.

These rather jejune considerations regarding 'right', 'correct', and 'just' were necessary to prepare the ground for a proper consideration of 'right' employed not as an adjective, as in 'right action', but as a noun, as in 'human rights'. Since in the English language the nominal and adjectival forms of the word 'right' are identical, the two uses are easily confused. Consider, for example, how the following questions, "What constitutes a *right* action?" and "What does it mean to possess a *right*?" can be a source of confusion. This terminological ambiguity can obscure both the similarity between the meanings of the two terms as well as their differences. It will be helpful to bear this in mind as the question of animal rights is considered.

If one inquires further *why* one should act justly, Aquinas's reply is the same as when answering the question why one should be courageous, temperate, or prudent, i.e., why should one lead a virtuous life. The criterion by which all free human actions are judged to be fitting or not is to be sought in their reasonableness. If an act is judged to be reasonable, it is morally good; if unreasonable, it is morally evil. The reasonable act in turn is judged to be such if it is understood to be in accord with man's nature; that is to say, if it is judged to be authentically perfective of the human as a human being.

This in bare outline is the core of Aquinas's natural law teaching. As is evident, it rests firmly on a metaphysics according to which beings are differentiated by their natures or essences, which makes them to be the kinds of beings they are. Ideally, actions are for the perfection of the agent performing them. And, paradoxically, it is the human alone, among all creatures, who is able to act in a manner contrary to his or her own nature; that is, is capable of performing acts that do not promote his or her own development and well-being as a human (cf. St. Thomas Aquinas, *ST* I–II, q. 71, a. 1).

This rather puzzling state of affairs is traceable to the fact that the human is free, able to choose any value, regardless of whether it represents an authentic human good or not. Acts are thus termed disordered or vicious if they do not promote the human's well-being as human, for they then vitiate the whole purpose of the activity of any creature, which is that it be consonant with the nature out of which it arises. Immoral or improper behavior, therefore, is always self-destructive, amounting to a denial of the very well-being of the individual who performs it. As Aquinas comments: "The vice of each thing seems to consist in its not being disposed in a manner congruent to its nature" (ibid.).

Since, then, as seen in the earlier discussion regarding knowledge and reason, the will receives the object of its willing from the intellect, the true ground of the will's being free is the human reason itself. This is because the mind is able to present to the will what has been experienced and evaluated as good or desirable in some particular way, and not good or undesirable when viewed from another perspective. Thus, even the most heinous crimes, such as selling illicit drugs to youngsters, can be viewed by the one selling the drugs as 'good', e.g., as a lucrative source of income, or as a means of gaining control over another person. Such conduct can be deliberately, freely chosen, because it can be viewed as advantageous in some way, though it is manifestly not good in a true moral sense, and can be so understood.

The human alone possesses the intellective power whose act is inherently immaterial. Owing to the immaterial nature of the intellective act, the human is able to transcend the particularities of sensory knowing and to apprehend the nature of the act he or she is performing, and is thus capable of judging the extent to which it conforms to and affirms the very nature of the one whose act it is. This is the manner in which an act is grasped as good or not good. If it is authentically self-affirming, it is good; if not, it is evil. Therefore, the moral quality of acts emerges from their intellective base, for one becomes responsible for a particular act—that is answerable—to the extent that the act is one's own, and this it can be only if one fully realizes what one is doing, and that it is done freely. For this, intellective awareness is indispensable.

Though the animal does possess sensory powers of knowing, these are not wholly detached from the singular, and hence cannot be fully reflective, with the consequence that the animal is unable to grasp fully the nature of its own acts; it cannot, therefore, ascertain to what extent what it is doing is self-affirming since the indispensable measure by which it can contrast the nature of its acts with its own nature is lacking. It is, then, precisely for this reason that the nonhuman animal cannot be held responsible for the acts it performs in the same sense that we attribute responsibility to the human.

Now it may reasonably be asked how it is that we are able to reach this conclusion. How do we really know whether or not the animal lacks the power of intellection, and hence of free choice? The response must be the same as that already given in our previous discussion of the question of freedom in Chapter Three. We are able to apprehend from the uniform nature of its behavior that the animal is not free. The lack of variability found in its highly predictable, patterned behavior is a telltale sign of this same behavior's being controlled by its own nature. Were the acts of the animal free, its behavior would be notable for its randomness and hence unpredictability. As it is, the lives of animals are strictly stereotyped, and hence, to the conscientious student of their behavior, an open book.

Because the human is by nature an intellectual being with the power of self-determination, the human actually controls its own behavior, thus personalizing the direction its own life takes. Consequently, the human can be *aware* of whether or not it is acting *justly*, that is, whether or not the actions it performs are compatible with what it itself is as a human being. Hence, when we speak of 'rights' in the contemporary sense, such as of having 'a right to life' or a 'right to privacy', we are speaking of a moral power which legitimizes, and thus provides moral protection for, those actions needed for the maintenance of life or the safeguarding of our privacy. A 'right', therefore, specifically demarcates an area of behavior that one is *morally* free to incorporate into one's life-plan, i.e., with the assurance that in so doing one is acting justly, which means, in a manner consonant with one's nature. A 'right', then, serves as both moral guarantee and protection of our freedom.

Correlatively, as a human I am aware of others who are likewise self-reflective and self-aware and who are also antecedently undetermined and free. I thus recognize that they must also possess the same basic moral powers that I do. What is fundamentally fitting for me to do is, given similar circumstances, also fundamentally fitting for them. I thereby implicitly acknowledge that I am obliged to respect their freedom. Otherwise I am acting contrary to their nature, which is to be free and self-determining, and, in so doing I act unreasonably. If I deny to others precisely that which I see to be altogether essential to myself, if I am to act as a human, then, since we share the same nature, I am acting in a way contrary to my own nature as well, and this is to act unjustly. In this way, I necessarily recognize my own obligation to respect the freedom of others, for to fail to do so is to deny authentic freedom to myself, since others possess the same free human nature as I.

My own freedom, then, which is the immediate source of my right or moral power to act, is the source of my duty or obligation to respect the freedom of others. In short, it is my nature as an intelligent, free being that grounds my rights to act in a certain way, and which at the same time limits my rights to act in other ways that would involve a denial of the nature of being human in my dealings with others. Thus the moral power to act in a certain way imposes on others the correlative obligation of respecting my freedom. This moral power which legitimizes my acting in a certain way is what we designate as a "right."

RIGHTS AND PERSONS

Rights and duties, then, are corollaries of freedom, and all those who are the subjects of rights are persons, for a person is "whatever subsists in an intellectual or rational nature" (St. Thomas Aquinas, *Summa Contra Gentiles* IV, q. 35). It is because the nonhuman animal does not have an intellective or rational nature that it is not a person and cannot, therefore, be considered the subject of rights. To apply the term 'person' to the nonhuman animal, as Singer and others do, on the basis of its being conscious, is to play word games, since it undermines the true, authentic meaning of personhood and of consciousness. If a living thing is a person simply because it possesses sensory consciousness, then there is little point in referring either to animals or to humans as 'persons' in the first place, since 'person' would add absolutely nothing not already contained in the term 'animal'.

RIGHTS AS A MORAL SHIELD

It is of the nature of a right that it serve as a moral shield; this is its basic function. Freedom is a commodity so valuable that it is without price, and at the same time it is a commodity most delicate in nature. Its exercise is very often dependent on the forbearance and acceptance of others who are also free and empowered to chart the direction of their lives through a process of self-programization. Rights have as their fundamental purpose the protection of freedom, guaranteeing that the individual will be allowed to make his or her own decisions without interference, provided only that the use being made of freedom is responsible, that is, reasonable, not infringing on the ability of others likewise to make responsible use of their freedom. The protection the right provides is not physical in nature but rather moral. Rights therefore, place restriction on others, for one person's right is another person's obligation. Consequently, rights and obligations are correlative notions. One does not exist without the other. There can be no obligation where there is no right, and there can be no right where there is not a corresponding obligation. For this reason rights may fittingly be likened to a moral shield, for they provide protection to those beings who possess the power of making free choices, i.e., are self-directing. The right protects one person from another person's unjust employment of freedom in disregarding the freedom of others.

The rights spoken of here are natural rights—inherent and inborn, not acquired or positive rights. Hence they are inseparable and thus inalienable from the person possessing them. Those who are subjects of rights possess them precisely because they are persons, and not because of anything they have specifically done. As natural rights they are not owing to any particular skills that have been acquired or significant accomplishments achieved. They were not acquired through some fortuitous set of circumstances, but rather 'inherited'. Moral rights do not, therefore, arise out of a 'contextualist environmental ethic' as Karen J. Warren proposes in suggesting that not only human responsibilities but the human nature itself on which these are grounded emerge from the interplay of environmentalist forces (cf. "The Power and Promise of Ecological Feminism," in *Earth Ethics: Environmental Ethics, Animal Rights, and Practical Applications*, ed. James P. Sterba, p. 240).

Since the purpose or nature of a right is the safeguarding of the responsible exercise of freedom by placing moral restraint on the consciences of other free, acting beings, the notions of rights, freedom, and intelligence are intertwined and form a single thread from three separate strands. This is why the call to consign rights to nonhuman animals entails a serious and fundamental misunderstanding regarding the nature of rights.

Though animals possess inherent value, they are not legitimate subjects of rights, for they do not share in a nature that is intellective, self-appropriating, and self-determining. At the same time, however, the human has an obligation to respect animals, not because they have rights, but because the human as a rational being is obliged to act humanely, that is, intelligently and responsibly, being aware of the nature of the acts it performs and their impact upon others, including, of course, not only nonhuman animals but the world of nature as a whole.

Owing to the fact that humans are free, they are capable of disregarding their obligations, misusing the power of choice conferred upon them by nature. In so doing humans act irrationally and unjustly by misusing the power conferred upon them. Unfortunately the nonhuman will often be the victim of such irrational behavior, of human selfishness and cruelty. Many of the causes actively supported by animal liberationists to enact legislation to protect the animal from such offensive treatment are certainly praiseworthy by nature, and deserving of the support of all fairminded humans. It is not, therefore, the legitimate protection of animals to which one can be opposed, but rather the philosophical basis upon which that protection is often alleged. To seek the enactment of protective legislation based on the assumption that animals have inherent 'rights' to such protection is seriously to misconstrue the meaning of 'rights' and to risk seriously eroding the meaningfulness of the distinction between the human and the nonhuman animal.

Finally, the pursuit of such a goal can only contribute to undermining the very meaning of 'right' itself. As has been often repeated throughout the course of this study, it is above all the Darwinist explanation of the origin of the human and the exaggerated emphasis on the alleged biological continuity between the human and the higher animals and, indeed, between all forms and levels of animal life, that set the stage and provided the impetus for the campaign of the animal liberationists to view the human as just another animal. But to move in this direction is wantonly to neglect the profound significance of the human's unique ability to understand, to speak, to build and make things, and to distinguish right from wrong.

ANIMAL LIBERATION AND SPECIESISM

It is often objected, of course, by animal rights supporters that for humans to refer to themselves as *persons* and to the nonhuman animals as *things* is to assume an untoward attitude. By so doing we are said to view ourselves as superior kinds of beings. "Speciesism" is the term animal liberationists often use to characterize such group 'arrogance', implying that such an attitude parallels those of racism and sexism (cf. Singer, "Animals and the Value of Life," in *Matters of Life and Death*, ed. Regan, p. 233). Yet, the parallelism in question is falsely alleged.

Racism is wrong because by it some humans regard other humans as inherently inferior by reason of race, when indeed all peoples of all races share, as humans, in one and the same nature and hence are, as humans, metaphysically equal. Sexism on the other hand, is wrong because by it some humans regard other humans as inferior by reason of gender differences, denying that men and women both share equally in one and the same nature, and hence are equally human. But 'speciesism' is alleged to be wrong because, even though the human and the nonhuman animal differ in nature and not merely according to race and/or gender alone, animals are thought of as though they were not humans.

The underlying fallacy is not hard to uncover. Racism and sexism deny equality of nature where equality of nature actually exists, and are hence based on an untruth. But what is considered 'speciesism' affirms inequality where equality does not exist. The fallacy feeds on the false supposition that what race and gender are to the human, species is also. Yet race and gender are accidental or incidental characteristics that modify one's humanity; they do not constitute one a human. On the other hand, 'species' is not an accidental characteristic, but a substantial determinant of the kind of being one is. Hence, if there are many kinds of beings, then they must differ in an essential way. Some beings must simply be ontologically more than others in order to be different from them. There are no other possibilities. And if one existent differs in being from another, it profits one nothing to deny this difference in kind. It is not being arrogant to bear witness to the truth, provided one's motive in doing so is to allow the truth to appear. For humans to regard themselves as superior to the nonhuman animal for reasons already alluded to, is no more worthy of condemnation than is the individual who thinks that he or she plays the piano better than someone who does not play the piano at all. Those bringing a charge of 'speciesism', then, against those who regard humans as ontologically 'superior' clearly argue fallaciously. Those leveling this charge, by the very fact that they recognize the nonhuman animal as sentient, and, to that extent, conscious beings, doubtless acknowledge the animal to be superior to plants, and both of these to be superior to all inorganic beings. Hence, if the human errs in asserting his superiority over the animal, he will similarly err in claiming the animal to be 'superior' to non-sentient, living things. Indeed, the charge of 'speciesism', as directed against those who consider humans to be superior to nonhuman animals, is as flagrantly illogical as the Parmenidean claim that all 'beings' are one.

Further, there is a quiet irony attached to the epithetical charge of 'speciesism', for animal liberationists making the charge do not themselves appear to admit to the underlying presupposition upon which such a criticism must rest. They do not, that is, grant that there are essences or natures at all, by which things are distinguished one from another, not only individually but by kind. Essence or nature for them is, as discussed earlier, merely a name for individual things. There is for them, therefore, no objective reality underlying the term 'essence'. Given this reading of the world of things we experience, any talk of *specific* differences among things is by definition 'speciesist' laden, since whatever differences there may be perceived to exist in nature are merely accidental or superficial differences. But in such a scenario it becomes meaningless as well as elitist to speak of 'rights' of any kind. It is on such unstable metaphysical underpinnings that the view that the nonhuman animal is not 'inferior' to the human rests. The hidden philosophical assumptions underlying the animal liberationist position run very deep indeed.

UTILITARIANISM AND ANIMAL RIGHTS

One might well surmise that those condemning the view that humans are ontologically superior to the nonhuman animal would straightaway affirm, as a consequence, that animals have rights as well as humans, since they are, after all, really equals. The rationale supportive of rights for the human could, in this view, also be favorably applied to the nonhuman animal. Logical as such an assumption might appear, it is however, not universally the case, for there are some animal liberationists who favor discarding the notion of 'rights' altogether. Peter Singer, for example, one of the first to use the term 'speciesism' to designate the human's natural superiority over the animal, does not himself put much faith in the concept of rights even for the human

("Animals and the Value of Life", in *Matters of Life and Death*, ed. Regan, p. 238). The same can be said of R. G. Frey, who in his article, "Why We Would Do Better to Jettison All Moral Rights" (in *Ethics and Animals*, ed. Miller and Williams, pp. 285–301), sees no purpose being served in speaking about moral rights at all, even a right to life.

Both of these supporters of animal liberation are simply unwilling to accept the basic presuppositions as well as consequences of any natural law theory. For them, all 'moral' issues are to be adjudicated solely on the basis of utility. Since, in their view, things do not really have natures, there could be no such thing as an inherent principle within living things which could provide any kind of measure against which their conduct could be ruled to be fitting or unfitting, proper or improper. Rather, all human actions are to be judged in terms solely of the benefits they might bring—not, be it noted, to the individual alone, but to all of mankind and the whole of nature. In this view, no form of behavior is ruled out of court as a priori wrong, that is, before the human calculus of reason has had the opportunity of measuring that behavior's cosmic fallout. Then and only then can the behavior be ruled either acceptable or unacceptable. This view is commonly referred to as 'act utilitarianism', for according to it, there are no other general principles or rules for evaluating behavior save the degree of utility it contains.

This is a strictly 'consequentialist' position, which means not merely that the consequences of an act play a significant role in determining its acceptability, but that they constitute its *only* relevant criterion for so determining it. It should be apparent, therefore, why Singer and Frey find the notion of 'rights' to be superfluous, for the only admissible criterion for judging whether a particular action is acceptable or not is whether, all things considered, the benefits resulting from it would outweigh the benefits that might accrue were the action not performed. There is in this view no guarantee, therefore, that in certain circumstances any act whatever could be judged to be morally unacceptable. In other words, for the utilitarian there is no such thing as an act that is inherently good or evil, befitting or unbefitting. Under given circumstances any form of behavior can or could be justified.

REGAN AND NATURAL RIGHTS

It is for this reason that Tom Regan, who might well be the most philosophically sophisticated advocate of 'rights' for the nonhuman animal, rejects the utilitarian position as inadequately providing the kinds of guarantee he feels are necessary for the protection of animal rights. For Regan, only if the animal has *inherent value* as an animal, independent of any external utilitarian calculus by which one concludes whether it would be preferable to allow it to continue living or not, is it possible to construct a consistent argument for respecting and sparing its life. In short, the animal must possess something within itself which requires that its continuing in existence be respected, if one is logically to consider it a subject of rights. Regan first examines, as a possible basis for the animal's having such rights, the mere fact of its *being-alive*. But this he finds inadequate to provide grounds for fully supporting such a claim (*The Case for Animal Rights,* pp. 242–43).

Interestingly, Regan readily grants that the animal is not a moral agent, that is, he does not argue for the existence of animal rights on the grounds that animals are responsible and accountable for their actions. He nonetheless considers them to be moral 'patients' in that they have rights, which others, e.g. the human, are obligated to respect. In Regan's view these rights rest on the animals being *subjects-of-a-life.* What he understands by subjects-of-a-life comes down to this: a being possessing some level of consciousness, capable of initiating activity *somehow* dependent on that consciousness, and having a sense of self-identity, at least to the extent that it seeks to protect and preserve its own life. It seems clear that what Regan understands by 'subject-of-a-life' is simply a being that is sentient, and therefore having some cognitive activity by which it acts as a partially independent agent.

In Regan's view it is, then, because animals are subjects-of-a-life that they possess an inherent value, which value is independent "of the utility they have for others" (p. 244). Regan believes that he has distanced himself from the utilitarian animal rights view, for the inherent value he is thinking of is categorical or absolute, admitting of no degrees (ibid.). By being the 'subject-of-a-life', the animal joins the family of moral beings in the sense, not that it is a *moral agent,* but in that it is a *moral patient* which other moral beings have an obligation to respect.

The distinction between moral agent and patient is of crucial importance for Regan's argument. To build his case in order to justify this key distinction, Regan draws upon the example of "young children and the mentally enfeebled of all ages," whom we recognize as lacking the power to make moral decisions, and hence to be moral agents, but whom we nonetheless view as retaining the right to be nurtured and provided for, and hence to be regarded as moral patients (ibid.). What Regan says about our obligations to young children and the mentally enfeebled is certainly true, but not necessarily for the precise reasons he alleges. The parallel Regan seeks to draw between the young, the mentally enfeebled, and the nonhuman animal rests on a fallacious assumption. In appealing to an inherent value residing within a living being which is entirely independent of the use to which it might be put or the actions that it might perform, Regan is aligning himself with a metaphysical view which in the end will bind him to conclusions he is unwilling to accept.

In admitting that living things have natures which, while subject to modification, do remain steadfast in defending their integrity, Regan is repudiating the nominalist view advocated by Hume and the utilitarians, and aligning himself with a natural law view akin to that defended by Thomas Aquinas. Things are what they are by reason of their essences or natures, which differentiate them from other beings, not as individuals, but as the kinds of beings they are. Hence a human *is* a human not because he or she is capable of thinking or choosing or acting in this or that way. Rather, he or she is capable of thinking or choosing or acting in this way because he or she *is human.*

Nature comes first, in the sense that it underlies all the activities a particular being is capable of performing. The activities do indeed provide evidence of the fact that this individual living thing is a human being, but they do not constitute this individual a human being. How we come to know something must be carefully distinguished from its metaphysical status as an existent being. The activities we observe a living thing such as a plant or animal perform do unveil to us the underlying nature of that organism, but these same activities are not the cause of its being the kind of being it is. Rather, the inverse is true. The organism does what it does because of what it is. Consequently, a young child deserves to be nurtured and cared for, not because it can presently give evidence of an ability to think or choose, but because it *is* human, even though it *is presently* incapable of manifesting through its activities its powers of thinking and choosing. The small child, therefore, is not said to possess rights *because* it is sentient and can feel pain and cry. And should one then ask: "But how do you know that this 'child' is human?" the answer can only be: "Because it is the offspring of parents whom we know by their behavior to be human." This is surely a presumptive conclusion that no one would reasonably deny.

On this point Regan seems ambivalent, for, while he convincingly rejects the manifest inappropriateness of the utilitarian view (p. 238), he nonetheless, in the examples of the young child and the enfeebled, allows them to be considered ontologically equal to the nonhuman animal. Thus he argues that since we are quick to grant rights to the small child and the enfeebled, we should logically grant similar rights to the nonhuman animal. But he has mistaken *why* the child is a subject of rights; it is a subject of rights, not because it actually thinks and chooses, but because it possesses the fundamental capacity to do so. It possesses rights because of its nature. Thus, by the fact that rights are accorded to the child, no justification is thereby given *ipso facto* for allowing the inference to be drawn that similar rights are owing to the nonhuman animal.

It is worth noting in this regard that the self-same argument Regan employs to support the rights of animals is often used by those promoting abortion rights to deny rights to the unborn. It is often claimed that the unborn is not a human precisely because it does not exhibit any of the acts normally performed by an adolescent or mature human. The pro-abortionist thus implicitly argues that, just as we are justified in taking the life of an animal, so are we justified in taking the life of the unborn, which is, at the fetal stage of its development, allegedly no more than a nonhuman animal.

H. J. McCloskey has even employed this very argument in support of active euthanasia in those cases where one is irreversibly enfeebled or terminally ill and/or suffering great pain ("The Right to Life," *Mind* 84, 1975, p. 420). If the criterion for granting rights is not *what one is*, but *what one actually does* or can do, then, clearly, all rights are irrevocably conditioned, and are not inalienable.

Though very young children and the enfeebled are not moral agents, since they are incapable of performing a moral act, yet they are moral *patients* and subjects of rights. This is because they are persons. Their personhood is a metaphysical reality directly dependent upon their rational nature. It is independent of their present ability or inability to act in a rational manner, that is, with reflexive consciousness and full freedom. Small children and the elderly enfeebled are, then, moral personalities even though they are not moral agents. In his widely discussed work *A Theory of Justice,* John Rawls grants that infants and children are entitled to "the full protection of the principles of justice" (1971, p. 509). Referring to infants and children as moral personalities, Rawls admits to a certain vagueness in his conception of such a personality, but is of the opinion that it agrees with our accepted and considered judgments, and is essential to avoiding arbitrariness in adjudicating matters of justice. He states:

> The conception of moral personality and the required minimum [for its establishment] may often prove troublesome. While many concepts are vague to some degree, that of moral personality is likely to be especially so. . . . In any case, one must not confuse the vagueness of a conception of justice with the thesis that basic rights should vary with natural capacity.
>
> I have said that the minimal requirements defining moral personality refer to a capacity and not to the realization of it. A being that has this capacity, whether or not it is yet developed, is to receive the full protection of the principle of justice. Since infants and children are thought to have basic rights (normally exercised on their behalf by parents and guardians), this interpretation of the requisite conditions seems necessary to match our considered judgments. (P. 509)

Regan's attempt to establish a basis for animal rights by claiming a parallel between animals, the young child and the enfeebled, whether young or old, does not succeed. Regardless of the fact that the youngster or the enfeebled are unable to act as moral agents, they are, nonetheless, still persons with an underlying, though inactive, capacity to perform as moral agents. This fundamental capacity the animal lacks.

This is not to deny that the animal possesses inherent value, since inherent value is the possession of every being, and not just those that are sentient. Inherent value is not a univocal property possessed by all beings in the same way. Where natures differ, there is by necessity a fundamental difference in the way things are. Consequently, some things simply *are* more than others, and their inherent value varies proportionately. Now Regan, rather than understanding 'inherent value' traditionally, has mistakenly construed it to mean only what Kant understood by it, viz. as applicable only to things that are ends in themselves and hence moral agents (*The Case for Animal Rights,* p. 239). Hence Regan's charge that "the attempt to restrict inherent value to moral agents is arbitrary" (ibid.) is quite justified as a claim, but only against Kant and those following his lead. It misfires when directed against the position outlined above.

MORAL AGENTS AND MORAL PATIENTS

In accepting that animals cannot be taken to be moral agents, Regan sees the need of establishing that they are, nonetheless, moral beings insofar as they are patients. This he claims to show by appealing to their ability to suffer pain, and this in the self-same sense that humans also experience it. "Some of the harms", he states, "done to these moral patients [animals] are harms of the same kind as harms done to moral agents. We cannot consistently hold, therefore, that moral agents and patients can never be harmed in relevantly similar ways" (p. 239).

But in so saying Regan apparently denies that the intellective act of the human is qualitatively distinct from that of the nonhuman animal, for he rejects the view that the intellect is an immaterial power, seemingly identifying that claim as inseparably one with Cartesian dualism. "There is a lesson," he says, "to be learned from Descartes's downfall. It is that viewing the mind as an 'immaterial something,' as a soul, is certain to land us in trouble" (p. 24). His objection to the Cartesian position is the familiar one, that it renders enigmatic the possibility of any true interaction between mind and body.

Regan's disclaiming of the Cartesian view is understandable enough, but he seems unaware that there are mind/body options other than the exaggerated form of dualism put forward by Descartes. Regan's entire case for 'animal rights' rests firmly on the continuist assumption inspired by evolutionary theory, that there is no qualitative difference between the human and the nonhuman animal. And so, finding the Cartesian solution to human consciousness wholly inadequate, Regan asks: "How else might this question be approached?" His response is immediate and direct: "Evolutionary theory provides a significantly different approach to the question of animal awareness than the one offered by Descartes. . . . Darwin, for one, is quite emphatic in denying a privileged status to human beings in this regard." "There is," he continues, "no fundamental difference between man and the higher mammals in their mental faculties." And, Regan further states, "The difference in mind between man and the higher animals, great as it is, certainly is one of degree and not of kind" (p. 18). Regan finally concludes: "One of the virtues of accepting an evolutionary view of the origin and development of consciousness is that it does not commit one to dualism regarding the mind and the body" (p. 24).

By denying that the human intellective act is inherently an immaterial act, and hence of an order different from any form of sensory activity, Regan has laid the groundwork for concluding that animal consciousness does not differ significantly from human consciousness. He thereby establishes what for him is truly the essential goal of his quest, viz., that the higher animals, at least, are authentically moral beings in some way, and hence possessed of rights. By minimizing the differences between intellective understanding and sensory perception, and by neglecting to set forth the rudiments of an epistemological theory to account for these phenomena, so sensitive to the question of 'rights' which he is addressing, Regan has set the stage for his unquestionably ingenious but still flawed defense of 'animal rights'.

Without a clear comprehension of the significant differences separating the two levels of consciousness, intellective and sensory, no

[handwritten margin note: animals seem just as conscious as humans and therefore have the same rights]

meaningful discussion of the issue of rights and the nonhuman animal can logically unfold. As our earlier discussion concerning the nature and origin of a 'right' hopefully made clear, the concept of right itself is a highly abstract notion, which rests on the threefold metaphysical base of intellective awareness, the power of self-determination or freedom of action, and an intersubjective community of persons. Rights obtain among equals, and they designate a moral power of an individual to be or act in a certain way without forceful hindrance. If the distinction between intellective and sensory awareness is overlooked or denied, it becomes superficially feasible either to deny moral rights to the human and the nonhuman animal, or to grant them to both.

[margin note: rights among equals.]

Regan chooses the second alternative, and this puts his case at risk, for it undercuts the validity of any rights argument. Since the animal is conscious, Regan argues, it has beliefs and desires, or, it is at least not unreasonable to assume that it has. Regan finds that "common sense and ordinary language support this, as does evolutionary theory" (p. 78). He regards this as a 'cumulative argument' providing the grounding for a 'burden-of-proof' argument which means, he claims, that "unless or until we are shown that there are better reasons for denying that these animals have beliefs and desires, we are rationally entitled to believe that they do" (ibid.). This passage provides a paradigmatic instance of the manner in which Regan reasons in order to build his case for animal rights. There is the joint appeal to common sense knowledge, which, he grants, is not conclusive (p. 25), and to the continuity among living things required by evolutionary theory.

[margin note: burden of proof: until we see that we don't, they must, we believe that they do.]

For the purposes of his argument, Regan again assumes the reliability of Darwinism to provide a creditable explanation of the origin of higher life-forms such as the human and nonhuman primates (pp. 18–19). By allegedly establishing that the more complex, higher animals possess a conscious life, Regan believes that he has laid the groundwork for concluding that they also possess rights. Yet, since it is doubtful that anyone would deny that animals enjoy some level of conscious life, the point is simply irrelevant as it refers to the rights issue.

Why? Because it is the quality of consciousness that counts here, and, as we have examined earlier, there is a significant difference separating the conscious life of the sense from that of the intellect, and it is precisely because the human is an intellective, and hence free being, that he is possessed of rights.

REGAN, AQUINAS, NATURAL RIGHTS, NATURAL LAW

All of which underscores a curious feature of the foundation upon which Regan builds his case for animal rights. Among the prominent moralists arguing in behalf of the liberation of animals and the recognition that they are deserving of our respect and concern for their welfare, he is, as far as I am aware, the only one committed to validating a 'natural rights' theory for animals.

As seen, Regan seeks to establish the existence of animal rights that are inborn, not acquired, and hence are independent of any positive act of human legislation. As conscious beings, having a life project and being capable of experiencing pain, animals, he concludes, are moral patients, i.e., full-fledged subjects of rights. To this extent Regan has made his own an important aspect of the natural law theory developed and explicated by St. Thomas Aquinas, for whom there are natural rights accruing to every human by the very fact that they are intellective beings. These rights can be reinforced by civil or positive law, but they are not dependent upon such law for their existence. Rather, they inhere in the nature of the human and hence universally possessed by them, including infants and the mentally enfeebled.

But, while accepting this part of the natural law theory of Aquinas and extending it (which Thomas did not do, indeed argued against) to the nonhuman animal as well, Regan rejects the very basis upon which Thomas's natural law theory is grounded. For Aquinas the natural law, which unfolds within human consciousness through the application of reason to those events and lived occurrences constituting one's everyday life experience, is itself in turn dependent upon a higher principle, namely, the eternal law. This latter is one with the rule of governance in God by which he rules the world (ST I–II, q. 91, a. 1).

The natural law Thomas defines as "the sharing by a rational creature in the eternal law" (ibid., a. 2). Without the higher paradigm of the eternal law, of which the natural law is a participated reflection, the latter would lack the necessitating quality that all law requires. Without eternal law, 'natural rights', which follow from it as a strict corollary, would fade entirely from view. Without the grounding of natural law in eternal law, no premise is provided by which one can move from a statement of simple, existential fact to one of strict obligation, i.e., from an 'is' to an 'ought'. The interminable discussions during recent decades which have centered upon the problem of how to move from the "is" to the "ought" bear ample witness to the irresolvability of this question, if the human's own desires and preferences must serve as the ultimate criterion for discerning which modes of behavior are morally acceptable and which are not.

The human cannot impose obligation upon him or herself. The supposition that one can is irremediably self-contradictory. Nor can the human alone determine what is right and what is wrong, or the kinds of behavior that are permissible and those that are not, for by what authority could one support such a claim? Merely by saying that this or that mode of behavior would be the 'reasonable' thing to do could not in itself suffice. On what grounds could one then conclude that one has an 'obligation' to be reasonable? Further, what criterion does one invoke to determine which form of behavior is reasonable and which not? Without a transcending standard by which the morality of an act is measured, one is left floundering about in a sea of incertitude and foggy probability.

Regan unambivalently affirms that what is morally acceptable is not a matter merely of personal opinion. There must be, he acknowledges, objective standards according to which a moral judgment is arrived at, for otherwise the judgment carries with it no more than the 'authority' of one person or another. A moral judgment, clearly, is a command, not a wish or suggestion. "Do not lie," is a moral judgment transcending the view of any single person or group. "I would hope that you would tell the truth," is not such a judgment, for it merely expresses the wish or preference of another, perhaps a parent or an acquaintance. To be sure, Regan is in complete accord with the need for an objectivist account of moral judgment. He is not an emotivist; nor is he a utilitarian, as is Peter Singer, and as the majority of animal liberationists appear to be. He readily grants that "our thinking something right or wrong does not make it so" (*The Case for Animal Rights*, p.

124), but he does insist that moral theory must derive its authoritativeness from 'human reason' alone. To ground it on God or a supreme moral authority would amount, in Regan's view, to basing morality on "an intellectually unsettled foundation" (p. 125), for "whether there is a god (or gods) is a very controversial question" (ibid.). Yet "the difficulties go deeper than this," for he is persuaded that even if a supreme moral authority were known to exist, "problems of interpretation abound." He wonders how are we to know precisely what such an authority means, and, when different interpretations appear, which interpretation is to be considered the authentic one?

Thus Regan seeks to circumvent the question of God altogether in his discussion of rights and morality by setting forth the conditions that we might reasonably assume should ideally be met, if we are to arrive at a correct moral judgment. "There are at least six different ideas that must find a place in a description of the ideal moral judgment" (p. 127), which are: (1) conceptual clarity (2) information (3) rationality (4) impartiality (5) coolness (i.e. emotional detachment) (6) valid moral principles. It is difficult to see how one could quarrel with these characteristics of an ideal moral judgment, if one grants at least that there are such things as objective moral standards transcending the likes and dislikes of individuals. And of the six requirements Regan lists, perhaps only two—rationality and valid moral principles—are likely to be sources of serious disagreement among ethicists.

Indeed, these two characteristics are all but indistinguishable, and may best be viewed as one requirement, for it is only through appeal to a moral principle that the moral judgement is judged rational, and it is only because it is 'rational' that the moral principle serves as a criterion of the ideal (correct) moral judgment. At any rate, to determine what these moral principles are, Regan, not surprisingly, calls upon an intuitive knowledge of some sort. I say "of some sort" because Regan's view here is not wholly unambiguous. What he calls 'reflective intuition' (p. 134) would seem to be the result of a kind of utilitarian calculus (for Regan, anathema); but we have already seen that utilitarianism ignores rather than supports any theory of inherent, objective rights.

Regan distances his position from that of G. E. Moore, for whom intuitions are incapable of proof, and from that of W. D. Ross, who, concurring in part at least with Aristotle and St. Thomas Aquinas, sees moral principles or intuitions as self-evident moral truths (p. 133). For Regan, intuitions are not 'prereflective', that is, self-evident, for they are subject to further testing against our moral beliefs and habitual ways of morally viewing situations. Thus the initial 'intuition', Regan grants, may have to be modified or changed depending on its ability to withstand the scrutiny of reflection. In this way the 'authentic' intuition, one might say, is for Regan 'postreflexive', that is, it has successfully measured up to the demands of the six criteria for effecting an ideal moral judgment (p. 134). Such an intuition is the same as a 'considered belief'. Though it still does not provide an absolute guarantee of correctness, it is highly reliable as a norm and should not be easily abandoned but requires a good argument for doing so.

Regan's concept of moral theory seems to bear some resemblance to Karl Popper's theory of falsifiability as applied to scientific laws or axioms. That is, no absolute claim to a judgment's reliability can ever be made. On the basis of experience it may seem to be true, and it may reasonably be assumed to be so until it is shown to be false or inadequate. This, then, is another way of saying that all moral judgments are at best conditionally true. If this is indeed Regan's intent, it seems difficult to distinguish his position—despite his protestation to the contrary—from that of utilitarianism, for his view seems equally subject to his own prior charge that the utilitarian position is reducible to moral relativism. His frequent denunciation of a relativistic view, as it pertains to moral theory, makes it clear that Regan does not see his own position in that light, for he could hardly be a firm advocate of 'inherent rights', as he most certainly is, nor a staunch defender of philosophical vegetarianism, which he also assuredly is, were he not convinced of the full correctness of the case he claims to make for 'animal rights'.

FAILURE OF REGAN'S COMPROMISE

In discussing these points Regan exhibits no hint of moral tergiversation. Yet his attempt to work out what appears to be a modified natural law theory of morality without grounding that theory on an authority transcending the human itself, seems irrevocably to have dictated the direction his own theory of rights would ultimately take.

Why should anyone, for example, feel obligated to act in a way conformed to the common judgment of others, for reasons other than those of personal utility and advantage? Why should the expressed will of others be seen as mandating one's moral decisions this way or that? What, then, becomes of the meaning of "I ought"? Do I really have to do what others command? Indeed, what is a command? Is a command really nothing more than an expression of preference on the part of the those commanding? Reductively, is not this view equivalent to the emotivist position regarding moral judgments, which relies on feeling as its definitive criterion? And if I seek to dissuade others from doing me harm by invoking my 'natural rights', why should they feel obligated by what for them might seem a mere expression of my own personal preference? What, then, do I ultimately mean by 'right', and whence does it derive? Why do I have a 'right' not to be mistreated, and others have a 'duty' to honor such a right? If one removes God from Regan's "subject-of-a-life" scene, the case for 'moral rights' is quickly dissipated, becoming little more than empty emotional talk or an ingenious way of seeking to secure my own will and advantage. In short, there can then be no convincing argument put forth in defense of the claim that "I, as a person, have a right to life and the pursuit of happiness." Under such circumstances no one is going to feel obligated, let alone motivated, to sacrifice their own will and pleasure for the good life of someone else. And why should they? Unless, of course, because they see some tangible value in it for themselves: But here again the 'morally commanded' act remains ultimately nonobligatory. Some form of utilitarianism, which, as seen, Regan strongly opposes, is the only remaining option for a "moral theory" arising from the pyre of ethical humanism, but it too, is a morality without an authentic, unconditional "ought." The expressions 'natural law', 'inherent inalienable right', and 'inherent duty' fall limp, devoid of meaning, becoming mere empty phrases.

The only kind of law left is 'law' that derives from human covenant, i.e. positive law, whose sole binding power rests in the arbitrarily threatened use of force, either to bring about compliance with the law or to punish the offender in some way should he transgress its prescription. This is the conclusion reached by Thomas Hobbes, for whom the natural condition of man is one of continuing conflict. Only for reasons of self-preservation are laws freely enacted to bring about an armed truce, and either a temporary cessation or a reduction of hostilities. The peace that results is a mere negative peace, being nothing more than freedom from war, but without the essential grounding of mutual respect and esteem for one's fellow man. Such esteem alone can secure an enduring peace, which fundamentally consists in a state of mind and a harmonious disposition of wills. Peace is, in Augustine's classic definition, *tranquillitas ordinis* (the tranquillity of order), and order can only be brought about through justice *(pax opus justitiae),* where the honoring of a right is seen as the object of justice, and law the means by which that right is secured and protected.

NATURAL RIGHTS AND GOD

Regan's theory of 'rights' fails, then, because it strives to establish the existence of rights and duties on an ephemeral foundation. He insists that it is only through "*a consensus among all ideal judges* concerning what principles are binding on all," that rights and duties are created and made known (*Case for Animal Rights,* p. 139). But, as just seen, there is no conceivable way by which one can move by dint of such premises from the 'is' to the 'ought' (and upon the success of such a transition all 'inherent rights and duties' depend) without first recognizing that the human is indebted to another Being for his or her existence and is, consequently, answerable to that Being for their behavior. With God eliminated from the scene, there remains no logical basis for deriving an obligation that is binding on all humans—that is, is inherent (and hence not acquired through the individual's personal effort or desire, or through fortuitous circumstance), while at the same time binding universally. Nor is it possible, Kant to the contrary notwithstanding, that the human can impose law upon him or herself. To think of oneself as originating such an obligation is to empty the 'ought' of its uniquely distinctive meaning. Such a 'personalized' law and/or right would have no more binding force than does a New Year's resolution. But a resolution is not a law, and imposes no real obligation on the one making it.

The point we have been arguing here has been made incisively by

Phillip E. Johnson. Paraphrasing a line of argumentation stressed in a 1979 lecture given by the late Yale University law professor Arthur Leff, Johnson argues that the heart of the problem of attempting to establish objective moral norms is "that my normative statement implies the existence of an authoritative evaluator. But with God out of the picture, every human becomes a 'godlet'—with as much authority to set standards as any other godlet or combination of godlets" ("Evolution as Dogma: The Establishment of Naturalism," *First Things,* March 1993, no. 6, p. 20). Johnson further affirms that all attempts to replace the unevaluated Evaluator with mere human authority in order to establish an objective moral order have proved unsuccessful. Yet he finds that there are no other options available for grounding a moral order because human authority is all that remains when God is removed from the scene (p. 21). Quoting from Prof. Leff's lecture, Johnson finds telling reinforcement for this claim:

> "If the evaluation is to be beyond question, then the evaluator and its evaluative processes must be similarly insulated. If it is to fulfill its role, the evaluator must be the unjudged judge, the unruled legislator, the premise maker who rests on no premises, the uncreated creator of values. . . . The so-called death of God turns out not to have been just *His* funeral; it also seems to have effected the total elimination of any coherent, or even more-than-momentarily convincing, ethical or legal system dependent upon finally authoritative, extrasystematic premises." (P. 20)

Johnson also points out that those who attempt to base moral theory on human authority alone manifest a reluctance even to mention, let alone discuss, questions relating to natural obligations. Such ethicians are, he states, "much more comfortable with the ideal of natural rights than with natural obligations" (p. 21). In fact, Johnson characterizes the impasse at which modernist moral theory has arrived as precisely consisting in its failure to justify the imposition of obligations (ibid.). This critique scores heavily against Tom Regan's theory of rights, which, while granting rights to animals, imposes upon them no obligations. Regan's theory also struggles, unsuccessfully, to justify the imposition of obligation on the human to respect the alleged 'rights' of animals.

The source of the pervasive nihilist atmosphere of the contemporary world is to be traced, Johnson goes on to say, to the fact that
"modernist thinking assumes the validity of Darwinian evolution,
which explains the origins of human and other living systems by an entirely mechanistic process that excludes in principle any role for a Creator" (p. 23). Appropriately, Johnson suggests as an epitaph for the
modernist view the prophetic words of King Lear: "Nothing will come
of nothing" (p. 25).

We have already pointed out the implausibility of Regan's position
where he speaks of inherent or inborn natural rights while denying the
only possible metaphysical justification for them, namely, the existence
of a Being who is the author of the natures of things as well as of persons, to whom one would owe the correlative obligation of acting in
accordance with those natures. What further adds to the incongruity of
Regan's bizarre case in support of animal rights we have hitherto alluded to only briefly. It is now time, finally, to look more attentively at the
effort to derive animal rights from an unlikely source, evolutionary theory.

EVOLUTIONARY THEORY AND NATURAL RIGHTS

On numerous occasions in his *The Case for Animal Rights* Regan
refers approvingly to Darwin's evolutionary view of the origin of living
things, the human in particular. He even concludes in at least one instance that his case enjoys a high level of credibility because it is fully
congruent with evolutionary theory. Regan finds most congenial to his
purpose Darwin's contention that man enjoys no privileged status over
the animal with regard to his level of consciousness, since, as Darwin
alleges, man's overall mental faculties do not differ fundamentally from
those of the higher mammals. The overall difference between the human and the higher animals is not one of kind but of degree only (p.
18). Darwin's contention is much to Regan's liking because it smooths
the way, as he thinks, toward establishing the credibility of his central
thesis—that, just as the human has rights, so also do at least the higher
animals.

The position he espouses, however, is a conglomerate of two incompatible views. Because of the 'continuity' principle, which requires Darwin to minimize the difference between all sentient beings, including the human, while maximizing their similarities, the concept of nature is no longer found to be relevant. Species for Darwin, as discussed earlier, is really merely a name. Evolutionary theory does not refer to differences in nature in any truly significant way, but merely to differences that may be considered strictly incidental. All living things have, after all, according to Darwin, descended from one original living cell, and the subsequent working out of sexual selection accounts for the multitudinous variety of living things on our planet. To speak, then, of essential differences among organisms is, for Darwin, arbitrarily to assign such difference to them, thus assuming a position wholly incompatible with evolutionary theory as well as with the findings of science.

Now Darwin's position cannot help but have far-reaching consequences, as metaphysical views are wont to do. First, there can be no such thing as a natural law by which all things are governed and according to which they develop. As a consequence, there can be no natural or inherent rights. Whatever rights there are must be the result of free human covenant—that is, be positive rights and not natural, inalienable ones. Furthermore, moral theory itself is profoundly affected. We note here that this point did not at all escape the quick mind of Darwin. In his *Descent of Man* Darwin acknowledges that "virtuous tendencies" are "more or less strongly inherited" (Pt. 1, ch. 4, p. 318a). He adds that moral qualities have their foundation "in the social instincts" (Pt. 3, ch. 21, p. 592a). Who, then, possesses these "virtuous tendencies"? Darwin does not hesitate to ascribe them to the lower animals as well as to the human: "These instincts are highly complex, and in the case of the lower animals give special tendencies towards certain definite actions; but the more important elements are love, and the distinct emotion of sympathy" (ibid.). He goes on to say that these social instincts have "in all probability been acquired through natural selection" (ibid.). Then, by a rather deft shift in terminology—which seems somehow symptomatic of the ambivalence underlying evolutionary theory—Darwin ascribes *moral qualities* to the lower animals, while not, however, expressly referring to these animals as *moral beings*. Here Darwin displays a certain hesitation, as though he is not altogeth-

er sure whether or not one ought to consider animals to be moral beings. A moral being, as Darwin defines it, is one "capable of reflecting
on his past actions and their motives—of approving of some and disap-
proving of others" (ibid.). Shortly thereafter, Darwin enlarges upon this
definition, indicating that there are three characteristics attaching to the
moral sense—social instincts, desire for the approval of others, and
highly developed mental powers. It is this last, he reasons, which is crucially important for distinguishing between the human and the lower
animals, since it provides the human with a superior memory, enabling
him to retain past thoughts with great vividness. This in turn, makes
possible the "looking both backwards and forwards and comparing
past impressions." In Darwin's own words:

> I have endeavored to shew that the moral sense follows, firstly, from the en-
> during and ever present nature of the social instincts; secondly, from man's
> appreciation of the approbation and disapprobation of his fellows; and
> thirdly, from the high activity of his mental faculties with past impressions
> extremely vivid; and in this respect he differs from the lower animals. Ow-
> ing to this condition of mind, man cannot avoid looking both backwards
> and forwards, and comparing past impressions. (P. 592a and b)

The manner in which Darwin understands conscience and his accounting for the experience of the "ought" in the human is noteworthy.
He clearly views the awareness of moral responsibility as emerging
from a kind of blindly working calculus of social instincts and past impressions, with the stronger instinct prevailing. Through the inner affirmation, "this ought to be done", the dominant instinct first finds its expression.

> Hence after some temporary desire or passion has mastered his social in-
> stincts, he reflects and compares the now weakened impression of such past
> impulses with the ever-present social instincts; and he then feels that sense
> of dissatisfaction which all unsatisfied instincts leave behind them, he there-
> fore resolves to act differently for the future,—and this is conscience. Any
> instinct permanently stronger or more enduring than another, gives rise to a
> feeling which we express by saying that it ought to be obeyed. (Ibid.)

It is clear, then, that for Darwin it is the human's superior mental
powers of reflection and moral being that most distinguish him from
the nonhuman animal and that ground the recognition of the human as

a moral being. As he states: "The fact that man is the one being who certainly deserves this designation [of moral being], is the greatest of all distinctions between him and the lower animals" (p. 592a).

Darwin, then, restricts the application of the term "moral being" to the human alone, even though he also attributes "moral qualities" to the lower animals. Among the latter he numbers the social instincts of "taking pleasure in another's company," "warning one another of danger," and "defending and aiding one another in many different ways." Such actions he deems as expressive of love and the emotion of sympathy, and these qualities, though found in some of the lower animals, have still not reached as high a state of development as that found in the human (ibid.).

So the difference between being a moral being and possessing moral qualities, is, for Darwin, shadowy at best. To rule out moral qualities for the lower animal would certainly present very real obstacles to his maintaining a firm line of continuity between the two 'species'; granting full moral status to the animal, on the other hand, would strain to the point of incredulity his ability to explain the undeniable differences between the behavior patterns of humans and those of animals. This nebulous characterization of his own position seems endemic to it. There appears to be no real prospect for its ever being successfully overcome.

For Darwin, then, the 'greatest happiness' principle of the utilitarian seems fated to become the ultimate or near ultimate criterion for distinguishing right from wrong. Accepting as true the general statement that "all humans desire their own happiness," Darwin reasons that "praise or blame is bestowed on actions and motives according as they lead to this end; and as happiness is an essential part of the general good, the greatest-happiness principle indirectly serves as a nearly safe standard of right and wrong" (*Descent*, Pt. 3, ch. 21, p. 592b). It also appears that, in the case of the mature individual, public opinion serves as a reliable and adequate standard for distinguishing right from wrong, if it is the opinion of a civilized state or populace (ibid.). Yet Darwin offers no explanation as to how one arrives at the conclusion that the oral convictions of a civilized populace are superior to others, and hence constitute a normative standard for forming correct moral judgments.

Though Darwin is silent on this matter, it is difficult to ascertain how he might arrive at such a discernment other than by again employing the utilitarian calculus. If this be so, Darwin still fails to provide us with the presuppositions necessary to support the claim that "all men desire happiness", and why, further, they then 'ought' to seek it. He appears to accept it as self-evident that the seeking of happiness is either an instinctive quality, and hence not acquired, or a simple factual statement of what indeed all humans happen to desire. Whichever it might be, it presents us with an anomaly of significant proportions within the context of Darwinist theory, given the assumption, that is, that the human descended randomly from lower life-forms. For this view represents these same life-forms as essentially discontent with their present life status. For whatever reason, lower forms reach out blindly to become something they are not, and without the 'slightest idea' of what it is they want to become, or how they are to achieve the 'unimaginable'. This inner urge to transcend and go beyond themselves must be indistinguishable from their pursuit of happiness itself, since both are simultaneously present within the being. Living life-forms long to become what they are not, so that, in effect, what they are truly pining for is their own demise. The desire for happiness turns out to be a 'built-in' tendency which is uncompromisingly suicidal. Such a scenario leads one well beyond the acceptable limits of credulity.

If, on the other hand, happiness for Darwin were to be thought to consist in each organism's retaining its own life status, while developing itself to the full limits of its special capabilities, then one is deprived of any coherent means of explaining how the emergence of higher from lower forms ever occurred in the first place or even why the lower would have wanted to mount 'higher'. The dilemma is very real.

But moral theory does really seem, finally, to entail an 'unnatural' appendage of sorts to Darwinism, even though it is quite understandable that the majority of Darwinists should, like Darwin himself, have turned to some form of utilitarian view in seeking to integrate the moral dimension of human life within the dynamic parameters of a self-energizing world. As we remarked earlier in discussing Regan's theory of rights, it is a hopeless venture to attempt to pry an 'ought' from an 'is' if there is no higher form of being than the human himself to account for the transition.

Only because humans, through their intellective powers, are able to transcend themselves—both in knowing themselves as they know the other and likewise in grasping the inevitability of a correlation between their actions, their own nature, and its fulfillment—can they be conscious of the presence of moral obligation of any kind. Hence, where there are no 'natures', there can be no grounds for forming a moral judgment. As neither possessing this level of reflexive knowledge nor freed from the encumbrance of the sensible world as spatially-temporally conditioned, the nonhuman animal is condemned to remain unaware of the aforementioned correlation. Hence, the animal can never attain to a conscious awareness of the 'ought'. In other words, it is plainly not a moral being. As a consequence, we do not, of course, hold the animal responsible for its actions. But for the Darwinist there are no true perduring natures, so he cannot logically maintain that this or that behavior is befitting a human (or a nonhuman animal) because it promotes the overall well-being of that human (or nonhuman).

DARWIN AND UTILITARIANISM

All of which certainly helps to explain how Darwin recognized a notable degree of affinity between his evolutionary theory and utilitarianism, since for the latter no act is ruled out a priori as unbefitting. Whatever promotes the well-being of the individual or the species can be adjudicated as morally acceptable behavior. Thus, in theory at least, any form of behavior could be declared acceptable or morally good if only more good might likely accrue from it than evil, whatever in this context 'evil' might mean. How one actually arrives at such a judgment, however, is never very clearly explained, as one could with good reason have suspected. Though the utilitarian view has retained some of the aspects of the natural law theory, it has cast aside the latter's most distinguishing and essential element, viz., its reliance on the nature of the human itself as the criterion by which an act is judged to be acceptable. Accordingly, whatever casuistical 'resemblance' between natural law theory and utilitarianism remains is resemblance in appearance only.

So, although the utilitarian view retains some of the vestiges of natural law theory, it is fully in truth a theory of a quite different sort, and it is understandable that it be widely accepted by Darwinists as the most satisfactory manner of explaining the phenomenon of moral discourse. Together with Darwinism, utilitarianism is nominalist in its roots, as it repudiates the concept of nature as being anything more than a name or mental fiction, and does not, therefore, experience any great difficulty in accepting Darwin's view that the human and other animals, while said to differ in species, differ nonetheless only in degree and not in kind. Consequently, the utilitarian experiences no inconsistency in arguing the appropriateness of the human's showing special regard for the nonhuman animal, particularly the primates, since humans, too, are 'animals'.

We have already noted how Darwin seeks to narrow the gap between the human and the non-animal by asserting that, although the animal is not a moral being, it does have moral qualities which at least the higher among them exhibit in caring for and defending their young, showing solicitude and compassion for other members of their own species, as well as for the affection and loyalty they display toward humans who befriend them, etc.

Now, interestingly, the distinction Darwin makes between moral beings and moral qualities is paralleled rather closely by Regan's differentiating between moral agents and moral patients as we have examined earlier in this chapter. It was then also seen, however, that Regan, though accepting of the Darwinist account of the evolution of species, strongly criticized the utilitarian code of ethics, considering it unsuitable for presenting a coherent case for the defense of animal rights, since the moral strictures it imposes turn out, in his view, to be subject to the charge of ethical relativism.

We have also seen how Regan strives to go further than other Darwinists and animal liberationists by arguing for *inherent* animal rights. Not only do animals possess moral qualities, but they are also subjects of rights, which then entitle them to be respected, not for utilitarian purposes alone, but for what they are, viz. conscious, emotional creatures, capable of suffering and subjects-of-a-life. For Regan, the rights of which he speaks are seen as inherent, entailing obligations on the part of the human, not because of what animals do or might do, but simply because of *what* animals are.

But such a view cannot find support from a Darwinian account of the human's moral life, for any talk of inherent rights must necessarily run afoul of the evolutionary conception of natures as emerging. Nor does it win the support of natural law theory, according to which mutual rights or duties obtain only between equals who are at the same time fully responsible for the mode of conduct they pursue. It is not because they are subjects-of-a-sense-life only that humans have rights, as Regan would infer, but because they are subjects-of-an-intellective-life which renders them self-conscious and hence free, responsible subjects-of-a-life. Nor is there need to further explain that this condition applies to babies and small children as well as to human adults, since they too possess a human nature, even though they have not yet reached the stage in their individual development where they are able fully to exploit its natural powers and hence to act in freedom and responsibly.

ANIMALS AND INDIRECT RIGHTS

For the above reasons I find the position advanced by Regan regarding animal rights to be basically incoherent and ultimately unavailing as regards the attainment of its alleged goals. Yet this is not to say that I am in disagreement with or unsympathetic to the motives which inspire him to lay out his case in defense of animals. What Regan wishes to achieve, i.e. a more humane treatment of animals, can, I believe, be attained without an appeal to an inherent rights theory for animals. By the application of the theory of indirect rights, which are a part of traditional natural law theory, coupled with the establishing of 'positive' rights through the enactment of restrictive human legislation dealing with the treatment of animals, his basic aims can, I believe, be successfully realized.

It is wrong to mistreat animals, not because they themselves have a direct right to have the integrity of their lives respected, but because they share, though not consciously, in the natural law. Since the human has the obligation to respect the will of the author of all nature, he is forthwith obligated to treat it respectfully and to make use of it in a reasonable way. As an intelligent being and as the conscious subject of his own actions, as well as the ultimate determinant of what those actions will be, the human ought always to adhere to the rule of right reason. To mistreat animals and wantonly destroy the flora and majestic beauty of our planet; recklessly to consume the minerals contained within it and heedlessly to pollute its rivers and oceans, is, clearly, not to act according to right reason. It is rather to act in a most unreasonable way, and therefore directly entails a denial of the law of nature of which the human, and the human alone among earthly organisms, is fully consciously aware.

Because, however, the natural law to respect nature is applicable to all humans in a most general way, there is still further need for a more detailed specification of the law to assure its reasonable application to the special conditions obtaining in different human societies and at different times, detailing how one ought concretely to treat the nonhuman animal. Such legal enactments are complimentary to the natural law, not opposed to it. It is through such positive, humanly enacted laws, with specific sanctions to support them, that further obligations are incurred by the human. Through such laws the natural law is both buttressed and further modified to suit particular situations, thus aiding in securing compliance with its general strictures.

Through the exercise of their legitimate right to legislate special restrictions governing the human's use of the environment, human societies have at their disposal all the powers that could reasonably be desired to protect the natural environment at all its levels. There is no need to resort to a doctrine of 'inherent rights' for the animal in order to provide adequate safeguards assuring that animals will be treated in a compassionate manner. But of course, as seen, natural law theory, by which the animal world is the recipient of and protected by indirect rights, obligates the human being, monitored by his own power of reason, to act in a responsible way, rests on the shared conviction that there is a supreme, intelligent being who governs the world with wisdom and justice. This rule, be it noted, is not arbitrary, for God cannot be unfaithful to Himself, He who is the ultimate source of truth and justice.

But, as noted, Regan elects not to allow God to enter into his universe of inherent rights, at least not into his philosophic universe, so he seeks to build his case for animal rights exclusively on the limited intuition that as sentient, emotional beings capable of experiencing pain and suffering and resembling humans in many ways, animals possess the same fundamental rights as do the humans to life, liberty and the pursuit of happiness. Whether or not Regan is familiar with St. Thomas Aquinas's theory of natural law, I an unprepared to say, but in his major work *The Case for Animal Rights*, I find no clear indication that he is.

FURTHER WEAKNESSES IN REGAN'S 'RIGHTS' CASE

Perhaps it is only within such a context that the task Regan has set for himself, namely, that of establishing rights for the animals, can be truly understood, for it then becomes comprehensible why he must narrow the gap between the human and the nonhuman animal in order to secure for them the protection natural law theory would afford them. Deep within Regan's mindset there seems to lie (and this appears to be true of other animal liberationists, as well) a lingering sense of resentment against a perceived human hubris which he dubs "anthropocentrism." When one inquires dispassionately into what the latter term is meant to signify, it is found in effect to refer to the view that humans consider themselves as existing on a higher plane than does the nonhuman animal.

There is, as a consequence, a seemingly general reluctance on the part of animal liberationists to acknowledge an irreducible difference between sensing and understanding, and between free and spontaneous activity. This seems traceable in the main to a pervasive conviction that Darwinism has given us the last word on the origin not only of the nonhuman animals but of the human as well. As alluded to earlier, to say that the human is a higher or more perfect kind of being than the nonhuman animal is not to deny that both share many qualities, both biological and cognative. They are both, after all, animals *generically speaking*. But to say that the human is an animal, and that the nonhuman is an animal is not to say that they share the same species. It is, then, illogical to claim that, because humans and nonhumans are both animals, they are therefore equals. The predication of 'animal' of the human and of the nonhuman is not univocal but analogical, signifying two quite different meanings which are, however, proportionately related. The human is an animal whose animality is transfused with rationality. It entails a logical fallacy, therefore, to conclude, as Regan does, that the human and the nonhuman are both animals without qualification.

To further infer that the nonhuman animal is a moral patient simply because it is able to suffer pain physically in a manner similar to the human, is, again, to be guilty of incurring the logical fallacy just alluded to, namely of confounding univocal with analogical predication. Though animals are clearly capable of suffering, they do not suffer as humans do through a consciously reflexive awareness of their pain, reading their suffering against the backdrop of their past and the horizon of their future life. Nor are they capable of the psychic pain the human experiences in recognizing, for example, that what they suffer is owing to an unjust act on the part of someone else, or that their illness or painful condition is terminal, etc. By equating the 'harms', as he terms them, which the human and the nonhuman experience, Regan is able to express considerable indignation at humans' possessing rights to protect themselves when the animal is accorded none directly. His words merit quoting:

> When these common harms are at issue, to affirm that we have a direct duty to moral agents not to harm them but deny this in the case of moral patients is to flaunt the requirement of formal justice or impartiality, requiring, as it does, that similar cases be treated dissimilarly. (*The Case for Animal Rights*, p. 189)

But one does not "flaunt the requirement of formal justice" in treating the human differently, because the nonhuman animal is *not a moral patient*, despite Regan's best efforts to render it so. It is not a human animal. It would be interesting to pursue further Regan's understanding of the phrase, "the requirement of formal justice". One receives the clear impression that Regan is here trading rather liberally on natural law theory, either without recognizing it fully or, perhaps, without willing to acknowledge it. Why, otherwise, should one feel obliged to act "justly and impartially" toward animals unless it be for mere utilitarian reasons. Regan himself offers no better reason than that this view reflects the consensus conviction of sincere, virtuous people. But though such reasons may indeed suffice in a pluralistic society for the enactment of a human law supportive of a like popular conviction, they will not suffice to establish rights and duties which are inherent, and hence simply beyond the reach of all forms of democratic processes. Regan's 'rights theory' proves, then, to be chimerical, for, like the Sphinx, though its head is the head of a human, its body is the body of an animal.

In concluding his critique of indirect duties, Regan announces with a faint hint of triumphalism that, in accord with the 'harm principle' just mentioned, it is clearly arbitrary to deny that animals are moral patients, and hence equally unjust to treat them as though they were not. "The harm principle," he continues, "undermines the credibility of any indirect duty view" (p. 192). Yet Regan fails to see that the indirect duty view he is thinking of here is that laid out by John Rawls, which restricts rights and duties to those capable of freely entering into covenant. Rawls is, understandably, hard put to find a basis for maintaining that infants, small children, and the unborn have any rights at all. But Rawls's "indirect duty" views are not those of natural law theory, which grounds rights, not on the ability of the individual to enter into covenant with another, but rather on the far more stable base that they possess natures which are intellective.

It is on the personhood of the individual that the rights of humans ultimately must rest, if we are speaking of natural or inherent rights, and not on biologically or historically determined factors such as gender, race, age, ability, or accomplishments. And thus, Regan not to the contrary, to affirm the existence of rights for small children, infants, the unborn, etc., and to deny them to animals (who, recall, do not qualify as persons simply by reason of their being capable of suffering, which constitutes a necessary but not a sufficient condition for personhood), is not at all "a symptom of moral arbitrariness" (p. 193). Rather it is but a recognition of plain ontological fact, viz. that nonhuman animals are not persons. Regan's parting claim, therefore, that "Since by definition, *all* indirect duty views deny that we have any direct duties to moral patients, no indirect duty view can provide us with an adequate moral theory" (ibid.), is factually in error.

VEGETARIANISM AND ANIMAL LIBERATION

Many of those sincerely interested in changing, and hence improving, our attitudes and practices toward the nonhuman animal, also actively promote vegetarianism as an expression of the appropriate moral response to what they view as the shameless exploitation of our less gifted "brethren". Vegetarianism, which in its strictest form mandates abstention from all forms of flesh meat (fish, fowl, reptile or mammal), is not a recent phenomenon. It has been for centuries not an uncommon practice in the Far East. Vegetarianism has also had its practitioners in the West, beginning with the early Greek period, as Daniel A. Dombrowski shows in his provocative work, *The Philosophy of Vegetarianism,* in which he briefly traces the history of vegetarians in the West. Among the Greeks the best known of the philosophers who professed and advocated vegetarianism were Pythagoras, Plotinus and Porphyry.

Vegetarians are distinguished one from another both qualitatively and in kind. They vary in the motives for their abstinence, as well as in the definition of "flesh meat" used. Some, for example, abstain from eating most or all varieties of flesh meat strictly for religious motives. Among these can be numbered the Hindus, for whom cattle are sacred animals. This would also seem to be true of other sects that accept some form of reincarnation, and therefore believe that the souls of their ancestors dwell within at least some species of animals.

Other religiously motivated vegetarians abstain from meat, not because their religious beliefs require it, but because they wish this gesture to serve as a forceful sign of their religious commitment. There have always been Christians who have been vegetarians for this reason. Among the more penitential contemplative orders of the Roman Catholic Church, e.g., the Trappists, Cistercians, and Discalced Carmelites, it remains to this day a universal practice.

Others practice vegetarianism for what can be called aesthetic and health reasons. They view abstinence from meat as promotive of an overall state of physical fitness, and/or contributing to their bodily comeliness. This form of vegetarianism has become much more common in recent decades.

Yet another form of vegetarianism is best qualified 'philosophical'. Closely allied to the question of animal rights, this last form presently commands our attention. 'Philosophical' vegetarians abstain from eating meat not for religious or health reasons, at least not primarily, but rather because they feel ethically obligated to do so. To use animals as a food source is for them to violate the animal's right to life as moral patients. Many of the animal liberationists are practicing philosophical vegetarians. They feel an obligation not only to abstain from meat themselves, but to seek converts as well among those not yet formally committed to the avoidance of flesh meat in their diet. The philosophical vegetarian regards abstinence from flesh meat as a serious moral issue.

From what has been said previously it should come as little surprise to learn that Tom Regan ranks among the most outspoken of the philosophical vegetarians. Vegetarianism (philosophical vegetarianism) as Regan appears to understand it (his position is not free of ambiguity) is obligatory on all, because the modern 'animal farming' system is unjust. It is unjust because, Regan claims, it "requires treatment that violates the respect principle, requiring, as it does, that individuals with inherent value are to be treated as if they lacked any independent value of their own and instead had value only relative to the interests of those who engage in the practice or to the preferences of those who support it" (*The Case for Animal Rights*, p. 343).

This position entails ambivalency owing to the fact that, though it is allegedly because animals have inherent value that they have a right to respect, the continual focus of Regan's polemic against nonvegetarians is *the unjustness* of the contemporary method of animal farming. Granted, it is unrealistic to assume that any system involving the deaths of large numbers of animals could be so changed as to remove all of the injustices Regan and others object to in contemporary animal farming practices, it is still meaningful to raise the question on the purely theoretical level. That is, what if these alleged injustices were removed? What would Regan's position toward the meat packing industry then be? Would it then be permissible to use the animal as a food source? I find in Regan's extended treatment of the evils of meat eating no direct response to this question.

Yet answers to several related questions might help to clarify Regan's position. If Regan were, for example, hiking in a remote, mountainous area, would it be morally permissible for him to fish in lakes or streams along the way, or to hunt for game such as rabbits or ducks to provide for his evening meal? There would be in this instance none of the excesses of neglect or cruelty associated with the slaughterhouse. If Regan responds negatively, then clearly his opposition is not limited to current animal farming practices but extends to killing animals for food. The suffering the animal undergoes would not, then, be what formally renders their death an immoral act, but rather the fact that they are deprived of rights they possess as animals in having their lives terminated merely for reasons of satisfying human hunger.

For Regan, the animal's right to life is a *prima facie* right and not an absolute one (*The Case for Animal Rights*, p. 330). There are instances, he grants, when the taking of the life of an animal can be justified, although such life-taking can never be just if it is administered as punishment, since the animal is not a moral agent (p. 331). If the animal's presence, however, poses a threat to a human, one would be justified in defending oneself, even to the point of killing the animal, but Regan sees this as merely an extension of the right of self-defense, which includes the animal as well as the human. According to this principle, even the taking of innocent life is permissible, provided the innocent party, without intending it, poses by its action, real or threatened (but not merely by its presence), a serious threat to one's life or bodily integrity. But apart from such situations or others in which one is faced with the option of either killing an animal or of starving to death, the animal's inherent right to life nullifies in advance, according to Regan, any alleged right the human might be thought to have to kill the animal for food.

If, however, the killing of animals for food in ordinary circumstances is always wrong, why does Regan find it necessary to restrict his criticism of animal farming with the added qualification, "as currently practiced"? This qualification seems to grant that, if certain aspects of present day animal farming were changed, it could, at least in theory, become morally acceptable practice. Regan introduces the above qualification in at least two separate instances. He states:

> Since, for the reasons given, the *current practices* of raising farm animals for human consumption fails to treat these animals with respect, those who support this practice by buying meat exceed their rights. . . . Since . . . the practice of raising farm animals, *as presently conducted,* routinely treats these animals in ways that are contrary to the respect they are due as a matter of strict justice, that practice violates the rights of these animals. (P. 346; italics added)

Whatever the precise meaning Regan attaches to these statements, elsewhere his condemnation of animal farming is sweepingly categorical, concluding that the alleged injustice endemic to it demonstrates the need to recognize vegetarianism as morally mandated, and not merely as a showcase gesture of consolidarity with the lot of the exploited animal. "Vegetarianism," he states, "is not supererogatory; it is obligatory" (p. 346). This conclusion can rest only on the alleged inherent dignity of the animal, and not on a complex calculus of the type advocated by the utilitarian, whereby the good achieved by the vegetarian's choice is finally seen to outweigh the negative consequences accruing to the rancher, the meat packer, and other allied industries. Such calculations inevitably bog down in a veritable sea of economic factors, rendering it all but impossible to reach a definitive conclusion one way or the other. "The case against the animal industry," Regan argues, "does not stand or fall, according to the rights view, on any individual's knowing the aggregate balance of good over evil for all those affected by allowing factory farming or by not allowing it" (p. 351).

It is ultimately, then, because the animal industry 'violates' the rights of the animals, that, for Regan, it is wrong to purchase their products and "why vegetarianism is morally obligatory" (ibid.). Regan calls for nothing less than the "total dissolution of commercial animal agriculture as we know it, whether modern factory farms or otherwise" (ibid.). This sweeping condemnation of animal farming would seem to settle in a definitive manner Regan's earlier-noted ambivalence concerning this issue, for by condemning as inherently wrong all forms of animal farming for purposes of providing food, Regan *is* affirming that it is not only because of the alleged flagrant evils attendant upon much of modern factory farming that this practice is rendered immoral. Rather, *it is a practice that is inherently evil because it always involves a violation of an animal's rights.*

In presenting his case for animal rights and, as he sees it, the concomitant obligation to adhere to a vegetarian diet, Regan makes it clear that he is speaking only of animals that have reached the age of one year. He needed to make this concession because his theory of animal rights was to rest ultimately, not on metaphysical but on what are properly psychological grounds, since, as we saw earlier, animals have rights because they are *conscious subjects* of a life. It is because the animal is the actual subject-of-a-life that it has a right to pursue that life to its natural conclusion without interference from members of the human species. But newborn mammalian animals are not subjects-of-a-life as Regan understands that term. Consequently, they are not candidates for the rights he ascribes to adult animals, i.e., those one year or more old. Yet, in order also to accommodate the young animals under the umbrella of rights that shields the adult animal from abuse, Regan quietly seems to play down the psychological dimension of the subject-of-a-life basis for rights, appealing more now to their natural physical endowments. Here Regan, whether consciously or not, is edging closer to grounding animal rights on an inherent nature, although he seems not to acknowledge this expressly.

Regan grasps the incongruity in his granting rights to parent animals and refusing them to their offspring. He is willing to grant, then, that in the case of the immature animal, "Because we do not know exactly where to draw the line, it is better to give the benefit of the doubt to mammalian animals less than one year of age who have acquired the physical characteristics that underlie one's being a subject-of-a-life" (p. 391). Indeed, he goes even farther, for he feels that we ought to extend the benefit of doubt to *unborn mammalian animals* as well, treating them 'as though' they had these rights. He draws an express parallel between the unborn animal and the unborn human. Regan feels that he must make this concession in order to root out the many forms of injustice the animal is subjected to through animal experimentation. He reasons that:

> To allow the routine use of these animals for scientific purposes would most likely foster the attitude that animals are just "models," just "tools," just "resources." Better to root out at the source, than to allow to take root, attitudes that are inimical to fostering respect for the rights of animals. Just as in the analogous areas of abortion and infanticide in the case of humans, therefore, the rights view favors policies that foster respect for the rights of the individual animal, even if the creation of these attitudes requires that we treat some animals who may not have rights as if they have them. (*The Case for Animal Rights*, p. 391; italics added)

Regan's prohibitions against the use of animals for experimentation and against their use as a food source obviously have the same fundamental ground, viz., the animal's inherent rights. In his view, as seen, animals have moral rights because they are moral patients, and they are moral patients because they enjoy sentience and are thus conscious beings capable of suffering, all of which constitutes them subjects-of-a-life. Regan and other animal liberationists, including those who do not accept his position regarding rights, build their case for respect for animals by emphasizing the similarities between them and the human. Placement of the animal within the elevated circle of personhood causes the specific differences between the human and the nonhuman to all but fade from view.

This leveling process, however, quite ironically, decidedly separates the human and the nonhuman animal one from the other. Because the nonhuman animal has a life of its own to lead, and built-in goals to attain which are proper to it alone, it should be allowed to go its own way and not have to suffer interference from its more gifted, perhaps, but nonetheless related cousin, the human animal. As parallel lines that never meet, the lives of the human and of the nonhuman animal are viewed as ends in themselves, following separate courses. Their worlds need never intersect in ways that are equally beneficial to both.

Though it is undeniable that there have been and continue to be instances of shameless cruelty to animals which no sane person would attempt to justify, animal liberationists, in seeking to rectify these abuses, have charted a course which can only have serious negative consequences for the human and for the nonhuman animal, as well as for the ecology of our world. To flatten out the differences between living things, especially between the human and nonhuman animal, is nothing but an exercise in futility, for it is to fly in the face of fact. Things *are* differently, and among living things and the environment there is the most intricate pattern of inter-dependency which ecologists are now so painstakingly and dramatically helping to draw to our attention.

In the world of nature specific living forms are obviously not ends in themselves, and predation, "the law of the jungle," is the law that makes possible the very existence of manifold species of living things. Clearly, many species of animals are dependent basically on another or several other species for their sole or main food source, and hence for their survival. Bats, e.g., depend on insects, penguins on krell, whales on plankton, lions on antelope and other quadrupeds. Should the food source of any of these predators be depleted, they are simply finished as a species. The ecological food chain is hierarchical, and it makes no sense to affirm that any species of animal is an end in itself, for in the wild all but the herbivores feed off of members of other species, and there are very few of the latter that are not in turn hunted as prey by other, usually larger, carnivores.

To claim, as Regan and other philosophical vegetarians do, that it is wrong for the human to kill animals for food when animals killing animals daily is a practice essential to the survival of the predator animal, is to exceed the limits of serious dialogue. Even the two most common of domesticated animals retained as pets worldwide, the dog and the cat, are by nature carnivores! How ironic that some contemporary vegetarians should chide Christians for not exercising charity toward animals, refusing to use them for sustenance, while themselves emphasizing that *we too are animals* (cf. Dombrowski, *The Philosophy of Vegetarianism*, p. 17).

Indeed we *are* animals, and precisely as a consequence we have need of interacting in a very physical, down to earth way with our environment in order to sustain our own human life. Each day we must convert some small portion of our surroundings into ourselves through the metabolic process common to all living things, but most particularly to all sentient beings. Since many insects are carnivores, those animals that do not feed on other animals as a matter of course do not represent an overwhelmingly large proportion of the total species population of animals. And the reason many animals are carnivores is clear: nature has not endowed them with the capability of feeding on grains and grasses nor with the tools needed successfully to digest them.

Just as the feline species is incapable of chewing and digesting the grasses that the herbivore feeds on, so the herbivore is generally incapable of tracking down game and subduing it. Without sharp fangs and canines it is unable to tear away flesh meat in small portions and hence ingest it. But nature has endowed the human with a balanced admixture of the abilities of both the herbivore and the carnivore animal. Like some species of animals the human is all but omnivorous, being able to eat and digest grasses, grains, and flesh meat. As an animal, the human is constrained, as all living creatures are, to eat in order to survive. Why, then, it should be forbidden to the human alone, among the myriads of species of carnivores, to eat the flesh meat of other animals, must rank as a conundrum of the first order. If humans ought to be vegetarians, then they ought not to have been constituted such as they are. Physiologically, the human becomes a biological anomaly if condemned to be a herbivore. Running through the philosophical vegetarian attitude is a hidden strain of gnosticism and latter-day Cartesian rationalism, since, paradoxically, it seeks to remake the human in the image of a diminished animality.

Contemporary vegetarianism thus completes the circle of paradoxicality by returning to the negation of the very premise from which it originally began. Arising out of the cosmic soup of Darwinism, which had declared all animals to be 'created' equal, it concludes its *danse macabre* on the prohibitive note that, of all animals, to the human alone is it forbidden to indulge in eating flesh meat. Under the cloak of vegetarianism, philosophical puritanism lies concealed here—a gnostic elitism that would seek, again ironically, to de-materialize the human (cf. *The Medieval Manichee: A Study of the Christian Dualist Heresy,* by Steven Runciman, pp. 171ff.).

Yet, at the same time, it must not be denied that the human is an animal unlike any other animal, for it is an animal that is intelligent. As discussed earlier, the human is the only animal that employs tools in the formal sense, and is the only animal that has mastered the use of fire, and therefore cooks its meals. The human is a subtle synthesis of the physical and the nonphysical or psychical, and the tension between these 'two worlds' is ever reflected by what the human does and what he says, as seen earlier in the chapters on human behavior, language and knowing. To deny this tension and the constant interaction within the human of this 'two-worldedness' is to deny to the human the uniqueness of his nature.

Though an intelligent animal, the human is not an angel (which is pure intelligence, without body); neither is the human an animal without qualification, since, though sentient, the nonhuman animal lacks intellective reasoning. The human shares similarities, therefore, with both the angel and the animal. In assessing the contemporary movement of animal liberation and its concomitant push for vegetarianism, it is important to keep before the mind's eye what its underlying assumptions are. At stake in this entire discussion is unmistakably a philosophy of the human person. The response one gives to the question of animal rights and the alleged obligation to abstain from flesh meat is inevitably and directly related to the way one views the human, as has often been reiterated during the course of this study.

One will note once again that those pressing for a reexamination and change in the manner in which we are to look upon and treat the nonhuman animal, strive so to emphasize the similarity between the human and nonhuman animal as to establish a glaring inconsistency in one's viewing the animal as something less than human. In short, since both are animals, they ought both, it is alleged, to share in part at least in a common ethical code. But this conclusion can be reached only if the distinction between intellection and sensation is trivialized to the point of effectively denying it—if the origins of both the human and the nonhuman animal are explained in essentially the same way.

This point is clearly illustrated by the bold contention of Dombrowski, who flatly asserts that no "rigid distinction should be made (à la Aristotle) between sensation and reason" (*Philosophy of Vegetarianism*, p. 97). Of course one might perhaps quibble over the precise meaning to be given the term 'rigid' in this context, but the tenor of the claim seems to make clear that what is objected to is the assuming of a true, essential difference between understanding and sensing. Once this distinction is understood in a Humean or Lockean or perhaps Whiteheadean sense, there is very little point in attempting to urge the matter further, for it is only through his rationality that the human can be held to differ specifically from all other animals. Yet Dombrowski further contends that "it is by no means clear why animals *must* be regarded as irrational" (p. 77). Apparently the term 'irrational' is construed to mean 'deranged', when it merely means non-rational. No one views the animal—except the rabid animal—as 'irrational' in the sense of 'deranged'.

Nor do the repercussions of philosophical vegetarianism stop here, for the question of the rationality of the human is also inseparably linked to the question of God. This Dombrowski seems to acknowledge when he quotes R. L. Clark. who argues that, assuming that God is now dead, there no longer remains a basis for distinguishing, in an absolute way, between man and beast. "'Rationalistic attempts to find an absolute stability in the moral domain, now that God is reckoned dead, are at once implausible in themselves and ineffective to produce any absolute dichotomy between man and beast. They are really no more than desperate and often unintelligible reconstructions of the Stoic ethic'" (*The Philosophy of Vegetarianism*, p. 76, quoting from Clark's *The Moral Status of Animals*).

Dombrowski readily acknowledges that it was Darwin's theory of the origin of man that started a "revolution in our perception of animals" (p. 17). If man derives from animals as a result of sexual selection alone, as Darwin theorizes, then there is no firm basis for arguing that the human is worthy of a special respect not accorded the nonhuman animal. This represents the double irony attached to the position that animals possess rights: elevating the animal to the level of the human jeopardizes the fundamental goal of securing more humane treatment for animals. The scales can be made to tip either way. If the human is seen as only *incidentally* superior to the animal, then rather than awarding rights to the animal, one might with equal consistency deny rights to both, arguing that, 'human rights' is nothing more than an empty phrase, having as little value for the nonhuman animal as for the human. As will be recalled, Peter Singer favors eliminating the term 'rights' altogether. In short, it is might that now makes a right. In pursuing the goal of liquidating all forms of anthropocentrism, the actual result of the vegetarian movement could be the elimination of any basis whatever for showing respect either to the human or to the nonhuman animal. Such an attitude could theoretically lead to a repeat of Auschwitz and Belsen on a global scale.

The critical attitude animal liberationists often reflect toward those humans who include flesh meat in their diet, suggests that they have failed to recognize that their argument cuts both ways. If one collapses completely the distinction between the rational and the nonrational, one could consistently argue that the so-called "human" animal need have no inhibitions in doing what "all" other animals do, i.e., in exploiting the weaker animals whenever that is seen to be to their advantage. Indeed, even cannibalism need not then be struck from the human's list of legitimate culinary options. The law of the jungle does, after all, accord the right of predation to all without prejudice.

On the everyday level, bypassing for the moment the claim that a vegetarian diet is more healthy, let us ask how realistic the universal strict vegetarian project might be. Many peoples worldwide, and especially within the third world, depend upon fish to provide the major source of protein in their diet. Many regions have neither large tracts of fertile land nor a climate favorable to the growing of grains. Fish and numerous other marine organisms constitute the staple food of peoples living in frigid climates and of many island peoples, such as those in the South Pacific and Carribean regions. Without fish they would be deprived of a nutrtionally balanced diet.

But the greatest inconsistency with the argument advocating equal treatment for humans and the nonhuman animal is, as before mentioned, that any theory of natural rights and duties must rest on a foundation more stable than that offered by the classical Darwinist theory of human and animal origins. Why is one beholden to nature at any level? Why should I show respect for any other individual, whether human or not? In short, given a Darwinian scenario, there remains no logical basis for anyone's claiming that one ought to behave in such and such a way toward animals or that one ought not deprive other animal organisms of life, nor indeed avoid doing anything that would cause them pain or suffering. One cannot just produce moral obligation out of social custom, tradition, or thin air. Moral strictures have many significant assumptions upon which they depend for their foundation, and without them they are mere *flatus vocis*.

The Darwinian explanation of the origin of living things, if left to its own devices, cannot provide the needed justification upon which moral theory must build. Why ought I do anything at all, if I am merely the fortuitous result of natural and sexual selection, and if every other living thing I experience has equivalently the same status as I as regards its origin? How applicable here is Dostoyevsky's arresting remark, "If God does not exist, then everything is permitted!"

The stark incompatibility of Darwinian theory with theism is underscored by Phillip Johnson, who views theistic naturalism (a view seeking to incorporate the evolutionary views of Darwin with traditional theism) as "ultimately incoherent" and "an intellectual strategy for coping with a desperate situation" ("God and Evolution: An Exchange," *First Things*, July 1993, p. 40). These-well intentioned theists, he maintains, want very badly to win acceptance from the scientific community.

Yet in Johnson's view, many of the scientists they cater to, physicists and evolutionary biologists such as Richard Dawkins and the late Carl Sagan, are busy issuing pronunciamentos of all kinds which go far beyond scientific and empirical evidence. Though they present their ideas allegedly as scientists, they are really speaking and thinking as philosophers or metaphysicians. "It may be," Johnson states, "these physicists—and evolutionary biologists who talk just like them—are no longer practicing 'science' and have become metaphysicians. What is important is that they mix metaphysics and science together and present the whole package to the public with all the awe inspiring authority of science" (ibid.). His final indictment of much of the contemporary scientific community is that claims are advanced as "purportedly factual ones—like the power of mutation and selection to create complex organs—based upon philosophical reasoning rather than on empirical investigation" (p. 41).

Situating Dostoevsky's trenchant comment, then, within the context of our present discussion, we might transpose it as follows: "If God does not exist, then why should I not be allowed to make use of animals as a food source?" A utilitarian response in some form seems the only conclusion remotely consistent with the underlying premise supporting secular humanism, but it is powerless to explain the origin, i.e. the reasonableness, of the moral 'ought'.

It seems, therefore, not merely coincidental that many who promote animal liberation and subscribe to Darwin's theory of man and nature openly profess either agnosticism or atheism. Darwin himself acknowledges in his autobiography that toward the end of his life he had lost his belief in a divine being. His theory of the origin of all living things left no place for a source of life that was not identified with nature itself.

Yet the human spirit cannot live without an absolute of some kind, for without it life is a meaningless adventure. Aristotle was clearly aware of this need, as one can gather from his *Nicomachean Ethics*. The absolute can be wealth or honor or pleasure, or it can be nature itself. It is this latter which appears to be the driving force behind most animal liberationists, philosophical vegetarians, and radical environmentalists. Nature becomes endowed with quasi-divine characteristics, idolized as the supreme value to which all else must be subservient. Nature becomes the true source of obligation; it is to be preserved and fostered at all costs; everyone must above all strive to protect it, help maintain its balance, and aid in its development. The streams and mountains, oceans and islands, lakes and meadows assume a parasacral character. The rich and varied beauties of nature are its face, making known some of the grandeur and power of the forces that lie beneath.

In this view the human is a part of nature but occupies no special place of honor. This is so because there are no fixed standards by which comparisons can be made. Hence it becomes improper and meaningless to speak of higher and lower in the scheme of nature. This, at least, is the view taken by many Darwinists and environmentalists, for whom, as seen earlier, the shrub, not the tree, provides the image-symbol most accurately reflecting the real nonhierarchical status of nature. Doubtless this is too extreme a position for many, who, like Mary Midgely, share the vision of Darwin but who, contrarily, see the human as the crowning achievement of evolutionary development. Among these also is, assuredly, Richard Leakey (cf. *Origins,* p. 189b). Yet Darwin would not be consistent with his own theory were he to maintain, as he does not, that the march of evolution might not well move beyond the human at some future date. He expressly alludes to such a possibility in the closing lines of his *Descent of Man,* where he surmises that man might hope for a still higher destiny in the distant future.

But whether or not one sees the implications of Darwinian theory, as Darwin himself did, as leading to ever higher and higher levels of development, it is unmistakably clear that the distinction between the human and the nonhuman animal becomes through it all, notably diminished. Unquestionably, the Darwinian vision of the human plays up man's role as a sentient being and downplays that dimension which sets him apart from the rest of nature, and, in a real sense as well, sets the human, in some respect, above physical nature.

The entire line of criticism directed against the human's higher status by animal liberationists, charging the human with the sin of speciesism in his dealings with the nonhuman animal and his own assessment of the human's place within the natural scheme of things, rests on the Darwinian perception of man as just another 'species' of animal, where the term 'species' has now lost its original meaning of nature or essence.

Much of the tortured reasoning which one finds throughout a good portion of Regan's *The Case for Animal Rights* results directly from his having to struggle with grounding the inherent-value concept, somehow merging it with a Darwinian theory of origins. Regan realizes only too well, and for this he must be commended, that an authentic moral theory has to be marked by stability and not be subject to the changing tides of historical events and human opinion. It is for this reason that he rejects all forms of consequentialism and affirms the postulate of inherent value, which all moral agents and patients possess. Regan does not claim that such value is sufficient in itself to constitute one a moral subject, but it is a necessary condition.

As we have also seen, Regan invokes, as a further distinguishing criterion of moral subjectivity, the notion of subject-of-a-life. But the only way in which Regan can satisfactorily support these claims is by accepting the more basic Aristotelian view that all things are good by reason of the nature they have. Goodness in its most fundamental sense is metaphysically grounded in being, so that every being, precisely to the extent that it is, is good in a metaphysical sense. But, unfortunately, Regan cannot have it both ways. He cannot be a Darwinian and an Aristotelian at the same time, for these positions rest on incompatible assumptions. It is for this reason that Regan is all but alone among animal liberationists in simultaneously embracing the Darwinian account of the origin of the human, and holding that the human and the nonhuman animal possess an inherent value inseparable or inalienable from their very being. The vast majority of Darwinians are consequentialists or utilitarians in some form; the rest turn to process philosophy in an attempt (futile as it turns out) to ground their moral theory.

Yet, essential as the claim that the higher animals possess at least some inherent value is for grounding his case for animal rights, Regan is not about to mortgage the family farm in defense of it. In a revealing passage wherein he discusses the grounds upon which this claim rests, he grants that its truth-status does not exceed that of a mere postulate or theoretical assumption (*The Case for Animal Rights,* p. 247). It is, in his view, to be preferred to other options advanced in support of animal rights, in that it avoids the extreme consequences he finds unacceptable. As Regan puts it: "To postulate that moral agents have equal inherent value provides a theoretical basis for avoiding the wildly inegalitarian implications of perfectionist theories, on the one hand, and, on the other, the counterintuitive implications of all forms of act utilitarianism (e.g., that secret killings that optimize the aggregate consequences for all affected by the outcome are justified)" (ibid.). By conceding that the 'inherent value' position is for him no more than a postulate, Regan shows that he is aware of the problem of trying to bring together under one roof Darwinian evolutionary theory and Aristotle's teleological view of nature.

As stated earlier, the notion of rights makes sense only within a context of free agents who are responsible for their actions, and hence it must be limited to *moral agents alone,* i.e. to humans. Trading on the sympathy most humans display for various species of domesticated animals, and the pervasive conviction that it is wrong to mistreat them, Regan has sought to revamp the definition of 'right'. He extends this revised definition to animals in an effort to provide reasoned, stable support for the above-mentioned near universal sentiment. But, as we have indicated, one does not need to attribute rights to animals directly in order to provide the latter with a shield that will morally safeguard them from maltreatment.

The protection Regan seeks to secure for the animal is in good part already provided for by the natural law, which requires that the moral agent act responsibly and intelligently in all his or her dealings with other humans and other creatures. To show cruelty to animals is simply to act inhumanely; to misuse creatures of any kind is a mindless, thoughtless form of behavior and hence a plain repudiation of the intelligence one has. But using creatures intelligently does not mean treating them as though they were human.

But why it is allegedly wrong for the human to do what is common practice among nonhuman animal species becomes unavoidably problematic, especially so when viewed within context, since many who would impose the vegetarianist prohibition on mankind are professed Darwinists. As a religious or aesthetic practice, vegetarianism is not without intelligible and justifying grounds, but as a prohibition emerging directly and uniquely from the diversity of existing organisms, it lacks foundation.

Defended and touted as a much-needed antidote to the rationalism of Descartes, which rent asunder the world of man and the world of nature, philosophical vegetarianism succeeds only in contributing to the spread of the same rationalist virus it seeks to contain. This it does by so absolutizing species that any meaningful interface between various species becomes unintelligible. Animal species become, in effect, ends in themselves, atomized realities having no other purpose than to seek their own particularized fulfillment, fleeting and purposeless as that may be. The vision of hierarchical arrangement of living things, of higher and lower levels of life, with the lower contributing to and forwarding the development of the higher, and all of nature culminating in the promotion of that life-form whose highest vital acts transcend mere sentience, is replaced by the vision of a world of myriads of individual species of organic beings frenetically striving to attain their own private 'goals' of self-fulfillment. The latter vision of utter disorganization and chaos offers no hope of a unified synthesis; the former vision of purposefulness provides a reasoned explanation of how unity underlies the rich manifold of diversity inhabiting the human's world.

The former vision is of course anthropocentric in the sense that the human is taken to be the highest life-form in our world. Through their power of reason, humans are the great gatherers or collectors of nature, for whatever they are able to survey, and their vision extends to all things, they can somehow understand. And understanding is the supreme unifying act, for through it alone the many can become one. It is the sole unifying act that is capable of grasping, and hence unifying, all that is. The human intellect can accomplish this because it is not bound by the limitations of time and space, since its operation transcends the realm of the particular.

Through the act of understanding, the human functions as the unique catalyst of all of nature. It is the human, alone among living things, who is capable of putting together the countless pieces of the puzzle of our universe and envisioning it as a single, seamless garment. It is the human alone who understands what it means to be an animal, and perceives the manner in which the different species of animals are interrelated. It is the human alone, therefore, who understands the difference between himself, as a self-understanding animal, and the other animals, whose knowledge, because it fails to divest itself fully of the temporal and the material, is unable to attain to the level of self-understanding and self-awareness. All histories of animals have, as a consequence, been written and compiled by humans.

Though the animal possesses inherent value, that value is wholly conditioned by the temporality of its life-span. That value, therefore, is not absolute but relative, for its value can be calculated only according to the contribution it makes to the well-being of the whole of nature. For this reason the animal cannot be said to have, in the proper sense, rights, which are reserved to the human alone, for only the human's lifespan transcends in a significant manner the temporal limitation of all vegetative and sentient living things. Not that the human does not die, for obviously all humans do. In this sense the human too, is subject to the law that all animals, indeed all flesh, are heir to, and hence is a temporal being. Yet the human's death, though it uniquely involves a change in life-form, does not entail an end of this inner, transcendent life, which, because of its immaterial nature, is permanently undying.

As we argued earlier, the concept of natural human rights is inseparably attached to a theistic view of the world. All creatures reflect, each in its own way, the light and beauty of the divine nature. As reflections of the divine nature, faint and imperfect as these may be, all created things are in themselves good; they positively contain inner perfection which is willed and loved. But to possess a 'right' is not synonymous with having inherent value. All rights indeed flow from inherent value, but only from a unique kind of value which transcends the limitations of the merely temporal. Everything that exists has inherent value, yet not everything has inherent rights, for the inherent value of each thing is itself directly proportioned to the nature or quality of being that a thing possesses.

The human is made in God's image in that he or she is an intelligent being capable of self-reflection, and hence open to the entire universe of

The Right to Life   311

being. Because of their uniquely spiritual quality, humans are persons, and person, as St. Thomas comments, designates that which is most perfect in all of nature (*ST* I, q. 29, a. 3). It is, then, on account of personhood that the human, most perfectly of all earth's creatures, reflects, however dimly, the creative power of God—as the most gifted of all God's creatures.

Through the human's ability to transform the world of nature by instilling within it a new vision and a new order, the world is brought to a higher and more perfect unity. A world without the human would be a world lacking ultimate meaning. It is in this that the world as authentically viewed is indeed profoundly anthropocentric. Though planet earth may not be the cosmological center of the universe in the sense in which Ptolemy thought it to be, yet it appears to be nonetheless the world's center in a truer, more significant way. This conclusion, daring or hyperbolic as it may seem, is still the only one that accurately reflects the present findings of science. Despite the most cherished expectations of astrophysicists and astronomers, all attempts to uncover signs of life elsewhere have thus far come up empty. Presently, the most advanced state of our knowledge of the solar system, of our own galaxy and the universe beyond, fully supports the astonishing conclusion that tiny planet earth is the *life-center* of the entire universe.

CRANIAL CAPACITY AND ANTHROPOLOGICAL CLAIMS

The world in which we live *is* anthropocentric, and not because the human has made it so, nor even because the human is the highest, most complex form of existing life on planet earth, but because the human is, while an animal, also an intellective animal, a person, who is the only organic being made in God's image.

This view of the human and the world is not, as many have claimed, anti-scientific. There is no way in which the evidence of the scientist can be judged as providing definitive support for the Darwinian claim with respect to man's origin, any more than science can show that there is not a difference in kind between sensory and intellective activity. The question of the origin of the human lies beyond the horizon of the scientist's vision as a scientist.

The paleoanthropologist is able to identify a fossilized skeleton or part of one as that of a human or gorilla or chimpanzee, only by comparing it with previously known skeletal fossils. Though the skull of a human has a much greater cranial capacity than that of any of the other primates, anthropologists are able to say that this ancient cranial specimen is humanoid only because they observe that it has the physical properties of other cranial specimens they know *in fact* to have been human. In other words, the anthropologist makes an inference that all cranial fossils having these particular specifications are human. But the paleoanthropologist 'sees' (observes) *nothing* in the fossil that tells him it is *human,* and without the ability to compare it with other fossils already known to be human, he is incapable of reaching such a conclusion. The upshot is that the paleoanthropologist judges the intellective capacity of the human by the physical size of the cranium, i.e. by its capacity to house a larger brain. The intellective capacity as such, of course, is wholly unobservable. It is not something that can be quantitatively measured, but only understood.

Consequently, there is nothing inherently contradictory in one's affirming that a cranium fossil of such and such a size normally interpreted as human is not human at all, but that of a nonhuman ape, for example. Could not a nonhuman primate have a brain the size of a human and still not be a human? Granted, this would not follow the regular pattern of experience the anthropologist is accustomed to observing, but he would be hard put to provide a cogent reason why this *must* be so. In short, in dealing with intellective capacity the inquiry is on a level beyond that of merely physical dimensions and of direct sensory observation. Only mind or reason can uncover and recognize the humanity of the human. All strictly 'scientific' knowledge in this regard is merely conjectural. The differences between the human and the nonhuman, being qualitative rather than quantitative, transcend the discerning capabilities of physical measurement and sensory observation alone. Indeed the very practices of the paleoanthropologist themselves reflect this, in that they admit as authentically human only those fossils unearthed in the proximity of various incontrovertibly human artifacts such as pottery shards, primitive tools, arrowheads, fire-ash deposits, etc. For the self-same reason, if one is prepared to grant that the human is ultimately the result of a creative act of God, no traces of such an act could be 'scientifically' uncovered. The most that could be thus affirmed is that during a particular era 'animals' appeared on the scene whose behavior differed markedly from that of other animals. Beyond this, nothing more could be affirmed on the basis of empirical data alone, obtained through a study of 'animal' fossils and various artifacts implying intelligent design not exhibited by other animal species. From the strictly anthropological point of view, something immediately unexplainable would be known to have occurred, but the why and how of this occurrence would remain merely a matter of conjecture.

But of course no scientist, indeed no thinking human, can long live without desiring a fuller explanation of their experiential data, for the human mind abhors an unexplained event as much as does nature a vacuum. A serious inquirer would doubtless be inclined to theorize that the human descended from ancestors morphologically more primitive, yet, for all that, in many respects identifiably similar. The anthropologist would, we may further suppose, not claim to know precisely how this all came about. The explanation of the 'how' would, assuredly, require further study and might, indeed, never be satisfactorily attained.

It is of considerable importance to note, however, that in insisting that the only acceptable evidence for the validation of truth claims be restricted to that capable of corroboration in quantifiable, measurable terms, the scientist is in that very assessment operating within a methodological framework which is itself subliminally transphysical. He then is proceeding on the assumption of a principle that is not itself sensibly verifiable, namely, that every event is explainable in physical terms alone, an assumption that clearly eludes the reach of any merely sensory or physical verification method. As a matter of course, then, the scientist, relying on the above alleged first scientific principle, will eschew any explanation that is not, whether directly or indirectly, sensibly verifiable—rejecting, as did Immanuel Kant, any transphysical mode of explanation as 'metaphysical' and thereby incurring the fallacy of "transcendental illusion." Now there is nothing illogical, of course, about employing the scientific method, accepting as validly established only what can be sensibly verified, as long as this manner of proceeding is not alleged to be the only authentic way of uncovering the truth of things. But where the scientist often errs is in making the further claim that the scientific method is the only viable pathway to truth. In so assuming he is forthwith guilty of the fallacy of scientism, of proclaiming out of hand that science, as he understands it, is the highest form of knowledge the human can attain.

Because of the notable differences, then, in method and approach employed by the anthropological and the philosophical inquirer, it would seem to be not altogether unlikely that the paleoanthropologist might one day announce that the date and place of the hominid has been discovered. In so claiming, the scientist might conceivably be stating the truth as regards the date and place, but could at the same time be erring in ascribing to natural selection the transition from a nonhuman primate to homosapiens. Yet, were we to assume that this transition had been effected by a divinely creative act, there would be no physical evidence by which this could be scientifically discerned. It is of further interest to note that, even apart from the above imagined scenario, the anthropologist has already gone beyond the strictures of the scientific method in identifying certain fossils as human, for in differentiating human from nonhuman fossils, the scientist must have had recourse to the act of intelligence by establishing the existence of artifacts of various kinds that could "only" be the products of an intelligent primate.

Such a creative act would have left no visible trace. What it would have left, however, would be gaps in the fossil record, which would remain inexplicable on the merely anthropological level. This would be true even if one were to suppose that the scientist, through an exceptionally good piece of luck, had uncovered significant skeletal remains so as to be able to identify them 'scientifically' as the likely remains of the first authentic human. Yet even so, it still seems highly questionable whether such a find could ever be accurately identified as such. But supposing this to be the case, the transition between the humanoid fossils and those of its nonhuman 'ancestors' would still forever remain mysterious, and quite beyond scientific explanation. Indeed, only as a theologian, if such matters were divinely revealed, could one truly know these skeletal remains to be those of the first human, for, short of this, one could never be sure that there did not exist still older but as yet undiscovered 'human' fossils.

The 'impossible' scenario just sketched may not be as many light years away from the contemporary anthropological scene as one might assume. The now-renowned American paleoanthropologist, Donald Johanson, who rocked the scientific world almost two decades ago with his best-selling book *Lucy: The Beginnings of Human Kind*, wherein he describes his remarkable fossil finds in Ethiopia, is refreshingly candid about what paleoanthropologists know and do not know concerning the origin of the species *homo sapiens*.

As he himself admits, he had the great good fortune to discover the oldest near human fossils on record anywhere in the world. The skeletal remains he uncovered, which are remarkably well preserved considering their estimated age of nearly three-and-one-half million years, are those of a female who cannot, according to Johanson and fellow anthropologists, be viewed as being those either of an ape or of a human in the modern sense. Johanson considers her to have been in some way an ancestor of contemporary man. Since Johanson is clearly committed to evolutionist theory, and his remarks on the question of the human's origin as well as on the present status of paleoanthropological studies are highly relevant to our present discussion, they are quoted at length:

> The only thing we had not discussed up to this point was the biggest enigma of all: how did it all get started? What pushed those ancestral apes up on their hind legs and gave them—some of them—an opportunity to evolve into humans?
> *That question is basic to the entire story of hominoid evolution.*
> But the Why of it was something else. *Why, of all the mammals that have ever walked the earth, did only one group choose to walk erect?* That tremendous enigma stumped us. Paleoanthropology alone could not solve it. *It would need the wisdom of disciplines that had nothing to do directly with fossils....* (*Lucy: The Beginnings of Humankind*, p. 306; italics added)

Johanson goes on to summarize the views of Owen Lovejoy, a specialist in locomotion, regarding the bipedal phenomenon so important to the human's unique mode of behavior. According to Lovejoy, whose views do not coincide with the traditional paleoanthropological explanation of how man came to walk erect, bipedalism was acquired before man became man, and certainly not as a result of his having to adjust to a new mode of life on the savanna (pp. 329 and 339). Lovejoy emphasizes how extremely complex the question of bipedalism is, since it is made possible only by a total restructuring of the entire skeletal frame. The foot, the leg bones, the pelvis and vertebrae must all be radically realigned to permit erect posture and moving with a striding gate as a matter of course. As Johanson puts it: "A chimp can walk erect, and often does for short distances, but it fatigues quickly because the abductor muscles leading down to the leg from the pelvis blade are very badly placed for that kind of striding . . ." (p. 348).

When one compares the bones that compose the retrieved skeleton of Lucy with those of other primates, however, one notes a decided difference. What one sees immediately, Johanson states, is that "the blade of the ilium has turned in just the way one might expect if the mechanical requirement were to be to provide better muscular attachment for erect walking" (ibid.). For such a change to take place, other significant adjustments of the skeletal frame must accompany them. Johanson argues that if the shape or the position of one part is to change, other parts are necessarily affected. When the blades of the ilium are moved forward, they put pressure on the entire lower abdomen. Hence they must become larger and longer (ibid.). These and other skeletal changes—not least among them being the restructuring of the foot and ankle (and the ankle is by far the most complex of all the joints in bipedal creatures)—are essential to permitting the human to walk erect as a natural mode of locomotion.

When it comes to offering an explanation as to how such changes might have taken place, Johanson informatively acknowledges that the present status of paleoanthropological enquiry is still far from free of nagging questions. The fossil-bearing deposits in the Afar region in Ethiopia do not tell us just *how* or *when* the crucial transition from ape to hominoid occurred. "This," he concludes, "is the *biggest remaining challenge* to paleoanthropology" (p. 375; italics added).

The origin of the human animal thus remains for the anthropologist exasperatingly puzzling, and, while at first glance this must seem a strange development, since the human is the most "intelligent" of all primates, yet one needs to recall that only humans seek to plumb the origin not only of themselves but of other organisms as well. The puzzlement stems from the human's imperfect knowledge. The human humbly needs to acknowledge that science alone can never succeed in providing the total, definitive explanation of the nature and origin of that which has always most intrigued him or her: namely, him or herself. Surprisingly, this assessment seems confirmed by Darwin himself, for in his *Descent of Man* he admits, "In what manner the mental powers were first developed in the lowest organisms, is *as hopeless an enquiry as how life itself first originated*" (Pt. I, ch. 3, p. 287b; italics added). Yet, however incongrously, Darwin is at the same time supremely confident in affirming that all life forms, including the human, randomly emerged.

6 Anthropocentrism, Biocentrism, Envirocentrism

REGAN AND THE RIGHTS OF NONSENTIENT BEINGS

A final question relating to animal rights but broader in scope, one that has attracted much contemporary interest, is the question of the human's relation to the environment. Is there such a thing as an environmental ethic, and, if so, how is it grounded? In his later writings Regan has given the question some attention. He was indeed all but constrained to address it, for many of those pressing actively for the implementation of an environmental ethic are sharply critical of his inherent rights theory. The reason for this opposition rests mainly on the fact, that, by his grounding of rights for animals on their ability to suffer, i.e., to be moral patients, Regan would seem to have systematically ruled out the possibility of there being an environmental ethic. The nonsentient world of plants and inanimate things would then lie outside the pale of ethical concern.

But Regan disputes the claim that his view of rights would automatically rule out the plausibility of an environmental ethic, although he by no means claims to have worked out a suitable explanation as to how this might be accomplished. For the present, however, he is satisfied with merely affirming that it is not obvious that 'rights' might not be extended to nonsentient and even inanimate things. In addition, he points out that by relegating the rights of individual moral agents and patients to a subordinate role within the environment as a whole, one

risks having moral agents and patients sacrificed for a greater environmental good. Quoting Aldo Leopold's holistic view of the environment
and its importance vis à vis the value of segments of the environment or
of individuals within it—"A thing is right when it tends to preserve the
integrity, stability and beauty of the biotic community. It is wrong
when it tends otherwise"—Regan comments:

> The implications of this view include the clear prospect that the individual
> may be sacrificed for the greater biotic good, in the name of "the integrity,
> stability, and beauty of the biotic community." It is difficult to see how the
> notion of the rights of the individual could find a home within a view that,
> emotive connotations to one side, might be fairly dubbed "environmental
> fascism." (*The Case for Animal Rights*, pp. 361–62)

Regan goes on to add, however, that the rights position he espouses
would not rule out the possibility of inanimate objects having rights, if
one is speaking not of *individual*, inanimate objects but of *collections*
or systems of them (p. 362). He has in mind such collective objects as
forests, mountains, rivers, lakes,etc. At the same time, Regan does not
grant that a convincing position in support of such rights for collections of objects has to this point been forthcoming. Indeed, he is clearly
skeptical that such a view would ever muster convincing support,
though this is of no consequence regarding the issue of inherent
'rights'. "What is far from certain," he says, "is how moral 'rights'
could be meaningfully attributed to the *collection* of trees or the
ecosystem. Since neither is an individual, it is unclear how the notion of
moral rights can be meaningfully applied" (ibid.).

Nonetheless, if a rights view for nonsentient and inanimate objects
could be worked out, Regan is confident that such a position would
provide all the protection environmentalists presently demand for the
environment. "Assuming [rights] could be successfully extended to
inanimate natural objects, our general policy regarding wilderness
would be precisely what the preservationists want—namely, let it be!"
(p. 363). So, even though Regan sees as dubious the possibility of establishing at some time in the future a viable argument supportive of
environmental rights, it is equally important to note that he does admit
to the reality of its possibility, however remote. In other words, Regan

sees nothing inherently contradictory in affirming that nonconscious living things such as plants and trees, as well as inanimate objects of the ecosystem, might possibly possess rights.

After meticulously developing his case for animal rights, and grounding it on what he chooses to call "the inherent value of the animal," and initially limiting his argument to include only those animals that might be said to be subjects-of-a-life, Regan now grants the possibility that other living things, including animals who are not-subjects-of-a-life, and even some nonliving things, might truly possess rights. By so doing, he is granting that 'being the subject-of-a-life' while a sufficient condition for having rights, is nonetheless not a necessary condition for having them. He also allows that the only necessary condition for rights may well be that the thing itself in question have inherent value. If this be his thinking, it is hard to see how Regan could logically deny rights to any object whatever, unless he were to admit that the value in question depends upon the manner in which the object is assessed.

But if this be the case, his position must be seen as sliding into some form of utilitarianism, whereby value would be granted to an object only to the extent that it somehow appears to contribute to the overall good or well-being of the universe. In this case the value would appear to be not an inherent, but rather an external one only. Moreover, utilitarianism in any form is a view Regan repeatedly rejects as an adequate support for his 'rights' case.

I have emphasized Regan's position regarding environmental ethics because, peripheral as it may be to his primary intent of establishing objective rights for animals, it does lay bare with signal clarity the internal inconsistencies of his argument. If one extends the meaning of rights to include moral subjects, as Regan understands that term, as well as moral agents, then one has stepped out onto a very slippery slope from which there is no safe, logical return. Lacking firm and precise criteria by which one might judge whether this or that object is the subject of rights, the meaning of the term 'rights' becomes so extenuated as to lose all identifiable meaning. It becomes nothing more than a thoroughly hollow expression, serving no useful purpose, other than perhaps a political one. And this seems to be the very reasoning behind the conclusion, as noted earlier, at which animal liberationists such as Peter Singer and R. G. Frey arrive—that in this matter we would be better off abandoning all serious talk of 'rights'.

PLURALISTIC STATE OF ENVIRONMENTAL ETHICS

Regan's position with regard to the rights of nonhuman animals, and particularly his tentative acceptance of the 'possibility' of there being a viable ethical theory extending also to the inanimate world, has been met with scepticism and opposition by perhaps the major portion of philosophers today who define themselves as environmental ethicists. Some environmentalists are critical of Regan's talk of natural rights, since it seems to embrace still a view singularly anathema to them, "anthropocentrism". Others, such as Peter Singer, find no real advantage in retaining the word 'rights' in our philosophic vocabulary, since it is misleading, having no clearly identifiable meaning, and is hence better abandoned entirely.

Notwithstanding their inability to formulate a commonly acceptable definition of 'rights', contemporary ethical environmentalists do find common cause in many of their practical goals. All share a deep concern for our environment in a technological age, and are keenly intent on slowing, and even possibly halting, the continuing rate of its exploitation. Practically, they agree that a change in our attitudes and practices regarding our relation to the environment must be brought about, and they likewise are of one mind that a philosophy of the environment that will provide a theoretic blueprint for such action is essential to the achievement of this goal. What is needed, they agree, is a viable environmental ethics which will provide guidance on the level of the individual citizen, but more importantly will eventually be inscribed in the laws of nations. Agreement pretty much stops here; there is a remarkable pluralism of views as to the nature and content of a satisfactory environmental ethics.

This state of affairs should cause no surprise, for a holistic ethical theory is not and cannot be (Kant's demurral notwithstanding) unrelated, as we have resolutely argued, to one's more basic views regarding knowledge, freedom, rights, and the human. In other words, one's philosophy of nature and metaphysics inevitably serve as a basis of one's ethical synthesis, and this extends to environmental ethics.

BIOCENTRISM

And thereby hangs the tale, for a very high proportion of contemporary environmental ethicists seek to build their ethical theory on the precarious structure of Humean metaphysics and the natural philosophy of Darwin. On this point we have their explicit acknowledgment. In his influential book, *Respect For Nature: A Theory of Environmental Ethics* (1986), Paul Taylor provides the following favorable appraisal of Darwin's evolutionary theory: "Evolution, as biological science has come to know it, offers a unified explanation for the existence of both *human* and nonhuman forms of life" (p. 111; italics added). Shortly thereafter he adds: "We can consider the theory of evolution to be a framework of thought that underlies one part of the biocentric outlook on nature. We understand ourselves as beings that fit into the same structure of reality that accounts for *every other form* of life" (pp. 112–13; italics added). And earlier Taylor had stated that "from an evolutionary point of view we perceive ourselves sharing with other species a similar origin as well as an existential condition that includes the ever present possibility of total extinction" (pp. 49–50). In an article appearing in *The Monist* a few years ago James Rachels approvingly acknowledges that for Darwin there are no fixed essences but only singularly existing things whose groupings into "species, varieties and so on—is more or less arbitrary" ("Darwin, Species, and Morality," *Monist*, 1987, p. 109).

The term commonly employed to designate in a generic manner the position of leading ethical environmentalists such as Paul Taylor, James Rachels, Baird Callicott, Holmes Rolston, Laura Westra, Richard Sylvan, and others, is *biocentrism*. Although the detailing specifications of this term vary from individual to individual environmentalist ethician, the core of the definition differentiates the biocentrist from all anthropocentric views. For the biocentrist the human is simply a 'part' of the environment; for some, the human is admitted to be the 'highest' form of living things, but possesses thereby no special rights or privileges. The human species is simply viewed as one among many, with no niche of privilege. The biocentrist view, then, is opposed to the anthropocentrist view, which does consider the human as representing the highest life-form, possessed of a dignity unambiguously and inherently superior to all other species of living things. It can be further noted that some ethical environmentalists distinguish between a hard and soft biocentrism. The former (hard) would maintain that from an ethical standpoint all living species are effectively equal in value, and that it is toward the preservation of the total complexus of living species that our concern should be directed. The soft biocentrist position, on the other hand, acknowledges that some species are of greater importance than others for the maintainance of a healthy biosphere, and hence can and should on occasion be awarded special privileged consideration. In a somewhat parallel way, the hard anthropocentrist views humans as so far above other living forms and the ecoworld as to have the right to make use of it in whatever way is seen to have most value to themselves. In the soft anthropocentrist view, however, while the human is seen as the highest of all living forms, possessed of both intellective and volitional powers. it is recognized nonetheless that the human still has the obligation of making use of the ecoworld in a manner consonant with his nature as an intelligent, responsible being, treasuring it as a gift to be used, not abused.

It can be mentioned here that two respected figures in the area of environmental ethics, Bryan G. Norton and Don E. Marietta Jr., question the correctness of the assumption made by many environmentalists that *anthropocentrism in any form* is inimical to the environment and indistinguishable from *egocentrism*. Marietta argues that an ethic which is truly anthropocentric will put humanity first and not the individual human, and hence should not be described as egocentric. He asks: "Do we need to keep *anthropocentric* as a term of reproach?" He answers his own question by asking further: "But what reasonable grounds can there be for preserving *anthropocentric* for an egoistic and ill-founded moral position?" (*For People and the Planet*, 1995, p. 75). That, however, the term continues to be employed as synonymous with an egoistic and ill-founded moral position will become evident from what follows.

BIOCENTRISM AND INHERENT WORTH

Of those who are critical of both hard and soft anthropocentric views, Paul Taylor ranks near the forefront. His opposition to such a position is uncompromising. The view that human life represents the highest type of existence is on his account "a totally groundless assumption." "We shall find," he continues, "that there are reasons for considering it to be nothing more than an unfounded bias in our own favor" (*Respect for Nature*, p. 113). Elsewhere, Taylor states: "To assume it [human superiority] without question is already to commit oneself to an anthropocentric point of view concerning the natural world and the place of humans in it, and this is to beg some of the most fundamental issues of environmental ethics. . . . It is not inconsistent for a human to believe in all sincerity that *the world would be a better place if there were no humans in it*" (p. 52; italics added).

Taylor's biocentric view, however, not only recognizes that humans have rights but also maintains that they alone possess them. "It is best," he affirms, "that the original idea of moral rights be accepted in its full, uncompromised meaning as applicable to humans alone." As a consequence he will argue that "assertions of moral rights would be made and understood within the domain of environmental ethics" (p. 226). Here he disagrees with Tom Regan and Joel Feinberg, since he denies rights to the nonhuman animal (p. 254). At the same time he will still argue that humans have duties toward the nonhuman animal, since these possess inherent worth. Their *worth* derives ultimately from the supportive but indispensable role they play in the universe of biota. Further, for Taylor the human's worth is *no greater* than that of other biota, since he concludes that "we owe duties to them that are *prima facie* as stringent as those we owe to our fellow humans" (pp. 151–52). Thus Taylor's biocentrism expressly commits him to assuming a hands-off attitude with regard to animals in the wild. Each species has its own inimitable worth, and is, consequently, as deserving of respect as is the human species itself.

In this view there can be no real evil in the universe, for everything follows set "laws" [sic]. The human should not interfere with nature, even by trying to protect nature in the wild or with a view to saving endangered species (p. 177). It should thus be noted that Taylor's environmental thinking does not admit of a distinction between the status of the wild and the domesticated animal. Both are equally worthy of respect, for all animals have inherent worth.

What has led Taylor to espouse what surely must be judged a radical form of environmentalism, would appear to be his conviction that it is the only way by which the anthropocentric view can be successfully countered. In the concluding chapter of *Respect for Nature,* Taylor reasons that by adopting his view of "inherent worth" for all living things, one can be assured that one then has "a solid basis for rejecting any human-centered view point that would justify an exploitative attitude toward the earth's wild creatures" (p. 226). As will become clearer in the analysis which follows, what is evidently the moving force behind the environmentalist ethic adopted by Taylor is his firm conviction that the prevailing contemporary ethical views laying out the direction a fitting interface between the human and the environing world ought to take are lacking in logical consistency.

Taylor and many of his co-environmentalist colleagues view with increasing alarm the common disregard of the technologically developed nations to exhibit provident concern for the well-being of the planetary environment.

They are thus striving to justify substantive changes in legislation and public policy pertaining to human use of the environment. While such motivation seems at first laudable, their resolve to retain the present environment intact is questionable. How does the desire to maintain the status quo accord with the Darwinian theory as it relates to the origin of all of the very species they seek to preserve? If the various levels of biota found in our contemporary world are indeed the result of random mutation, orchestrated by the pervasive lottery of natural selection, then how explain their dedication to preserving the world unchanged—a world that on this account has fed exclusively on the phenomenon of circumstantial change. Is their program not a concerted effort to bring to a halt the onward march of evolution? If what has occurred during the past "millionennia" through the process of natural selection is so worth treasuring and preserving, by what logic can one question or deny that the best is likely yet to come? Ought we not simply "let nature take its random course"? And if one advocates "letting species be," as does Taylor, then why bracket out the human species from this dynamic equation? Is not the very suggestion an attempt to reverse the evolutionary process? Why not let the humans be; why not allow them to be as they "want to be," if their very wanting to be is itself the result of blind natural selection? Teleology is, after all (as Baird Callicott emphasizes), representative of "the scientific anathema of explaining things in terms of final rather than efficient causes" (*Earth's Insights*, 1994, p. 40). Yet how can one (humans) destroy or interfere with the plan of an "unplanned" universe? Why is the human singled out from all other species as the one species that is the "unthinking upsetter" of an "unplanned environment"? And who or what are the ones, and by whom authorized, to pass judgment on what is fitting or unfitting, if everything that occurs is and has always been itself the result of natural selection?

One suspicions that these are not questions the evolutionarily oriented environmentalist is anxious to address; one seeks in vain for evidence that they have addressed, or even raised, such questions. Is not this the "new" dogmatism? The present ecological crisis that unfolded during the latter half of the twentieth century seems in great part owing to the fact that humans had lost a sense of their own definition and meaning. Without self-knowledge it becomes impossible for us to understand our relationship to other organisms and to the universe of which we are a part. Immersing the human in a cosmic biotic stew does not augur well for our approach to understanding who we are, or who the animals are, or what the universe is, and how all three relate one to another.

HOLISTIC ENVIRONMENTALISM

But it is time to explore further the subtleties of the key term employed by most environmentalists—biocentrism—for all biocentrists are in agreement that the universe is a massively complex, intricate, organic unity. In this sense, the biocentrist view extends beyond the biota

themselves and includes indirectly the environment in which these latter exist, since they cannot exist without those other elements of nature
which are not themselves living. Whence it is that the biocentrist position entails the inclusion of the biosphere, or circuit of all living things,
in its considerations. Accordingly, the adjective "holistic" is frequently
added to emphasize the all-inclusiveness of their concerns. They see the
universe as interdependently networked, so that in order for any particular grouping of biota to manage survival and remain healthy, all occu-
pants of planet earth, together with its nonvital components, must
flourish. Thus Don E. Marietta, Jr., writes in his recent book *For People and the Planet: Holism and Humanism in Environmental Ethics:*
"What we need to promote is a holistic understanding of what humans
are as part of the system of nature" (1994, p. 79). He obviously understands by "holistic" a view that includes all elements of the environment, both organic and inorganic. His position does not much differ
from Taylor's, though he does emphasize the need of contextualization
in making ethical assessments. Marietta defines his ethical theory as
that of "a context oriented humanistic holism" (p. 153). Yet, contrary
to Taylor, he does not see any impediment in considering the human
species as superior to other living things, for this need not of itself
"lead to a lack of concern for ecosystems on which human life most depends" (p. 79). He sees values as falling into clusters, as "one value
leads to another, calls for another, takes strength from another" (p.
139).

Marietta is also in agreement with those environmentalists who
grant that humans have obligations to nonhuman animals, and he
specifically agrees with Taylor that all members of the land community
should have moral standing (p. 164). Along with most other environmental ethicists he sees no necessity for grounding one's ethics on a
metaphysical system; rather he opts for a contextualist ethics which
rests on two underlying principles, namely, that of utility and that of
justice (pp. 170–72). He provides no realistic support of the claim that
we are 'obliged' to protect the environment. As we have seen, mere
utility cannot adequately support an ethical claim, and it is difficult to
envision how Marietta might successfully derive a viable principle of

justice from a context-oriented, humanistic holism, which will in fact differ from a so-called purely utilitarian ethic. Marietta does, however, recognize that "one of the most difficult tasks of the moral philosopher is addressing the logical issue of justifying moral principles" (p. 7). He also denies that there exists a real or unbridgeable "gap" between an "is" and an "ought" (p. 8). With this one can agree, as is evident from our earlier treatment of the psychogenesis of the first moral principle, "Do good and avoid evil," for the "ought" cannot be so "derived" unless it is already present in the primordial awareness of things as existing. Further, such recognition of obligation requires that what is experienced have a determinate nature which is itself an expression of order and hence a sign of intelligence. But this explanation is not truly available to Marietta, for whom all experience is contextual only. As he argues, "if we talk about knowing things as they really are, we must keep in mind that we are talking loosely. . . . The concept of 'objective' knowledge of 'brute' facts lost its credibility and usefulness long ago" (p. 66).

Marietta goes on to characterize his moral theory as 'pluralist', because he believes that it is often necessary to apply more than one moral principle in order to ascertain what is the right thing to do. He considers these principles to be irreducible to only one common principle, and hence inherently inconsistent one with the other. He cites the principles of utility and justice, and points out that determining the right course of action often requires more than simply applying the principle of utility; often one must also inquire into the fairness of such a course of action.

In many respects the view just outlined by Marietta seems to receive support from Richard Sorabji, who, while in some ways sympathetic to the thinking of Singer and Regan on animal and environmental issues, is still not completely satisfied with their positions. He feels that a fitting solution to fundamental questions of environmental ethics has yet to be found (*Animal Minds and Human Morals*, 1993, p. 216). Specifically, Sorabji thinks that the formulation of the right rationale for engendering a greater respect for animals has yet to be worked out (p. 8). He leans in the direction of placing greater emphasis on the singular cases of animals and less on what he refers to as abstract theory or principle, calling for an ethical theory that is more pluralistic in the sense of embracing considerations focusing especially on the more concrete aspects of problems. He sees his position as primarily opposed to what he terms "one dimensional theories." The phrase he particularly employs to describe his ethical theory is that it is one of "multiple considerations" (pp. 218–19). I take this to mean that one ought to approach ethical questions employing different methods, investigating them at different levels: that Sorabji opts for a pluralistic view somewhat akin to the view of Marietta. He will himself refer the reader to the thought of Charles Taylor, Tom Nagel, Bernard Williams, and Isaiah Berlin as representative of a similar approach to grounding ethical theory.

Yet what Marietta will call a pluralist ethic differs little, it seems, from a traditional natural law ethic, as long as one grants that the principles of utility and fairness are hierarchically ordered and that they are merely more specific determinations of the first and ultimate moral principle: "Do good and avoid evil." In this way the two principles need not be irreducibly incompatible, and conflicts experienced in attempting to employ both simultaneously can be resolved (cf. *For People and the Planet*, pp. 169–73).

This is a highly important point to emphasize, since one of the most common misapprehensions many environmental ethicists seem to labor under is that of assuming that the natural law tradition involves the unenlightened claim that "ready made solutions" to complex moral questions are always to hand. Nothing could be less true, and there are perhaps few areas where more complex moral problems can be encountered than in the field of environmental ethics. For the natural law ethicist, there are, indeed, many moral principles, differentiated one from another by their specificity or levels of generality or particularity, but all derive their moral quality from the fact that they are traceable as further determinations of more and more general principles, all of which rest finally on the one ultimate principle of total universality as regards human free behavior. How these various principles are to be applied, as well as which of them in given instances is relevant to a particular ethical question, is often extremely difficult to determine. Often-times one must humbly recognize that one cannot affirm with full certainty which course of action would be the most fitting overall; consequently, one will need to settle for living with the uncertainty of probability. In short, the natural law ethicist might often end up concluding somewhat as follows: "I consider this to be the best course of action, but I can't really say for sure." Sometimes that's the best one can do or say, for the sorting out and balancing of moral principles in complex situations can turn out to be a humanly insuperable task. This explains why the Supreme Court is more often divided than not, and why minority opinions almost always accompany the majority view.

VALUE, UTILITY, AND THE ENVIRONMENT

Another prominent ethical ecologist who has much influenced the development of an environmental ethics and has been actively involved in the various debates that have swirled around its proper identification and definition is J. Baird Callicott. His aim has been, as he tells us, "the construction of a postmodern, evolutionary and ecological environmental ethics" (*Earth's Insights*, p. 188). Callicott sees his ethical theory as universal in nature, and he intends that it serve "as a standard for evaluating the others [environmental ethics]" (ibid.). Though 'officially' Callicott sees his ethic as nonanthropocentric, he grants that it can be viewed as anthropocentric in the weak sense, since the anthropocentrism it supports is "thoroughly transformed by ecology" (p. 208). The environmentalism to which Callicott subscribes is, he says, that exhibited by the American Indian, which he sees as "identical to the ecological concept of a biotic community which is foundational to the Leopold land ethic" (p. 189). Callicott affirms that Leopold's Darwinian ecological world view "has served as a standard for evaluating the environmental attitudes and values associated with traditional cultural world views" (p. 189). The synthesis he envisages does not include an ultimate creative being as the origin of the universe, and Callicott readily admits that "one is hard pressed to find a course of value outside finite (human) consciousness" (p. 21). In an article written in 1980 he had stated: "It is my view that there can be no value apart from an evaluator, that all value is, as it were, in the eye of the beholder" ("Animal Liberation, A Triangular Affair," in *The Animal Rights/Environmental Ethics Debate,* ed. Eugene C. Hargrove, 1992, p. 48). In a recent editorial comment, responding to a critic whom he charges had badly misunderstood his position on the origin of value, he reiterated: "Regular readers of this journal know that I hold a subjectivist theory of value which I trace back to Hume. Time and again I have insisted that there is no value without a valuer; that value is a verb first and not only derivatively" (*Environmental Ethics,* Summer 1996, p. 219).

Callicott thus clearly rejects any attempt to find inherent value in the eco-world. At the same time, it appears that he has no intent of denying that said eco-world possesses inherent worth. Value he sees as being the human's subjective assessment of the worth of various objects experienced. Thus, while *value* is an anthropocentrically derived term, *worth* appears to be totally nonanthropocentric, or objective, for it exists whether or not humans are there to assize the value of nature's worth. He expressly affirms that it is through the human's sympathetic acceptance of natural objects that they become "ethically enfranchised." "Soils, waters, plants, and animals," he writes, "stimulate our social instincts and sympathies. They bring into play our moral sentiments. Accordingly, we extend these fellow members *moral consideration;* we grant them moral entitlement; *we enfranchise them ethically.* Individually and collectively they command respect" (*Earth's Insights,* p. 204; italics added). What Callicott, then, seems to support is a position similar to that of Paul Taylor, since the moral enfranchisement of the sentient and inorganic world follows upon humans having judged it as being deserving of respect.

Yet, further on Callicott will explicitly acknowledge that there is a decidedly anthropocentric dimension to his manner of regarding the ecology and promoting its conservation, for, in defense of the environment, he appeals to the human's sense of self-interest. He first notes that our survival as a species depends upon a functioning biosphere, and, second, he argues that "we ought to care about the natural environment for reasons of embedded self-interest, for our identity as individuals as well as our collective identity as a species depends on the integrity both of our local bioregion and of the whole biosphere" (p. 209). This seems to smack very distinctly of a utilitarian ethics.

Since, further, Callicott finds dubious the proposition "that a transcendent, personal God exists in fact as well as story" (p. 22), he must search elsewhere if he wishes to firmly ground his environmental ethics. To provide this he invokes the golden rule of authentic environmentalism, alluded to above, laid down by Aldo Leopold in his now classic work, A Sand County Almanac, namely, that "A thing is right when it tends to preserve the integrity, stability, and beauty of the biotic *community*. It is wrong when it tends otherwise" (quoted in Earth's Insights, p. 204; Callicott's italics). Callicott is unsympathetic to those who do not accept the Darwinian thesis, but "still cling to the dream of a special metaphysical status for people in the order of 'creation.'" He insists that "people are (and only are) animals" ("Animal Liberation: A Triangular Affair," p. 46).

Whatever the true intent of the author, the view defended seems difficult to distinguish from a pantheistic position, for the singular emphasis on the biotic community seems to make of this latter the supreme value, against which all activity is to be measured and evaluated. Indeed, the environment itself is now seen to take on a near-sacral quality, as the human is but one species among countless others comprising that community. Thus, according to this "land" ethic, "the moral worth of individuals [including, n.b., human individuals] is relative, to be assessed in accordance with the particular relation of each to the collective entity" (p. 51). Callicott himself characterizes his view as a "genuinely postmodern, scientifically informed and constructed theory for environmental ethics. It is genuinely postmodern because its conceptual foundations are not Cartesian" (Earth's Insights, p. 209). He goes on to comment that its intellectual content resembles more Eastern than Western cultures, as it supports a sense of unity between self and the environment which is alien to the Western world but which harmonizes with indigenous cultures.

From the above account of Callicott's philosophy of the human person and of the environment, one is well prepared for the deflating description (definition?) he gives of the individual human. "From an ecological point of view," he states, "oneself is a nexus of strands in the web of life. . . . Any entity from an ecological point of view, is a node in a matrix of internal relations" (*Earth's Insights*, p. 207). From the ecological point of view it is indeed difficult to understand just how and where the human fits in, and, if the basic ethical thrust is towards the maintaining of the present balance between the various biota and the environment, it is anything but clear what the role of the human might or ought to be. What, for example, does the human contribute to the "matrix of internal relations"?

Many of the environmentalists, including Callicott, warn of the imminent danger of overpopulation of the human species; that, indeed, even now planet earth is badly overpopulated, much to the detriment of numerous animal and plant species, many of these latter having become extinct. And, if the human is seen as an evil force ecologically speaking, and if the main concern of a viable environmentalist ethic is the conserving of the delicate balance obtaining among the many species of biota, and the flourishing of the environment, then might not draconian measures aimed at firmly reducing the number of human biota, or even terminating the human species altogether, be ethically warranted? If there is no transcending dimension to the human, how can one justify the human's being the recipient of special consideration among the other biota of the environmental world, save to the extent that the human contributes to the health and balance of the same ecological world? Finally, the reader's attention is once again drawn to the incoherency of a position which insists on maintaining the status quo in the earth's ecology when the very "balance" that is now experienced within nature is the unplanned result of a fortuitous interplay of forces following blindly upon the "law" of natural selection. Why should the present status of the ecology be considered the best ecology of all ecological worlds? Shades of Leibnitz?

ENVIRONMENTAL ETHICS AND THE PRINCIPLE OF INTEGRITY

A recent work on environmental ethics, which lays claim to providing an original basis for a biocentrist view which latter we have already examined at some length in discussing the ethical positions of Taylor,

Marietta, and Callicott, is of special interest to our study. In a work entitled *An Environmental Proposal for Ethics: The Principle of Integrity*, Laura Westra sets out to provide a more satisfactory basis for environmental ethics, acknowledging that previous efforts in this regard have fallen short of accomplishing this goal. As have a number of environmental ethicists who have preceded her, she is seeking to establish an ethics of the environment that is holistic—that is, which is all-embracive, inclusive both of all the species of living beings and of the inorganic world of nature as well. What Westra claims to be unique to her position is that it emphasizes in a formal way the primacy of the characteristic of "integrity" as the authentic leitmotif of a holistic biocentrism. She makes the broad claim that "present environmental problems require not only a new ethic, but also a deepened ontological or metaphysical grasp of human being, *who* they are and *where* they are" (1994, p. 8).

Westra's major complaint with regard to the more traditional approach to the human is that it sets the human apart from the rest of nature. She characterizes such a position as atomistic, divisive, and anthropocentric. It is clear that for Westra the latter term is also a term of reproach. In her view a symbiotic, holistic approach to the human is needed, which views the human as part of the whole, so intertwined with the other biota and nonorganic beings within the universe that our destiny as humans is inseparable from the health and well-being of the nature that encompasses us. "If our ability," she states, "to think rationally and to conceptualize has evolved from our interaction with our *whole* environment, when we allow it to become depleted we risk a corresponding depletion of the mental capacities of human being" (ibid.).

The extent to which the position she advances truly differs from other positions previously considered is debatable. Westra herself grants that her synthesis is similar in many respects to the environmental ethics of Aldo Leopold. It also bears resemblance to Paul Taylor's major focus on the inherent dignity and worth of nature, as well as to Marietta's holistic biocentrism and to the weak anthropocentrist view defended by Baird Callicott.

Anthropocentrism, Biocentrism, Envirocentrism

As is also true of the driving force behind these other pos: Westra openly allows that her ultimate concern is the establishm a stable, nonrelativistic philosophical base upon which deontolｏｇｉｃａｌ prescriptions regarding the human's use of the environment can rest. In short, there is a hidden political agenda, which aims at saving the environment from the destructive forces of "technocentrism" (cf. p. 103). The *principle of integrity* is for Westra that firm base, the "ultimate value," from which can flow the "deontological formulations" needed to protect the environment.

The integrity of the world of nature is thus the first principle of Westra's environmental ethics. Whatever contributes to the preservation of the harmonic unity of species, both in and among themselves and in their relation to the nonbiotic world, is good, while whatever militates against this holistic state is evil. Westra's position is not extremist. She does not insist, along with Regan for example, on mandatory vegetarianism, but merely requires that humans make use of nature according to the legitimate needs they have as part of the ecosystem, and that they respect the delicate balance of that system. Thus she says, "The first moral principle is that nothing can be moral that is in conflict with the physical realities of our existence, or cannot be seen to fit within *the natural laws of our environment*" (p. 92; italics added).

Two things, particularly, about the above moral principle are worthy of remark: First, Westra does not envisage environmental ethics as a branch of ethical theory in general, but rather, conversely, she sees interhuman ethics as a subdivision of a more generalized ethical theory, and environmental ethics as prior to it and foundational. Environmental ethics alone is thus taken to treat of and contain the *first* moral principle, all other divisions of ethics being subordinate to it. Second, it is to be noted that this first principle entails a clear reference to the natural laws of our environment.

Now in what sense might one speak of *natural laws* of the environment, if one supposes, as does Westra, that the world, for all its complexity, is the result of an evolutionary process that unfolds according to random mutation "guided" by natural selection? This is, of course, a point that has been emphasized before. But owing to its centrality in the whole issue of the origin of rights and obligation, and the proclivity on the part of environmental ethicists to overlook the incompatibility of both assigning laws to nature and assuming a Darwinian, nontheist view of the origin of the world and all that is within it, it seems to bear repeating. That the position Westra espouses does not necessarily include a theist view is made clear from her admission: "While I cannot claim to have proven beyond a doubt that the interpretation of 'dignity' I suggest is the correct one, it seems to me that it is the only possible satisfactory candidate, moreover, it is one which permits albeit without demanding it, a theistic view of the universe" (p. 101). The 'dignity' referred to is the same as 'fittingness', she argues, and is the quality or characteristic shared by all beings, regardless of 'species', conferring on them moral value.

In affirming that a theistic view is not a component part of the nonanthropocentric position she defends regarding the status of the ecosystem, Westra is of course simply saying that such a view is irrelevant to the whole question of the origin of moral worth and/or 'dignity' as related to both the bio- and ecospheres. But then, how is the "ought" in question—the human's obligation to avoid behavior that interferes with the balance of the ecology—anything more than conditional, and hence ultimately reducible to a utilitarian calculus, which in turn she finds inadequate and objectionable?

Put another way, and more directly, what is the ultimate ground upon which an ethically biocentrist view rests? Other than motives of enlightened personal self-interest, what other reasons can be advanced supporting the contention that the human is *obliged* to "respect" the integrity of the ecosphere? Further, in what, precisely, does this "integrity" consist? Does not the very notion of "integrity" include a teleological component? Indeed, is not purposefulness the very soul of integrity, guiding the various elements of the ecosystem to function in continuing harmony? But the evolutionary model of explaining the so-called origin of species shuns all mention of final cause or purpose, and there can be no question but that Westra views her "integrity" position as firmly within the parameters of the evolutionary model. Though she does attempt somehow to reconcile her position with what Aristotle "really meant" when speaking of the unity of organisms and their thrust toward full development, she nonetheless is critical of his science precisely because its "major flaw is . . . that he has no conception of evolutionary theory" (p. 136).

It is indeed ironic that Westra's very criticism of Aristotle's view— that (allegedly) it lacks a developmental dynamism—is the very argument she advances *for* her support of a "holistic position". This latter "seeks to reintegrate and restore to importance and value *all* parts of ecosystems, rather than exalt one group at the expense of others." To exalt one group would be to contradict "the principle of 'integrity', which seeks to revalue all individuals of all species, through membership in ecosystems" (p. 138). Clearly, Westra wants the ecosystem to remain just as it is. But how is one able to say that this is the "best of all possible ecosystems"? As times and situations change, might not human ingenuity assist the ecosystem to make adjustments that might prove beneficial to all "parties" concerned? And who is to say a priori that this might not be achieved through the reduction or, even, in critical situations, the practical elimination of some particular species of biota?

Five or so billion years ago, no ecosystem whatsoever existed on planet earth, for our solar system was not yet in existence. Within 'x' number of years this solar system will cease to exist as such. The life-forms on planet earth, both individually and specifically considered, will, at some unknown point in time, cease to exist. What, then, is the "inherent" value of the earth's ecosystem? Or, an even prior question can be asked: Why, to begin with, is there an ecosystem? And why should one strive to preserve it in its present state, if it just happened to be? If it has no teleology or purposefulness? If its total reality consists simply in its remaining as it now is? *Ad quid?* The ethic Westra proposes turns out to be self-refuting, for it depends on the gratuitous assumption that the integrity of the ecosystem, whose lifespan is temporal, is an unconditionally ultimate value, eliciting in the human a moral obligation to protect and conserve it.

Nor can she seek recourse by appealing to the value of the ecosystem as instrumental, in that it is vitally important for the survival of the human that the weather patterns on our planet, as well as other conditions of the planet's environment, remain very much as they now are. To argue in such fashion Westra must abandon her biocentric position by subordinating the environment to the well-being of the human, thus accepting an anthropocentric view of sorts.

To affirm, as Westra and numerous other environmentalists do, that the ecology of our planet has in and of itself the "authority" to oblige the human to structure his behavior in what relates to the environment in ways that contribute first and foremost to the maintenance of its health and viability as a system, amounts at once to an unceremonial dehumanization of the human, as well as to a divinizing of nature. Further, it simultaneously involves a signally fallacious leap from inherent worth or value to possessing the awesome power to impose inordinate moral constraint and obligation on the planet's only self-determining organisms. Westra awards to nature, which through time has randomly and fortuitously arrived at its present state of incalculable complexity and balance, the moral power to require of the human the pursuance of behavior that is altruistic, i.e., not calculated to enhance or promote the well-being of the individual human, nor even perhaps of the entire human race. In this way the human is subordinated to the environment. The latter is capable of imposing obligation on the human, but the human is forbidden to restructure nature in ways that promote the human good.

As discussed in some detail in the previous chapter, obligation and rights are correlative notions, the former being dependent on the latter. There can be no obligations that are not predicated on rights. Hence to claim that the ecosystem can constrain humans to act in accordance with what befits its own health and 'integrity', one must first explain how an amoral colossus can impose such obligation on a moral being. Whence does the power to impose obligation derive? Westra is surely aware that this is not in accord with Aristotle's own ethical view with regard to order and development in nature, but she incongruously suggests that one could argue "on Aristotelian grounds" that the environment could impose obligation on the human if one takes the goal of nature to be "not only 'good' but an ultimate value." In the latter case she concludes that "some obligation not to interfere with a naturally un-folding self-actualizing reality should be in order" (p. 139). But what does one really understand here by "ultimate value"? How could one consider integrity the *ultimate value* without at the same time implicitly committing oneself to a pantheist position? Otherwise, why should "some obligation . . . be in order"? Whence, precisely, does this obliga-tion derive? Further, what place does "a naturally unfolding self-actual-izing reality" have in a Darwinian universe?

Though it is true that some contemporary ethical environmental-ists, following the lead of Peter Singer and others, refrain from employ-ing the term 'rights', either absolutely or in reference to inanimate na-ture or to nonhuman biota, all, nonetheless, seek to impose upon the human definite obligations vis à vis the overall environment or seg-ments thereof. Hence the above strictures apply with equal force to any form of evolutionary environmentalist ethic as well as to the "integrity environmentalism" of Westra, who categorically maintains that "the ideal of environmental 'integrity' is the basis of obligation" (p. 116).

Further, Westra sees the human as profiting by being a member of ecosystems, for "understanding a human being as part of a valuable whole adds objective value to human beings *precisely* as parts of ecosystems . . . *a necessary premise for defending a biocentric environ-ment ethic*" (p. 122; italics added). Necessary perhaps, but far from sufficient! In support of these conclusions Westra cites a passage from Holmes Rolston III to the effect that genetic laws within organic beings constitute *"a normative set"* distinguishing "between what *is* and what *ought to be*." Rolston qualifies and defends this claim by denying that he holds organisms to be moral systems; what he does mean is that "the organism is an axiological system" (as quoted by Westra, pp. 139–40, from Holmes Rolston III, "Biology without Conservation: An Environmental Misfit and Contradiction in Terms," in David Western and Mary C. Pearl, eds., *Conservation for the Twenty-first Century,* 1989, p. 232; italics added).

But clearly, the above reasoning equivocates with regard to the term 'obligation'. *It unjustifiably equates obligation in the moral sense with physical necessity based on a teleological intent inherent in nature itself.* But his latter necessity is not what ethicists understand by obligation. Unfortunately, the fundamental theme of Westra's ethical project appears to rest exactly upon the confusion of obligation with nonmoral necessity. Only a behaviorist—one who denies that there are free acts—could argue in this fashion and still maintain a semblance of consistency. Yet, Westra does not seem to view her position as behaviorist. In concluding her study Westra lays out what it is she claims to have demonstrated. "It is perhaps ironic that a 'fascist', holistic approach can be used finally to readmit *all* biota to moral considerability. . . . When, . . . as it has been claimed here, no one understands the meaning, scope, and role of ecosystem integrity, and is prepared to adopt *the imperatives that follow upon the acceptance of the principle* [of integrity], then sustainability in agriculture and with it the elimination of world hunger will remain a mirage, rather than a goal" (p. 224; second italics added).

In the end the "obligation" to respect nature according to this "rationale" can only arise out of subjective prescription or from some form of utilitarianism, or both. This is not to deny that the human has obligations to respect nature and the environment; it is only to say that such obligations can not coherently be thought to derive directly and solely (as Westra and numerous other environmentalist ethicists seek to establish) from the environment itself.

THE GREENING OF ETHICS

Of the environmental ethical theories of which I am aware, the view presented by Richard Sylvan and David Bennett would seem to be the most radical. Authors of *The Greening of Ethics: From Anthropocentrism to Deep-Green Theory* (1994), they lay out their philosophy of the world with considerable candidness. They leave no doubt in the mind of the reader that they consider the human to be the quintessential enemy of the environment, describing him as a thoroughgoing "class chauvinist" (p. 104). They reject the distinction between human and nonhuman, holding that it is "not ethically significant" (p. 142). What then follows is merely the working out of the logical implications of this denial.

The name Sylvan and Bennett have given to their ethic of the environment is Deep-Green Theory. It is distinctive in that it "calls for a new nonhuman centered ethic with regard to the environment" (p. 90). It "sets anthropocentric concerns within ecocentric concerns. This amounts to an ethical Copernican revolution" (ibid.). Their position aims at replacing *actocentrism and homocentrism* with an ethics that makes "the nonhuman world a proper object of concern, either directly or indirectly" (p. 17). This view has much in common with deep ecology (the name given to an environmental ethical theory developed by Arne Naess). Rather than relying on the deep-ecology principle of "biospherical egalitarianism" (a radical equality of species), however, deep-green theory invokes the *principle of eco-impartiality,* whereby, though differences of species are acknowledged, it is argued that there should be no discriminatory treatment of any species without clear justification. Thus they claim: "The appropriate impartial treatment is substantially independent of the comparative value of the item treated. Most important, impartial treatment does not entail equal treatment, or equal consideration, and does not require equal intrinsic or other value" (p. 142). Here there seems to be agreement with Westra: the ultimate criterion of ethically acceptable behavior is whether it contributes to the integrity of the environment.

Another fundamental theme emphasized by the deep-green theory is "that a range of environmental items are valuable in *themselves,* directly and irreducibly so, so that their value does not somehow reduce to or emerge from something else, such as features of certain valuers or what counts for them" (ibid.). Thus value is taken to be completely independent of the valuer, whether the latter be human, sentient or pertaining to "other value-responsive classes" (ibid.). Such things, therefore, as mountains, lakes, rivers, seashores are for Sylvan and Bennett intrinsically valuable, independently of the use to which they may be put, or howsoever they may be perceived.

Yet, when it comes to providing a rationale for the value theory underlying deep-green ethics, one enters into an area that appears to be much grayer than it is green. One is told, for instance, that "goodness is not an objective quality (nor subjective either, but *nonjective;* it is not simply intuited, but appreciated through reflective methods based on emotional presentation; no sort of *consequentialism* is flirted with, not even ideal *'utilitarianism'"* (pp. 142–43; italics added). The reflective methods are reformed to include what is termed "enhancement methodology" through which one's procedure in evaluating goodness is reinforced by newly acquired data. The 'enhanced' information can then be further sharpened or improved "through presentational input and assessment for overall coherence" (ibid.). Also, the manner in which the evaluator may be affected emotionally by the information received must be added to the equation. As a concluding note, it is argued that an ethical theory must include an "axiological system elaborating deep environmental values and virtues, a deontic framework, supplying obligations, rights, taboos, and similar" (p. 146).

As it turns out, the authors of *The Greening of Ethics* are pushing a political agenda to protect the environment. Though they devote considerable space to seeking to distinguish deep-green theory from deep ecology, the theoretic difference between the two appears significant only in the application of the theory to concrete environmental problems. Deep-green theory finds remnants of anthropocentrism and pragmatic motivation in deep ecology. It views itself as the more radical of the two environmental theories, and the only one that can accomplish what supporters of both theories ardently desire, namely, the preservation and promotion of the earth's environment.

The proponents of the deep-green theory find "all standard ethics mired in heavy prejudice, a prejudice in favour of things human and against things non-human" (pp. 139–40). They consider their ethics to be "a new nonstandard ethics, superseding established ethics in order to further environmental causes" (p. 26). They therefore contend that human chauvinism is "a cardinal weakness" of practically all other ethical systems. By human chauvinism they understand systematic 'differential' (i.e. preferential) treatment of the human coupled with "inferior treatment of items outside the class, by sufficient members of the class concerned, for which there is no sufficient justification" (p. 140). Thus the deep-green theorists are pressing for an ethical theory purged of a specially privileged place for humans (ibid.). Their new ethics would allow no criteria of moral relevance to be based on differences of species or subspecies, or on single characteristic features, such as life or sentience (ibid.).

This, however, does not mean that "traditional ethical notions such as rationality, self-awareness, having interests" are rejected (p. 141). Apparently, the traditional ethical values may be actually applied only within the parameters of the human species. What is not acceptable is the application of human standards to other species, each species being considered as an encapsulated whole, operating, as it were, in a self-contained world. Allegedly, each has its own laws and statutes, which render it "untouchable" by any other species which, by definition (proclamation?) must be "judged" according to its own proper laws and statutes. Such an interspecies arrangement of course raises some interesting questions regarding the "unity" of the environment itself. One seems to be dealing here with a plurality of environmental "worlds" rather than "a world." Even more perplexing, deep-green theory considers "the human/nonhuman distinction not to be ethically significant" (p. 142). It would seem, then, that the single purpose of human existence is to contribute but another link in the chain of diversity of the earthly environment, thereby helping to achieve and maintain an equilibrium within the whole of nature (cf. p. 146).

Sylvan mentions three areas in which deep ecology and deep-green theory are in substantial agreement—ecocentrism, ecoregionalism, and human population. Ecocentrism considers the environment as the ultimate good to be conserved and enhanced, and each of the various species contains value accordingly as it positively contributes to the "organic" health of the environment. Ecoregionalism, on the other hand, recognizes that the needs of different areas throughout the planet are various, and that consequently different criteria and plans are to be enacted and implemented according to this regional disparity. With regard to human population, both theories agree that "population policies should be aimed at decreasing the human population and that a substantial, but carefully monitored, reduction in the human population is compatible with maintaining human values, especially cultural values" (p. 156). In short, the human does not occupy, according to supporters of deep ecology and deep-green theory, a place of greater dignity or value than do other biota or inanimate elements of the environment. One might aptly term this view 'envirocentric'.

CONCLUDING SUMMARY AND CRITIQUE

The foregoing reflections on ecological theory and environmental ethics should serve to bring into bold relief the profound importance of having carried out the investigations conducted earlier in this study, investigations of those areas of human activity which singularly distinguish the human from other living things as well as from the environment itself. Those areas included human knowing and understanding, human language, willing and choosing, and the nature and origin of rights and obligations. Even a cursory scanning of contemporary thinking regarding the area of environmental ethics will quickly persuade one of the crucial importance of a thorough and in-depth investigation of these fundamental aspects of the human person both for defining the human and for clearly delineating the human's relationship to the other living and nonliving beings that co-populate our planet.

One can, then, unquestionably admire the deep commitment environmental ethicists often display in seeking to raise the consciousness level of the contemporary world regarding the urgent need to take immediate steps to safeguard our earthly environment. The point they make—that the human's range of obligation extends far beyond the limited parameters of the human family—is certainly well taken. At the same time, however, one can hardly avoid concluding that their promotion of bio- and ecocentrism or envirocentrism is often carried to notable excess. It is truly regrettable that their eloquent appeal to protect the environment is often wasted through their failure to provide a viable philosophical grounding for their pronunciamientos, so often sweepingly moral. Not infrequently they incur the fallacy of invoking a noble end to justify, as means to achieve it, judgments and evaluations of the human that fall far short of accounting for the full extent of the horizon of human experience.

Thus, with seemingly few exceptions, those characterizing themselves as environmental ethicists insist on laying at the feet of 'anthropocentrism' the blame for the admittedly often gross neglect of the ecosystem on the part of the human. It is undeniable that the past record of the human regarding the treatment of the environment, particularly since the beginning of the industrial revolution, has been abysmal. It is, however, not convincingly evident that the manner in which the human has treated the environment is owing directly to an authentic anthropocentrism. Even less evident is truth in the assumption that anthropocentrism is inherently incompatible with a caring, respectful attitude toward the world of nature. The aberrational misuse of nature can more fittingly be traced, I believe, to two distinct but related causes. First, until well into the past century, humans little realized the limited capacity of planet earth's ability to supply the raw materials required to feed its industrial machine. Secondly, even as such knowledge became increasingly available, humans, overcome by greed, in large numbers continued callously to give scant heed to nature's warnings.

In good part, then, the problem is closely related to the phenomenon of human freedom. It appears that the biocentrist environmentalists, who identify "anthropocentrism" as the major or sole source of environmental neglect and abuse, fail to distinguish the two basic meanings of that term. Anthropocentrism can refer (and it is in this sense that the environmentalists often seem to intend it), to the specific nature of the human, so that what is criticized is the acceptance of humans as superior to other life forms populating planet earth. Taken in this sense, any human behavior judged to be destructive of ecological harmony is seen as endemic to the anthropocentric world view.

The term "anthropocentrism" can also refer to *an inflated attitude of superiority* which can strongly motivate humans to behave toward the environment in ways that are unreasoning and inappropriate.

Behavior arising from this second form of anthropocentrism entails a clear misuse of human freedom; as such, it is not necessitated. Further, since it takes its rise from the phenomenon of human freedom, there is no feasible way in which such behavior can be fully eliminated, or perhaps even adequately controlled, although measures can be taken through the enactment of laws and statutes by the body politic to discourage such misuse. The possibility of the misuse of the power of freedom is inseparable from that very power. One can always be tempted to pursue selfish goals at the expense of the rest of humankind and of the environment itself, and even the most stringent and detailed laws can never eradicate such a possibility. If 'anthropocentrism' is understood in this second sense, it can clearly be taken as a term of reproach. Yet this by no means justifies restricting the term to this latter sense only, using it as a critical canon whereby humans, by their very nature, are viewed as implacable enemies of the environment.

Furthermore, to maintain that the human is the highest life form on the planet is not to deny the obligations of humans to the environment. We have shown that humans have a real, though indirect, obligation to respect the environment. It is from the author of nature and not from nature itself that humans incur the responsibility of respecting the garden-world in which they live.

Further, the anthropocentric ethicist need not deny that the human is part of the ecology. What the anthropocentrist will deny—and this is what differentiates his position from that of the biocentrist—is that the human is totally submerged in the environment in which he lives, and that the ultimate human good is to be identified with the healthy, affluent status of the environment. This romantic vision of human happiness seems much akin to a terrestrial Garden of Eden. The human has indeed a deep interest in the environment, since the quality of human existence is very much dependent upon its healthy status, but this does not exhaust the human's menu of desirable goods, for his ultimate fulfillment is hardly irremediably identifiable with the well-being of the world in which he lives. As an intellective being, the human's horizon of interest is much broader, extending well beyond the seas and mountains that frame the ultimate reaches of his earthly vision. A citizen of two worlds, the sensory and intellective, the human sees the world through two different, though complementarily aligned lenses, focussing them to a single vision. This constitutes the uncompromisable uniqueness of the human experience. The beauty of the universe is not solely in the eye or mind of the beholder, but it is the human *alone* among earth's biota who 'lives' in a world of beauty, a world transcending all manner of sensory perception, who can appreciate and ad-

mire that beauty, imitate it imaginatively through creative art and re-
shape and order that world in limitless ways.

It ought also, of course, be emphasized that most if not all those
identifying themselves as biocentrists, also formally identify with a
monist view of the human, and are exceedingly ambivalent when discussing intellective and sentient knowing. Their cognitive roots seem
commonly traceable to the philosophy of David Hume, and, especially
with regard to moral theory, to Immanuel Kant. The basis for their
monistic position regarding the human is often openly acknowledged
to have its source in the evolutionary theory of Charles Darwin. As we
have earlier discussed at some length and retrospectively alluded to on
numerous occasions, one accepting the Darwinian theory of the emergence of species-forms is ultimately left with no other recourse than to
view the human as no more than an animal, all differences between the
human and nonhuman animal being reduced to a state of practical inci-
dentalness.

The term environmental ethicists continually use to give expression
to the above mentioned reality is 'biocentrism'. Yet biocentrism is fatally destined to fall short of providing an accounting of the human that
embraces all the facets of human experience to which we have previously alluded. Biocentrism, as commonly understood by ethical environmentalists, while focussing on the commonality of all living things,
strongly minimizes their differences. In this sense, it becomes transformed into a highly abstract term which can provide little basis for a
philosophy of the environment. Above all, it lacks the breadth and flexibility needed for coordinating the various life forms in a way that recognizes their obvious diversity while simultaneously acknowledging
their similarities. 'Biocentrism' can provide no overarching synthesis of
the human with other living things and the environment, nor support
an ethical theory that authentically coordinates their differences. By
leveling all living things to the anonymity of merely living entities, biocentrism throws away the only key that can unlock the secrets of the
plurality as well as the unity of the environment which constitutes our
world.

7 Animals in the Human's World

The preceding chapters have provided a detailed comparison of the human and the nonhuman animal, focusing on their salient differences while at the same time seeking to underscore the manner in which they share a life in common. In this, the concluding chapter of our study, we turn our attention to the unique role the nonhuman animal plays in the world of the human. We will explore the importance of the animal to the human and attempt to visualize what life for the human might have been were the human's world completely devoid of animals, particularly quadrupeds and birds. We will also reflect on how the sharing of our lives with animals contributes to a better understanding of ourselves and offers an antidote against loneliness and a welcome analgesic to psychic stress and pain. These considerations will be made against the backdrop of the reflections already made in the preceding parts of this study, and will, it is anticipated, serve to persuade the reader that those earlier reflections do not and need not exclude an attitude of concern and affective appreciation of the many species of animals that co-inhabit our world.

As suggested on several occasions earlier, there is much in the agenda of the animal liberationists deserving of thoughtful consideration and much with which one can agree. But one need not adopt a radical egalitarian stance vis à vis the human, the nonhuman animal, and the environment in order to oppose the clear excesses of some humans in their treatment of the nonhuman animal and the environment. One can

348

have salutary recourse to other options short of elevating the status of the animal to or near the level of the human, which, as we have seen, debases the quality of rights that humans enjoy and, in the end, renders 'rights' at all levels powerless to provide protection to the human or to the nonhuman animal, or to the environment.

Turning our attention, then, to the distinctly positive contribution the animal makes to our world, it will be helpful first to recall how much there is that animals share in common with us humans. We, too, are animals in the nuanced sense already painstakenly set forth, for we are animals who are at the same time rational and intelligent. As humans we are a unique union of two distinct elements, body and mind. This uniqueness of our nature, coupled with the fact that we combine the seemingly incompatible ingredients of sensation and intellection, leads to the inexorable conclusion that we are unable to find adequate reflections of ourselves deriving from any area of our world other than from that which is human.

As sentient and intelligent beings, we humans neither sense without understanding nor do we understand without sensing. This means on the one hand that we have no direct, Cartesian-like, pure awareness of understanding as completely isolated from our sensory activity. On the other hand, once we have developed our power to reason, we individually have no consciousness that is purely sensory. Epistemological theories, in attempting to account satisfactorily for human awareness, have historically oscillated between the two extremes of radical empiricism and absolute idealism. Because it is so difficult to integrate the mental with the physical in such a manner that components of each are congenially united—rather than opposing the two as alien, seemingly conflicting, worlds—we have repeatedly emphasized throughout this study that the human is an animal that is rational. Though the human possesses an intellective capacity, the exercise of this power requires that it remain in contact with the sensory world from which its intellective knowledge derives. It is precisely because of this dependency on sense knowledge that the human intellect must reason in order to unify its intellective knowledge which derives from insight into phantasm, in order that it might give formal expesssion to its truth through the complex act of judgment. Though truly intellective, the human soul occupies, as Aquinas states, the lowest rung on the ladder of intellectual substances, since it possesses no knowledge whatsoever prior to its exposure to the world of sense and since it is constrained to acquire its knowledge in this life through the intermediacy of the senses and, consequently, of the body with which it is united as its form (cf. *On Being and Essence*, ch. 4, # 10, and *ST* I, q. 84, a. 7). It can be seen, then, that this view is as opposed to the Kantian-Humean view of the human as it is to the Cartesian.

ANIMALS AND THE HUMAN'S APPROPRIATION OF SELF

In our encounter with the animal world, and particularly with the higher vertebrates, we come into contact with a consciousness that is sensory but lacks intellection. Only in nonhuman animals can we see the operation of animality as separate from rationality. Only in nonhuman animals do we experience beings that are sentient but nonintellective, that lack the higher level of consciousness made possible to us humans through intelligent insight and reflective understanding. We ought not underestimate the value of this encounter. The presence in our world of animals that are not rational beings allows us to draw back from ourselves and to imagine what it might be like to be simply sensory beings lacking the power of intellective knowing. Such an encounter thus provides us with a unique window on our own human behavior, opening up for us a clearer vision of the significant difference that intellective awareness makes in the realm of behavioral activity.

Our encounter does not, however, permit us to penetrate fully into the elusive inner state of consciousness of the nonhuman animal. Our everyday, common awareness of the nonhuman animal's interior sense-world becomes inescapably suffused with the transcending, nonsentient activity of human intellection. There is here a rather odd, and doubtless unexpected, partial convergence of the human's knowledge of animal consciousness and the general conclusion Kant has drawn regarding the whole achievement of the human intellective project. For Kant claims that what we can term authentic, i.e. augmentative, knowledge is strictly limited to what he terms synthetical judgments, which are always phenomenal in nature. That is, they provide knowledge not of things as they are in themselves, existing independently of human awareness, but only as they appear to us in consciousness. Kant's view finds a rather interesting parallel in the matter at hand, though to be sure not in the manner he himself would have foreseen. The parallel consists in the fact that our awareness of sensory activity is never an awareness of sensation as it is in itself, that is, as completely separate from an intellective awareness, although this is the manner that sensory awareness actually exists in the nonhuman animal. Thus our human psychological world does not and cannot fully coincide with that of the nonhuman animal, and conversely, the nonhuman animal is unable to cross the threshold between the world of strictly sensory consciousness and the world of intellective awareness.

Granted that our lived experience can never be the 'lived' experience of the nonhuman animal, it nonetheless is true that our contacts with the nonhuman animal do provide us with a uniquely valuable insight into our own sensory experience. But, even here, much of our experience of what it might mean for us to be without intellective awareness is present to us in a fashion more negative in character than positive. We are able only to conjecture imaginatively what the limitations of pure animal consciousness are as experienced by the sensing organism itself.

This means, then, that by our association with animals that are not human, our task of separating out the dual components of our own human consciousness is greatly facilitated. Consequently, our contacts with the animal world assist us in a most positive way to achieve a more precise awareness of who we are as humans. They thus provide us with valuable clues as to how the two sources for human consciousness, the sensory and the intellective, in their confluence constitute human consciousness. By our observation and monitoring of the behavior of the nonhuman animal (and concurrently by remarking the animal's inability to perform creatively and in freedom), we humans are better able to delineate the limiting parameters of sensory awareness *sans* intellective consciousness.

Finally, a comparative study of the nonhuman animal's behavior and our own provides us with a deepening realization of the profound importance of human intellective power to the process of distinguishing humans from other conscious earthly organisms. The conclusion that they are distinct is warranted, since the actions of nonhuman animals would otherwise not remain, as they clearly do remain, stagnantly invariant over millennia and beyond, were these same animals possessed of the equivalent of the human's intelletual acumen. During our earlier discussion of animal language the same line of reasoning was traced in detail by addressing the question of whether the nonhuman animal may indeed be capable of employing language after the manner of the human: "But how do you know what animmals are thinking, and how can you be sure the sounds they utter are not a genuine form of language?" "Is it not possible that they are indeed speaking, but that we simply cannot interpret the language they are using?"

There are of course many extrinsic similarities among the various species of higher animals. They provide the impetus for asking such questions. But the similarities are accompanied by significant, though often overlooked, dissimilarities. The tension that arises from the dialectic of similarities and dissimilarities has led to our rather lengthy and painstaking investigation of the comparative status of the human and the nonhuman animal. This has resulted in a more finely nuanced appreciation of the mutual complementarity of the two worlds of sensory and intellective knowing, and, particularly with regard to language, to our being able the better to recognize the unique constructive harmony achieved, within the single human organism, between sensory and intellective knowing.

Hence, by our interaction with and observation of animals, we humans become more acutely aware of the difference between our own acts of understanding and our acts of sensation, for we are then allowed to appreciate with a sharper clarity the constraining limitations that the sensory powers, when left to themselves, impose on the knowing subject. Negatively, we are led directly to the stark realization that there are many aspects of our environing world to which the senses fail to attune us, simply because the sensory act, when isolated from the universalizing dynamism of intellective awareness, is not up to the task of unfolding before us the plethora of riches contained in the world we inhabit.

At the same time, we are, conversely and on the positive side, led to a deeper awareness of what, precisely, it means to 'understand', for we are enabled more clearly to experience the difference that understanding makes and to experience its singular contribution to our heightened quality of awareness, which carries us far beyond the limiting confines of the senses. Stated somewhat paradoxically, and by way of summation, we may justly conclude that contact with animals that are not human is all but essential to our own full appropriation of who we are as humans. The presence of animals in our world helps us to understand and better come to terms with our own animality, while at the same time reflecting in the mirror of our inner consciousness faint traces of our ever-present, but so often frustratingly elusive, rationality.

ANIMALS AS FRIENDS

On a perhaps less obviously philosophical level there are other ways in which the nonhuman animal significantly impacts the human's world and notably enhances it. Many families 'adopt' animals into their homes as pets. Although dogs and cats are far and away the most common species that are so accepted as added 'members' of a family, they are not by any means the only ones, for there are many animals that are attracted by human 'companionship' and readily adjust to domestication. The few reflections that follow, however, will be restricted to the canine species, which is, from time immemorial, universally recognized as the most companionable of all animals that man has befriended. Nonetheless, many of the following reflections could, with almost equal validity, be applied to pets of other species such as horses, ducks, geese, song birds, squirrels, cows, pigs, deer, and, of course, the domesticated cat.

Doubtless, people have many different reasons for keeping pets, but high on the list must surely be the companionship they afford. It is remarkable the intimate bonds that can develop between the family dog and its adoptive family. The warmth and depth of the affection a dog can display toward its owners and all members of the family, especially to the children, is legendary. The dog is also extraordinarily trustworthy and faithful in its loyalty to its owners.

If we inquire into why it is that so many families retain animals as pets, especially dogs, I believe the answer is to be found mainly in the fact that the affection a pet animal offers the human is quite unique. One finds in the animal a spontaneous and, on most occasions at any rate, a full return of affection, which, assuredly, cannot always be said of one's relations with members of one's own species.

There are of course good reasons for this striking discrepancy between one's dealings with one's fellow humans, even members of one's own family, and one's pet. When contrasted with the human, the pet leads a comparatively simple life. It has few if any worries or anxieties. It is not troubled by what tomorrow might bring, nor preoccupied by problems of the past. It does not appear to experience the capricious mood swings commonplace for humans during the course of a single day. Consequently, the animal's disposition is far more constant, and its attention, for reasons discussed earlier in chapter two, is focused exclusively on the present. It does not ordinarily bear grudges, quickly forgetting a rebuke or scolding it might have received but a few moments earlier. Because the pet lacks an authentic inner world upon which to focus, it is always ready to give its full attention to whatever occurs at the moment, and, if what it experiences is pleasant, it reflects total, unremitting joy. Its emotional response is immediate and it is genuine. Its joy is not feigned; its affection not contrived.

Further, the pet is a good listener. We are usually able to speak to it without being interrupted or contradicted, and the very uttering of words aloud may act, as we hear ourselves speak, as a soothing balm to our own complex and perhaps troubled inner world. The affective exchange between the human and the animal 'friend' occurs in a climate of complete trust and assurance; the animal is not a threat to our psychic world, and we are fully enabled to relax and unwind in its presence.

And what may be the most important point of all to emphasize is that the pet is able to "knit up the ravelled sleeve of care" precisely because it is in its affectivity restricted to the world of sense-based emotion, lacking as it does an intellective experience behind its emotional world. The animal lives in a conscious world of the present only. It experiences no difficulty in making itself available without restriction to its 'friend' of the moment. And yet, because the outpouring of affection from the animal is so spontaneous and so unrestrained, we are easily convinced in anthropomorphic fashion that we are indeed, for the moment, the most important 'person' in its world, for our being the recipient of all that 'love' and attention.

ANIMALS AS GUARDIANS OF CHILDREN

One of the more important reasons many families have adopted a pet is to provide companionship and protection for their children. It is commonplace to observe that animals and children get along most famously together, almost as if they were emotionally made for each other. What might be the reason behind this remarkable phenomenon? The main reason seems not hard to find; identifying it will provide us with an opportunity of bringing more clearly into focus several points made earlier in our study of the animal and the human. These revolve around the immature status of the child. The very young child has not yet developed all of its mental powers, so its behavior displays the many ways in which it shares the emotional and psychic level of consciousness of the nonhuman animal.

As one reflects on the symbiotic relationship between the family pet and the small child, it becomes clear why the child and the animal get along so well together. Their respective mental states and basic interests diverge considerably less than do those of the animal and the adult human. As does the animal, the very young child of two to four years lives almost totally in the present. It recalls little of its past and the distant future offers no overriding concern. Thus the small child and the family dog make ideal companions one for another, for their communication level is very nearly the same. The child speaks to the animal, if it speaks to it at all, not in full sentences, but in one or two word phrases which are more signals for its emotions, than signs of an interior intellective awareness, and the pet is quite capable of handling that.

Thus the pet is able to provide valuable companionship for the small child or children and allows the mother or parents to attend to various other tasks without leaving the child completely alone and companionless, where it might soon become bored or lonesome and begin calling for attention. The domesticated pet, having little else to do anyway, with no pressing need of hunting for food, welcomes play and diversion as much as does the child. I am sure that all who had pets as a child are most grateful for the patient, forebearing companionship they provided during those long, often tedious years of growing up.

ANIMALS: COMPANIONS TO THE ELDERLY

In treating of this very issue of friendly interaction between humans and animals, Mary Midgley has suggested that we humans are by nature neotonous, that is, our personalities are deeply influenced by our childhood experiences, and these remain with us to some extent throughout all our adult lives. This very quality of retaining some of the characteristics of the child she points to as explaining our spontaneous attraction toward animals of various species, why we seek them out as companions. Furthermore, she views the companionship we form with animals not as a replacement of human friendship but rather as complementary to it. I believe that in this she is surely correct.

Yet Midgley goes on to surmise that it might have turned out quite otherwise; that we humans could well have been so programmed that we would experience no affective attraction toward any alien species (cf. "The Mixed Community," in *Earth Ethics,* ed. James P. Sterba, 1995, pp. 85–86). This precarious speculation disturbingly implies an internal contradiction in the assessment of human behavior: The attraction we humans experience toward other animals is natural, that is, it flows from the animal nature we share with them. Should we feel no attraction whatever toward what are in fact our "generic cousins," we would be repudiating an aspect of our very selves, the very self-image we observe as we gaze into the "living mirror" of nonhuman animals. An enduring internal conflict of this kind would, while involving an "ontogenetic" contradiction, also almost certainly have catastrophically negative repercussions in human psychology. Such a condition would drive an enduring wedge between the intellective and sensory dimensions of the human, thus falsely dichotomizing the human person and rendering human love for members of our own species all but purely Platonic in nature, since its sensuous dimension would disappear. Midgley's scenario can be entertained, welcomed even, in a Cartesian perspective, but certainly not in an Aristotelian one, which envisages the sensory world as the original seedbed of all human intellective awareness and aspirations. Doubtless the scenario would also be warmly welcomed by those onetime friends and cultist colleagues of St. Augustine, the Manichees.

In addition, and for analogously similar reasons, household pets provide much comfort and companionship to the elderly and infirm, who are now much less active than they formerly were during their youth, and who often have nothing, if not time, on their hands. Pets can bring joy and welcome diversion into their lives, which are often lonely, and help turn their attention away from themselves and from their own problems and infirmities. In recent years the affection and warmth the pet is able to bring into the lives of the sick and infirm has been discovered in rest home facilities and even hospitals, where puppies, have been brought in on occasion to be petted and held by the patients.

This practice has proved wonderfully effective toward brightening their spirits and their day. Again, as in the case of the small child, with the elderly and infirm the gap separating them and the nonhuman animal has been significantly narrowed. This time, however, not owing to the undeveloped state of the elderly patients' intellective powers, but because their exercise is noticeably impaired by the onset of old age and physical infirmity. The elderly are thus in a better position to enjoy the frolicsome antics of young animals, especially those of puppies and kittens, much as young children do, for with the advance in years they increasingly find enjoyment in the simple pleasures they enjoyed so much as a child. Intellective interests often gradually give way in the elderly and infirm to absorption in the everyday life of the senses.

ANIMALS: SOURCE OF ENJOYMENT AND BENEFITS

Finally, the contribution to the human made by animals in the wild ought not be overlooked. Apart from the many species of animals that have perennially provided a rich food source for the human, among which can be numbered the vast variety of fishes and marine animals, birds, chickens, water fowl, sheep, goats, antelope, elk, deer, bears, bison, cattle, as well as many others, we need also consider the aesthetic pleasure millions of different species of animals have collectively provided by their presence in our world. Anyone who has strolled through meadows or woods or along the shores of our lakes, rivers, streams and oceans, in the mountains or on the prairies or deserts of our planet, cannot but have experienced the enriching presence of many varieties of birds as well as terrestrial and marine animals.

Allowing ourselves for the moment to imagine our world *sans* non-human animals may serve to bring home to us their importance in our world. Even should we assume that the food supply would still be adequate to our needs, and that machines could successfully supplant animal power in countries not presently enjoying mechanization, yet what an overwhelmingly impoverished world that would be. A wholly precious dimension of our world would fall silent, much of its visual beauty and attractiveness gone; for, apart from plants and trees, our planet would be populated only by ourselves, by humans, and we would doubtless feel very much more alone, and longing for the return of our animal coinhabitants.

The conclusion that animals are essential to our lives on this planet seems inescapable, if one but quietly reflects on the myriad ways in which animals contribute to our needs, and our enjoyment. As reiterated at the outset of this chapter, the role animals play in holding up to us the mirror of nature, allowing us to better understand just who and what we are, we humans, is an indispensable one. The millions of diverse species of animals collectively help us to better understand and appreciate the great mystery of the life we share, recognizing with a deep sense of awe, that, though there are on planet earth countless life forms of every imaginable structure and configuration, there is none, save the one we call 'man', that possesses the power to ascend to the summit of the intellective mountain from whence a panoramic, all-encompassing vision of our world in all its grandeur fills our gaze.

Epilogue: The Animal and the Theologian

In what has preceded, the comparison between the human and the nonhuman animal was sedulously carried out on the philosophic level. Before concluding our study, however, it seems appropriate that some consideration be given to the theological dimensions of the place of the animal in the human's world. To the extent that the nature of the animal is viewed as related and similar to that of the human, the theologian must of necessity be concerned, for whatever intimately touches the nature of man will inevitably bring within its train notable theological repercussions, as should be evident from the fact that the central Christian mystery, the Incarnation, directly involves the union of a singular human nature with the second person of the Holy Trinity.

According to the Christian interpretation of the New Testament and the earliest tradition stemming from the time of Christ, and as expressly declared in the fourth century at the Council of Ephesus, Christ is, though one person, both God and man. He is not, however, two distinct individuals but one, a Divine Person having two natures, which makes Him both truly God and truly man. According to the Christian tradition the Second Person of the Trinity assumed to Himself a human nature, so that Christ is God become man. Though He is human, Christ is not a human person, since His humanity, although individual and perfect in the order of its nature, does not have an independent existence as a created person but exists as assumed into the Godhead by the Son, the Second Person of the Holy Trinity.

Now from this it becomes clear why any philosophical position which either directly or indirectly, intentionally or unintentionally, minimizes, perhaps even to the point of factually denying, a true distinction of kind between the human and the nonhuman animal, threatens seriously to compromise the Christian teaching whereby Christ is held to be both God and man. This is why Darwin himself acknowledges in his autobiography, composed toward the end of his life, that he is no longer a Christian believer, for seemingly he could not help but recognize that his theory of the evolution of man through sexual selection dealt a severe if not fatal blow to Christian belief—and, indeed, perhaps even to any form of monotheistic belief. This assessment of Darwin's is corroborated by what appears to be the consensus opinion of contemporary life scientists, particularly biologists and anthropologists, accepting the Darwinian explanation of the origin of species. Stephen J. Gould, Richard Dawkins, and the late Carl Sagan are three instances of prominent scientists who are self-defined atheists. As Darwinists they see the existence of a supreme being as unnecessary to explain the origin of the human and the universe.

CHRISTIAN THEOLOGY AND VEGETARIANISM

If, then, there is held to be no substantive difference between man and the animal, the Christian position regarding the Incarnation could imply that God, in becoming man, became not only man but a nonhuman animal as well, a clearly intolerable conclusion for a believing Christian. It is mainly to the question of philosophical vegetarianism, however, that I direct the reader's attention in this epilogue, since it is held by some of its proponents to be inherently compatible with Christian belief. Philosophical vegetarianism, as alluded to earlier, is the view that it is intrinsically, and hence morally, wrong to use the flesh of nonhuman animals as food.

In his book *The Philosophy of Vegetarianism*, Daniel A. Dombrowski adopts and firmly defends this view. He finds it not to be "at all clear that animals completely lack rationality" (p. 8) and further concludes that Darwinian theory renders "every earlier justification of man's supreme place in creation and his dominion over animals untenable" (p. 17). It is, then, not surprising that Dombrowski should be uncompromisingly critical of the Judeo-Christian tradition's treatment of the animal. Indeed, he does not stop short of characterizing this tradition as *speciesist* (p. 9), contending that "even if human beings do most closely approximate the divine likeness, *there is no reason to infer that animals are our slaves* and can be treated in any way whatsoever, as long as our treatment of them does not lead to cruelty to human beings" (ibid.; italics added).

There are two logical problems with this view. First, we have seen that if one accepts the Darwinian hypothesis regarding the natural origin of all life forms, man included, then one cannot logically contend that the human has any moral obligations whatever. Within a Darwinian context it may indeed be in an individual's best interests to act 'morally' by avoiding giving offense to others, but it would be logically unsupportable to claim that such behavior is *morally* necessitated or mandated.

Secondly, for a Christian (who perforce understands Scripture to be divinely inspired), it would be incongruous blithely to put aside the express views of both Old and New Testaments regarding the nature and place of animals in the divine plan for the world. In the Book of Genesis we read that the human is the climax of God's creative activity, and that God has given humans dominion over all living things on the face of the earth, including the animals of every species:

> Let the earth bring forth all kinds of living creature: cattle, creeping things, and wild animals of all kinds. And so it happened: God made all kinds of wild animals, all kinds of cattle, and all kinds of creeping things of the earth. God saw how good it was. Then God said: "Let us make man in our image, after our likeness. Let them have dominion over the fish of the sea, and the birds of the air, and the cattle, and over all the wild animals and all the creatures that crawl on the ground." (Gen. 1:24–26)

Surely God, the source of all living things, has not acted unjustly in subjecting animals to the dominion of man. Yet that appears to be the implication of Dombrowski's comments on this and other passages of the Hebrew Scriptures. With thinly veiled sarcasm he grants that the Scriptures advise some restraint in man's treatment of animals, but complains that in effect this is too little too late. "It is true," he states, "that men were given some guidance regarding animals; for example, they were not to boil a kid in its mother's milk [Exod. 23:19]. But this offers no solace to the kid, who could presumably be boiled nonetheless" (*Philosophy of Vegetarianism*, p. 5). In a similar tone Dombrowski adds: "It is also true that Isaiah predicted a time when the lion would dwell with the lamb; unfortunately, there is no clear indication that the lamb could dwell peacefully with *man*" (ibid.; italics added). His final comment on the Old Testament's teaching regarding the place of animals in the human's world is that "the Hebrews viewed man as the crown of creation, a status which denigrates animals" (ibid.). It is difficult to uncover in these remarks a pattern of logical consistency. May not the Lord of the universe judge what is fitting in dealing with creatures that His wisdom and power have brought into the world? St. Paul, a man certainly not unaccustomed to the ridicule of a hostile pagan world, responds forthrightly to those questioning the propriety of divine providence as it manifests itself in the world:

> Friend, Who are you to answer God back? Does something molded say to its molder, "Why did you make me like this?" Does not a potter have the right to make from the same lump of clay one vessel for a lofty purpose and another for a humble one? (Rom. 9:20)

In another place Paul drives home the same point: "Who has known the mind of the Lord so as to instruct him?" (1 Cor. 2:16). The conclusion Paul obviously expects his hearers to draw is that a God in need of counsel is a God in name only.

Nor is Dombrowski's critique of the scriptural views regarding the status of animals restricted to the Old Testament, for he is hardly less critical of the attitude toward animals shown by the writers of the New Testament. Indeed, he does not withhold his criticism even from the person of Jesus. In a passage that occasions Dombrowski's criticism, Matthew relates that two men possessed by devils approached Jesus. The devils entreat him, if they are to be cast out, to be allowed to enter into a herd of swine nearby. Jesus grants them their request, and Matthew then records: "they came out and entered into the swine; and behold, the whole herd rushed down the cliff into the sea, and perished in the water" (Matt. 9:32).

Dombrowski complains that Jesus showed 'indifference' toward the plight of the swine who ran headlong into the sea, drawing the further conclusion, "The New Testament seems to leave animals in the same situation as the Old. Jesus himself showed indifference (if not cruelty) to nonhumans when he unnecessarily forced 2,000 swine to hurl themselves into the sea" (*Philosophy of Vegetarianism*, p. 5).

Yet, in another instance, using a quite different tack, Dombrowski asks why Christians should not treat animals with charity when God himself shows so much concern for them. The particular passage to which he alludes is again from the Gospel of Matthew. It refers to Jesus' teaching that we should trust God, who has much more concern for us than He does for many sparrows, even though not a single sparrow dies without the Father's consent.

> Are not two sparrows sold for next to nothing. Yet not a single sparrow falls to the ground without your Father's consent. As for you, every hair on your head has been counted: so do not be afraid of anything. You are of more value than an entire flock of sparrows. (Matt. 10:29–31)

From this passage Dombrowski draws the following conclusion: "It is not clear why we cannot condescend to show charity to animals when even God can; it should be remembered that Jesus commended his Father for caring even for the fall of a sparrow" (*Philosophy of Vegetarianism*, p. 10). But to interpret this passage to mean that the Father is showing 'charity' to the sparrows is simply to misread it. Jesus is teaching His hearers that they as individual humans are of much greater worth than even a whole flock of sparrows. There is no mention of charity here with regard to the Father's providence over the sparrows, but simply the comment that not one of them dies without His consent. Indeed, Jesus' teaching in this passage must be taken against the backdrop of His entire gospel message. If His disciples but listen to His words, they have nothing to fear, for the Father dearly loves them. As proof of that love He has given His only Son as ransom for them. It is humankind that the Father so loves; not the sparrows. As a parenthetical aside to his interpretation of this passage, Dombrowski states: "Indirectly I am respectfully implying an inconsistency on the part of Jesus, who was no vegetarian" (p. 142, note 19). It is indeed remarkable to witness the lengths one is prepared to go in order to establish a basis for philosophical vegetarianism. He who is being criticized for inconsistency, because He was not a vegetarian, is none other than, in Paul's words, "Christ, the power of God and the wisdom of God" (1 Cor. 1:24), who said of Himself, "I am the way, the truth and the life" (John 14:6), and "Philip, he who sees me sees also the Father." (John 4:9). It is also not without historical interest to note that this same line of argument regarding the sparrows was used centuries earlier by Porphyry, an embittered enemy of Christianity.

St. Paul, in his turn, is brutally direct in his condemnation of those who forbid marriage and who brand the eating of certain foods as evil. His words undoubtedly refer to the Gnostics, forerunners of the Manichaeans, and, as regards the strictures on the eating of certain foods, also to the Jews.

> The Spirit has explicitly said that during the last times there will be some who will desert the faith and choose to listen to deceitful spirits and doctrines that come from the devils; and the cause of this is the lies told by hypocrites whose consciences are branded as though with a red-hot iron: they will say marriage is forbidden, and *lay down rules about abstaining from foods which God created to be accepted with thanksgiving by all who believe and who know the truth. Everything God created is good, and no food is to be rejected, provided grace is said for it: the word of God and the prayer make it holy.* (1 Tim. 4:1–5; italics added)

These passages make abundantly clear that the issue of philosophical vegetarianism is, from a theological standpoint, by no means a peripheral one. It sends roots deep into the foundations of Christian belief. Prescriptive vegetarianism is radically antithetical to the Christian Church's central teaching that Christ is, as savior of the world, both Son of God and Son of Man. "The word was made flesh and dwelt among us" (John 1:18).

THE LAMB OF GOD

The symbolism of the lamb is of central importance in both the Old and New Testaments. David and his ancestors, beginning with Abraham, the father of nations, were all shepherds. Sheep were a major source of their livelihood, providing wool and skins for clothing as well as milk and meat for the table. Since He was a descendant of David, shepherding was a major part of Jesus' heritage. His parable of the lost sheep and His frequent reference to shepherds illustrate in a very human, tangible way the care and affection the shepherd characteristically had for his sheep. It will be recalled how, in the parable of the lost sheep, the shepherd searches out the single lost sheep, and, finding it, returns it to the flock, carrying it on his shoulders. Christ often likened Himself to a shepherd; not to any shepherd, but to a good shepherd who shows genuine concern for his flock (cf. Matt 18:12–14; John 10:11–18).

Christ also makes full use of the sacrificial imagery of the Old Testament. In commemoration of the Hebrew's deliverance from the state of slavery to the Egyptians, the blood of a sacrificed lamb was sprinkled on the lintels and doorposts of the Hebrews' homes; the meal (called the Passover meal) was eaten according to the ritual stipulated by Mosaic Law, which included above all the sacrificed lamb. The meal was eaten hurriedly while standing, to symbolize the need to be prepared for an imminent departure (Exod. 12:21ff.).

In compliance with the Mosaic Law the Passover meal was to be reenacted each year. From the scriptural account we learn that during His public life Jesus gathered with his disciples each year to share this Passover meal (Matt. 26:17–19), the high point of which was always the consuming of the sacrificial lamb, of which no bone was to be broken. The last time Jesus celebrated this meal was the night before He died.

At the very end of the meal, Jesus formally substituted himself for the sacrificed lamb, thereby unveiling for the first time the full hidden meaning of the rich symbolism of the lamb whose flesh had been eaten. In instituting the Eucharist, Jesus revealed that He was Himself the sacrificial Lamb of God. Of all the incidents of Jesus' life, this Last Supper is, together with His death on the Cross of which it is the unbloody anticipation, held as most sacred and most revered by believing Christians. By this sacrificial act Jesus reveals Himself as the Good Shepherd who gives His life for his sheep, and He does so by becoming God's Lamb Who is sacrificed.

The core of Christian belief is inseparably related to the symbolic eating of the flesh of the lamb, the victim of sacrifice. By His own actions Jesus confirmed that practice most solemnly, for the eating of the Paschal meal provides the lasting symbol He chose to communicate to the world the true meaning of the mystery of His life as God become man. Just as the sacrificial lamb provided nourishment to those who partook of it, so the Mystical Lamb of God gives His life that others might live, not their own lives, but His.

In addition to the rich symbolism Jesus employed in referring to Himself as the Good Shepherd and in offering Himself for mankind as the sacrificial Lamb of God, there are numerous other instances when Jesus legitimized by His own actions the eating of flesh meat. One will recall among others the multiplication of the loaves and fishes which is recorded by all of the evangelists, and which Mark (6:34) and Luke (9:12) both indicate occurred on two separate occasions. In each instance several thousands of people were fed with a "few fishes" and loaves which, miraculously multiplied, Jesus personally handed out to His disciples to be distributed to the crowds.

Furthermore, the gospel narrative makes clear that many of the twelve apostles were fisherman by trade. Not only did Jesus select fishermen to be numbered among His closest followers—and indeed one of them, Peter, to be the very rock upon which His Church was built—but on two occasions recorded in the gospels He aided them in their efforts to catch fish. While standing on the shore He directed the apostles who were fishing just off shore, to cast their nets "to the right of the boat" and they would find a large school of fish (John 21:6). St. John relates that the disciples subsequently made a huge catch of fish. John further indicates that Jesus instructed the fishermen to bring Him some of the fish: "Bring here some of the fishes that you caught just now" (21:10). Jesus also invited His disciples to come and have breakfast, and personally handed to them bread and fish which He Himself had cooked over a fire. "Jesus said to them, 'Come and breakfast! . . . And Jesus came and took the bread [from the fire] and gave it to them, and likewise the fish" (21:12, 13).

St. Luke records Jesus' appearance to His disciples in the upper chamber a week after the resurrection. They were badly frightened, be-

ing convinced they were seeing a ghost. He asked them if they had anything to eat. Luke records that they gave Him some cooked fish, which
He ate in their presence to show that it was not a mirage they were seeing. "'Have you anything here to eat?' and they offered Him a piece of
broiled fish and a honeycomb. And when he had eaten in their presence, he took what remained and gave it to them" (Luke 24:41–42).

It is particularly noteworthy that these last two incidents occurred
after the resurrection of Jesus. Hence, clearly, even the risen Jesus approved of the eating of flesh meat, for He both presented fish to His
apostles with His own hands and actually partook of some fish himself.
These events thus make clear that the resurrected Jesus did not in any
way alter His practice of eating meat. He continued to eat and drink
with them just as He had done before.

By eating with them, He reassured them in the most persuasive
manner that He who now appeared to them was truly the same person
Who had eaten with them on so many occasions previously; he was not
a ghost. In short, Jesus' partaking of the broiled fish provided incontrovertible proof of His resurrection from the dead. The scene makes clear
as well that His followers' everyday habits of living and eating were to
remain as they had been before Jesus' death and resurrection. Jesus imposes no new dietary restrictions on His nascent Church that had not
been imposed on them during His years spent with them prior to His
death.

JESUS AND PHILOSOPHICAL VEGETARIANISM

As is well known from the Old Testament, animal sacrifice was a
common practice among the Hebrews and included the offering of various animals—goats and sheep, calves and oxen, turtledoves and pigeons. Ritual sacrifice of animals was, indeed, prescribed by the Mosaic
Law. Consequently, St. Luke relates that when the Child Jesus was
brought to the Temple by His parents to be consecrated to the Lord in
accordance with the Mosaic Law, they were required to bring either
turtledoves or pigeons to be sacrificed. St. Luke records: "and when the
days of her purification were fulfilled according to the law of Moses
they took him to Jerusalem to present him to the Lord . . . and to offer

a sacrifice according to what is said in the Law of the Lord, 'a pair of turtledoves or two young pigeons'" (Luke 2:22, 24).

Nor did Jesus at any later time repudiate the Mosaic Law as cruel or unjust; neither did He in His teaching avoid referring approvingly to the common Jewish practice of using animals for food. We have already noted His directing His disciples to make preparations for the Paschal meal, which included the sacrificing and eating of a young lamb. And in His parable of the prodigal son, Jesus relates how the father, overjoyed by the return of his wayward son, ordered his servants to kill the fatted calf, and to prepare a feast in his son's honor (Luke 15:11).

On still another occasion, Jesus and Peter, upon entering a synagogue, were asked to contribute the coin required as the Temple tax. Not having a coin, Jesus instructed Peter to go to the lake shore and cast a line into the water. A fish would take it, Peter was informed, and inside the fish's mouth he would find a coin which he was to bring and offer as the Temple tax (Matt. 17:23–26). These stories emphasize yet further that Jesus neither practiced nor taught mandatory vegetarianism; rather, by word and deed He clearly opposed it. Philosophical vegetarianism, whereby one abstains from the eating of flesh meat on the grounds that to eat it is inherently immoral, is not only not a Christian tenet, but plainly incompatible with Christ's teaching.

ORIGINS OF VEGETARIANISM

Prescriptive vegetarianism has its true roots in paganism and certain Eastern religions, which never succeeded in developing a clear awareness of the distinction between man and beast. Often lurking in the shadows cast by the vegetarian view is the belief in metempsychosis or reincarnation. According to this view the soul of the human, if not sufficiently purified at death, will be required to redescend, this time perhaps into the body of an animal of one species or another, to undergo yet another period of purification. Just such a view is presently lending impetus to the semi-religious advocacy of an exaggerated reverence for animal life, for it provides grounds for believing that the soul of the animal might possibly be the reincarnated soul of a human ancestor. Such a view appears to have originated in India, and many of those presently championing the cult of vegetarianism have been deeply influenced by teachings of religions from the East.

In the West, the doctrine of a morally obliging abstinence from flesh meat, or at least of strongly encouraging it, makes its appearance in the early stages of Greek philosophy. Pythagoras expressly taught vegetarianism; Plato seems at least to have recommended it. Later, after the beginning of the Christian era, it was strongly advocated by Plotinus, the Greek philosopher from Alexandria, Egypt, and by his gifted disciple, Porphyry. Porphyry's influential writings were singularly responsible for the public dissemination of vegetarianism during the early centuries of Christianity, especially in the West. As is known, Porphyry was deeply embittered against the Christian religion, vigorously opposing it in his writings, for reasons that are not fully clear. It has been suggested that he had once been a Christian but later defected. Whatever may have been behind his hostility to the Christian religion is of no real relevance here. Nor can it be claimed with any assurance that Porphyry became an advocate of vegetarianism because he considered it to be an effective way of attacking Christian beliefs, though he was perhaps the first to have used Jesus' teaching about the flock of sparrows in defense of vegetarianism.

Much more to our purpose, however, is the role strict vegetarianism played in the disciplinary teaching of many of the religious sects that appeared during this period. Indeed, vegetarianism was often a required or at least a strongly endorsed practice among many religious sects that sought to combine elements of Eastern religions with the teachings of Christianity. These sects proliferated, especially in those regions of Asia and Europe where Christianity had been recently introduced. Most of them can be subsumed under the umbrella term 'Gnostics' ('Knowers'), even though each was usually identified by a more particularized name, such as Manichees, Paulicians, Bogomils, Messaliens, Waldensians, Cathars, or Albigensians.

Common to all Gnostics is a dualist view of the origin of the world. According to this view the world drew its origin from two sources, a god of light and a god of darkness. The latter was the cause of the material world, which included the human body, while the former was the cause of mind and spirit, which inhabit body. Hence the god of darkness was the source of evil, while the god of light was the source of all that was good. Between the two gods of good and evil a state of incessant warfare exists, as each strives to overcome the power and influence of the other. The battlefield on which the conflict rages is the world, where body and spirit are joined.

This basic teaching of all the Gnostic sects is in effect a mosaic of various tenets of religious teachings found in Eastern religions as well as in Judaism and Christianity. The dualistic view results in the adoption of an attitude of pronounced pessimism regarding the body and material world, for matter is seen as inherently evil. The body belongs to the domain of the god of darkness, serving as the prison in which the life-spirit is held captive. Authentic life, then, consists in an ongoing struggle on the part of spirit to liberate itself from the clutches of body and hence of darkness. Freedom from dependence and submission to the powers of this world, in the Gnostic view, is the supreme goal of life.

In his scholarly study of the dualist sects of the early centuries of the Christian era, the respected historian Steven Runciman gives an insightful summary of the diverse elements contributing to the origin of these sects. He writes:

> But Greek philosophy in the later, wearier centuries wondered again about evil. Stoics and Neoplatonists each in their way condemned the world of matter; and Jewish thinkers of Alexandria began to face the problem, influenced by the emphasis on spirit that they found in the Hermetic lore of Egypt. Over the frontiers in Persia Zoroaster had taught long ago of the permanent war between Good and Evil, spirit and matter. Farther away to the East Gautama the Buddha told his followers that only by dissociation from the world could goodness be found, that peace came only with material annihilation. Such doctrines trickled through to the countries where Christianity was passing its childhood. (*The Medieval Manichee: A Study of the Christian Dualist Heresy,* 1961, pp. 171–72)

It can readily be seen how Gnosticism differs from the religions of the East and from Christian teaching. It differs greatly from Buddhism and Zoroastrianism in admitting that Jesus is God's own Son. But it often dilutes this teaching to such an extent that the Christian belief that Christ is God is effectively denied. It tends generally either to reduce Jesus' role to one of triviality, as Runciman states (p. 175), or it denies outright that Jesus was man in the true, unqualified sense. Rather, he *appeared to be* man, or was an angel appearing in human form. As Runciman concludes, "There is no room for Christ in a truly dualist religion" (ibid.), for there can be no doctrine of Atonement, since the material world is, in this view, inherently evil and hence unredeemable. For the Gnostic it is unthinkable that the God of Light could willingly have entered into union with the world of darkness.

It is precisely here that vegetarianism makes its entrance in a dualist system, since spirit as spirit is of one level only. Consequently, the only difference between the human and the nonhuman animal is traceable to the body in which spirit is imprisoned; as spirits they are equal. This view is consistent with the further teaching that an interchange between all creatures having 'animal' spirits can occur. This in turn entails the doctrine of metempsychosis or reincarnation, which appears to be a teaching common to all dualist systems. The eating of flesh meat is hence forbidden, since this could render more earthbound any fragments of spirit that the meat might still contain. This would occur as a result of the metabolic process. Thus to avoid any further contamination of spirit, the eating of meat is shunned (cf. *The Medieval Manichee,* pp. 151–52). It does appear that in practice this was often merely an ideal to be aimed at, for only those who had attained to a high level of purification appear to have followed a regimen of strict vegetarianism. The 'Perfects', as the Cathars called them, who had attained the highest level of spiritual perfection, lived on a strict diet of bread and vegetables with some fish permitted on occasion (p. 160).

Historically, then, we find a distinct link between the doctrine of the transmigration of souls and the practice of vegetarianism understood as a moral duty, and both are closely tied to dualist systems that derive from attempts to explain the origin of physical and moral evil in the world. Since vegetarianism, taken as a moral imperative, derives its binding power from the alleged assumption that the souls or spirits of the human and the nonhuman animal are fundamentally equal, one can appreciate why it is that vegetarians might subscribe to the Darwinian theory of the origin of the human about which much has already been said.

Because Satan is, for the religious dualist, the cause of the material world, that world is evil; to the extent possible it ought to be shunned by those who are seeking to be truly spiritual. For this reason the Dualists often, if not always, urged their followers to abstain from eating flesh meat, for, as Runciman states, quoting from Alain of Lille regarding the French Cathars, "the Cathar adepts abstained from meat because of their doctrine of metempsychosis. The meat might contain a fragment of earthbound soul which would thus become more earthbound by metabolism" (pp. 151–52). It was the Dualists belief in the transmigration of souls that led them to look upon the souls of animals as equal to those of the human.

We may conclude, then, that historically there is a distinct link between the theory of the transmigration of souls and the practice of vegetarianism as a moral imperative. Philosophical vegetarianism is incompatible with Christianity, for it renders impossible a coherent understanding of the central mystery of Christian belief, namely, the Incarnation. A thoroughgoing philosophical vegetarian position would have to hold that Jesus was unaware of the full compass of the moral law, having failed not only to condemn the 'sin' of using animals as food, but, even worse, having himself succombed to this evil practice. But if that were so, Jesus could not have been truly divine, or even an authentic prophet. Further—and this point addresses even more directly the doctrinal dimension of the mystery of the Incarnation—vegetarianism, in granting 'rights' to the nonhuman animal, so obscures the distinction between man and beast as to render arbitrary and meaningless the fundamental claim of Christianity that the Second Person of the Trinity became 'man', and that He entered this world to save mankind. Rather, it could then be argued, the Second Person of the Trinity also became 'animal' in order to redeem not just the human but the nonhuman animal as well. This would then render the lives of all sentient organisms sacred and worthy of the same respect as humans. These are of course heretical theological conclusions, utterly repugnant to the Christian conscience and profoundly altering the true nature and purpose of God's having sent His son into the world, as that son Himself testified. Through Christ all nature is redeemed, but precisely insofar as man's contemplation of the goodness and beauty of the world can now assist him in raising his mind and heart to their and his own Creator.

The early practice of the Church was never tainted with any false rigorism of vegetarianism. It is true that Christians were forbidden to eat the meat of idols, that is, meat that had first been ritually presented to one of the pagan gods as a sacrificial offering. But this stricture in no way proceeded from any form of vegetarian persuasion; it was solely to prevent Christians from implicating themselves in false worship, since the meat they would be purchasing for their meal would have been first consecrated to one of the pagan gods. The eating of such meat, furthermore, would provide the pagans a firm basis for judging that those calling themselves Christian converts were still at heart believers in pagan gods. Consequently, as is recorded in the Acts of the Apostles, the apostles and elders of the Church at Jerusalem sent word to the Christians of Gentile origin in Antioch "to abstain from things sacrificed to idols and from blood and from what is strangled and from immorality" (Acts 15:29).

Also recorded in the Acts of the Apostles is St. Peter's vision at Jaffa, of a large sheet coming down from the sky supported at the four corners, "and in it were all the four-footed beasts and creeping things of the earth, and birds of the air. And there came a voice, 'Arise, Peter, kill and eat.' But Peter said, 'Far be it from me, Lord, for never did I eat anything common or unclean.' And there came a voice a second time to him, 'What God has cleansed, do not thou call common'" (10:11–15). Now it is true that the full meaning behind this vision, as Peter soon learns, is that God is no respecter of persons, (10:35) and that the message of His Son is not intended merely for the Jews but is for the Gentiles as well; that they, too, are worthy of evangelization. Yet, the command given Peter, to "arise, kill and eat" could not have been given him were it wrong to kill animals for food, for this would render the meaning of the vision not only unclear to Peter but simply unintelligible.

THE NONHUMAN ANIMAL: A GIFT TO HUMANKIND

These few instances based on passages taken from the New Testament provide an indisputably clear summarization of (a) the Christian theology of man, (b) the theology of animals, and (c) the human's relation to the nonhuman animal. After man himself, and of course God's gift to man of His own Son, there is no greater gift that has been given to humans than the animals. Without the presence of animals in our world, human existence would truly be impoverished beyond all imagining. Animals enhance human life in so many different ways, providing companionship, service, and nourishment.

Though the animal is a living and sentient being, and thus capable of experiencing pain, and though it acts in a highly intelligent manner, the animal is *not* an intelligent being; it is not reflexively conscious of itself as a knowing subject or agent. The nonhuman animal is not made in God's image. Hence it does not formally possess 'rights', though God's law serves to protect it from abuse and misuse by man, for by that same law the human is required to make responsible use of God's gifts, employing them with gratitude and respect and by using them intelligently and reasonably as befits his legitimate needs. To act otherwise is to display disrespect primarily toward the Giver of the gift rather than toward the gift itself. But to treat the animal as though it were human would be to insult the One Who made both man and the animal, but Who made man alone in His own image; Who so loved the world as to send His own Son into it as man, that man might share His Life; and Who gave animals to man that, using them wisely, they might offer Him fitting thanks and praise (cf. Gen. 1:28).

The mystery of man—his singular dignity and his relation to the world—is sublimely heralded by the Psalmist David, who sings:

What is man that you should be mindful of him,
or the son of man that you should care for him?

You have made him little less than the angels,
and crowned him with glory and honor.
You have given him rule over the works of your hand,

putting all things under his feet:
all sheep and oxen, yes, and the beasts of the field,
the birds of the air, the fishes of the sea, and
whatever swims the paths of the seas.

O Lord, our Lord, how glorious is your name
over all the earth!
    (Psalm 8:5–10)

Bibliography

Adler, Mortimer J. *The Difference of Man and the Difference It Makes.* New York: Holt, Rinehart, Winston, 1967.

———. *Ten Philosophical Mistakes.* New York: Simon & Schuster, 1996.

Aiken, William. "Ethical Issues in Agriculture." In *Earthbound: New Introductory Essays in Environmental Ethics,* edited by Tom Regan, pp. 247–88. New York: Random House, 1984.

Aquinas, St. Thomas. *Summa Theologiae.* 4 Vols., Ottawa edition. Ottawa, 1941.

———. *In Aristotelis Librum De Anima Commentarium,* 3d edition. Edited by P. F. Angeli M. Pirotta, O.P. Rome: Marietti, 1948.

———. *Quaestiones Disputatae.* Turin: Marietti, 1949.

———. *In Aristotelis Libros Peri Hermeneias et Posteriorum Analyticorum Expositio.* Edited by Raymundi M. Spiazzi, O.P. Turin: Marietti, 1955.

———. *On Being and Essence.* Translated by Armand Maurer. Toronto: Pontifical Institute of Mediaeval Studies, 1968.

Aristotle. *The Basic Works of Aristotle.* Edited by Richard McKeon. New York: Random House, 1941.

Armstrong-Buck, Susan. "Whitehead's Metaphysical System as a Foundation for Environmental Ethics." *Environmental Ethics* 8 (1986): 241–59.

Augros, R., and G. Stanciu. *The New Biology: Discovering the Wisdom in Nature.* Boston: Shambhala, 1987.

Avicenna. *De Animalibus.* Frankfurt am Main: Minerva G.m.b.H., 1961.

Ayala, F. J., and G. L. Stebbins. "Is a New Evolutionary Synthesis Necessary?" *Science* 213, no. 4511 (1981): 967–71.

Barker, Eileen. "Scientific Creationism in the Twentieth Century." In *Darwinism and Divinity,* edited by John Durant. Oxford: Basil Blackwell, 1985.

Barnes, Joshua, Lars Hernquist, and Francois Schweizer. "Colliding Galaxies." *Scientific American* 265, no. 2 (August 1991): 40–47.

Bartolomei, Sergio. *Etica e Ambiente*. Milan: Edizioni Guerini, 1990.

Beardsley, Tim. "Smart Genes." *Scientific American* 265, no. 2 (August 1991): 86–95.

Becker, Lawrence C. "The Priority of Human Interests." In *Ethics and Animals*, edited by Harlan B. Miller and William H. Williams, pp. 225–42. Clifton, N.J.: Humana Press, 1983.

Behe, Michael. *Darwin's Black Box: The Biochemical Challenge to Evolution*. New York: Simon & Schuster, 1996.

Benjamin, Martin. *Splitting the Difference: Compromise and Integrity in Ethics and Politics*. Lawrence: University of Kansas Press, 1990.

Berra, Tim. *Evolution and the Myth of Creationism: A Basic Guide to the Facts in the Evolution Debate*. Stanford: Stanford University Press, 1990.

Bickerton, Derek. *Roots of Language*. Ann Arbor: Karoma Publishers, 1981.

———. *Language and Species*. Chicago: University of Chicago Press, 1990.

Birch, Charles, and John B. Cobb. *Liberation of Life: From the Cell to the Community*. Cambridge: Cambridge University Press, 1981.

Blackstone, William T. "The Search for an Environmental Ethic." In *Matters of Life and Death*, edited by Tom Regan, pp. 299–334. New York: Random House, 1980.

Bramwell, A. *Ecology in the 20th Century: A History*. London: Yale University Press, 1989.

Cairns-Smith, A. G., and H. Hartman, eds. *Clay Minerals and the Origin of Life*. London: Cambridge University Press, 1986.

Callicott, J. Baird. *In Defense of the Land Ethic: Essays in Environmental Philosophy*. Albany: State University of New York Press, 1989.

———. "Animal Liberation: A Triangular Affair." In *The Animal Rights, Environmental Ethics Debate*, edited by Eugene C. Hargrove, 37–70. Albany: State University of New York Press, 1992.

———. "Animal Liberation and Environmental Ethics: Back Together Again." In *The Animal Rights, Environmental Ethics Debate*, edited by Eugene C. Hargrove, 249–61. Albany: State University of New York Press, 1992.

———. *Earth's Insights: A Survey of Ecological Ethics from the Mediterranian Basin to the Australian Outback*. Berkeley: University of California Press, 1994.

———. "American Indian and Western European Attitudes." In *Environmental Philosophy: A Collection of Readings*, edited by Robert Elliot and Arran Gare, pp. 231–59. St. Lucia: University of Queensland Press, 1983.

Cassirer, Ernst. *An Essay on Man: An Introduction to a Philosophy of Human Culture*. New Haven: Yale University Press, 1944.

———. *The Philosophy of Symbolic Forms*. 3 Vols. Translated by Ralph Manheim. New Haven: Yale University Press, 1955.

Cavalli-Sforza, Luigi Luca. "Genes, People, and Languages." *Scientific American,* November 1991.

Cheney, Dorothy L., and Robert M. Seyfarth. *How Monkeys See the World: Inside the Mind of Another Species.* Chicago: University of Chicago Press, 1990.

Cheney, Jim. "Callicott's 'Metaphysics of Morals.'" *Environmental Ethics* 13 (1991): 311–25.

Chomsky, Noam. *Aspects of the Theory of Syntax.* Cambridge: M.I.T. Press, 1965.

———. *Chomsky, Selected Readings.* Edited by J. P. B. Allen and Paul Van Buren. London: Oxford University Press, 1971.

———. *Language and Mind.* New York: Harcourt, Brace, Jovanovich, 1972.

———. "Human Language and Other Semiotic Systems." Paper delivered on February 16, 1978, in the symposium "Emergence of Language: Continuities and Discontinuities," at the annual meeting of the American Association for the Advancement of Science.

———. *Knowledge of Language: Its Nature, Origin, and Use.* New York: Praeger, 1986.

Clark, Stephen R. L. *The Moral Status of Animals.* Oxford: Clarendon Press, 1977.

———. *The Nature of the Beast: Are Animals moral?.* New York: Oxford University Press, 1983.

———. "Good Dogs and Other Animals." In *In Defense of Animals,* edited by Peter Singer, 41–51. Oxford: Basil Blackwell, 1985.

Colinvaux, Paul. *Ecology.* New York: John Wiley, 1990

Collin, Remy. *Evolution.* New York: Hawthorne Books, 1959.

Collins, James. *Three Paths in Philosophy.* Chicago: Henry Regnery, 1962.

Comins, N. F. *What If the Moon Didn't Exist? Voyages to Earths that Might Have Been.* New York: Harper Perennial, 1993.

Crick, Francis. *What Mad Pursuit: A Personal View of Scientific Discovery.* New York: Basic Books, 1988.

Darwin, Charles. *The Descent of Man: And Selection in Relation to Sex.* Great Books, vol. 49. Chicago: Encyclopedia Britannica, 1952.

———. *The Origin of Species By Means of Natural Selection.* Great Books, vol. 49. Chicago: Encyclopedia Britannica, 1952.

———. *The Autobiography of Charles Darwin, 1809–1882.* Edited by Nora Barlow. London: W. W. Norton, 1969.

———. *Darwin: Selected Texts, Contemporary Opinion, Critical Essays.* Edited by Philip Appleman. London: W. W. Norton, 1970.

———. "Comparison of the Mortal Powers of Man and the Lower Animals."

In *Animal Rights and Human Obligations,* edited by Tom Regan and Peter Singer, 72–81. Englewood Cliffs, N.J.: Prentice-Hall, 1976.

Dawkins, Mary Stamp. "The Scientific Basis for Assessing Suffering." In *In Defense of Animals,* edited by Peter Singer, 27–40. Oxford: Basil Blackwell, 1985.

Dawkins, Richard. *The Selfish Gene.* Oxford: Oxford University Press, 1976.

———. *The Extended Phenotype: The Long Reach of the Gene.* Oxford: Oxford University Press, 1982.

———. *The Blind Watchmaker.* New York: W. W. Norton, 1986.

Denton, Michael. *Evolution: A Theory in Crisis.* Bethesda, Md.: Adler and Adler, 1985.

Descartes, René. "Animals Are Machines." In *Animal Rights and Human Obligations,* edited by Tom Regan and Peter Singer. Englewood Cliffs, N.J.: Prentice-Hall, 1976.

Dewey, John. *The Influence of Darwin on Philosophy, and Other Essays in Contemporary Thought.* New York: Holt, Rinehart and Winston, 1938.

Dombrowski, Daniel A. *The Philosophy of Vegetarianism.* Amherst: University of Massachusetts Press, 1984.

———. *Hartshorne and the Metaphysics of Animal Rights.* Albany: State University of New York Press, 1988.

Durant, John, ed. *Darwinism and Divinity: Essays on Evolution and Religious Belief.* Oxford: Basil Blackwell, 1985.

Dworkin, Ronald. *Taking Rights Seriously.* Cambridge: Harvard University Press, 1977.

Eimerl, Sarel, and Irven DeVore. *The Primates.* Life Nature Library. New York: Time Inc., 1965.

Eldredge, Niles. *Time Frames: The Rethinking of Darwinian Evolution and the Theory of Punctuated Equilibria.* New York: Simon and Schuster, 1985.

———. *Unfinished Synthesis: Biological Hierarchies and Modern Evolutionary Thought.* New York: Oxford University Press, 1985.

———. *Life Pulse: Episodes from the Story of the Fossil Record.* New York: Facts on File Publications, 1987.

Elliot, Robert. "Moral Autonomy, Self-Determination and Animal Rights." *Monist* 70, 1 (January 1987): 83–99.

———, ed. *Environmental Ethics.* Oxford University Press, New York: 1995.

———, and Arran Gare, eds. *Environmental Philosophy: A Collection of Readings.* University Park: The Pennsylvania State University Press, 1983.

Feinberg, Joel. *Rights, Justice, and the Bounds of Liberty: Essays in Social Philosophy.* Princeton, N.J.: Princeton University Press, 1980.

Finnis, John. *Natural Law and Natural Rights.* Oxford: Oxford University Press, 1985.

Flew, A. G. N. *Evolutionary Ethics.* New York: St. Martins Press, 1967.

Fossey, Dian. *Gorillas in the Mist.* Boston: Houghton Mifflin, 1983.

Fox, Michael W. "Philosophy, Ecology, Animal Welfare, and the 'Rights' Question." In *Ethics and Animals,* edited by Harlan B. Miller and William H. Williams, 307–16. Clifton, N.J.: Humana Press, 1983.

Fox, Michael Allen. *The Case for Animal Experimentation: An Evolutionary and Ethical Perspective.* Berkeley: University of California Press, 1986.

Frey, R. G. "On Why We Would Do Better to Jettison Moral Rights." In *Ethics and Animals,* edited by Harlan B. Miller and William H. Williams, 285–304. Clifton, N.J.: Humana Press, 1983.

Frey, R. G. "Autonomy and Valuable Lives." *Monist* 70, no. 1 (1987): 50–62.

Froman, Wayne Jeffrey. *Merleau-Ponty: Language and the Act of Speech.* Lewisburg, Penn.: Bucknell University Press, 1982.

Frankena, W. K. "Ethics and the Environment." In *Ethics and Problems of the 21st Century,* edited by K. E. Goodpaster and K. M. Sayre, 3–20. Notre Dame: University of Notre Dame Press, 1979.

Fuchs, Alan E. "Duties to Animals: Rawls's 'alleged dilemma.'" *Ethics and Animals* 2 (1981): 83–87.

George, Wilma. *Biologist Philosopher: A Study of the Life and Writings of Alfred Russel Wallace.* London: Abelard-Shuman, 1964.

Gillespie, Neal C. *Charles Darwin and the Problem of Creation.* Chicago: University of Chicago Press, 1979.

Gilson, Étienne. *The Unity of Philosophical Experience.* New York: Scribners and Sons, 1937.

Godlovich, Stanley and Roslind, and John Harris, eds. *Animals, Men, and Morals: An Enquiry into the Maltreatment of Nonhumans.* London: Gollancez, 1972.

Gomez, Juan Carlos. "Are Apes Persons? The Case for Primate Intersubjectivity." In *Etica & Animali,* edited by Paolo Cavalieri, pp. 51–63. Milan: 1998.

Goodall, Jane. *In the Shadow of Man.* Boston: Houghton Mifflin, 1971.

———. *Through a Window: My Thirty Years with the Chimpanzees of Gombe.* Boston: Houghton Mifflin, 1990.

———. *The Chimpanzees of Gombe: Patterns of Behavior.* Cambridge, Mass.: Belknap Press of Harvard University Press, 1986.

Goodpaster, K. E., and K. M. Sayre, eds. *Ethics and Problems of the 21st Century.* Notre Dame: University of Notre Dame Press, 1979.

Gould, Stephen Jay. *Ever Since Darwin: Reflections in Natural History,* London: W. W. Norton, 1973.

———. *Ontogeny and Phylogeny,* Cambridge: Harvard University Press, 1977.

———. "Is a New and General Theory of Evolution Emerging?" *Paleobiology* 6, no. 1. (1987): 119–30.

———. *An Urchin in the Storm: Essays about Books and Ideas.* New York: W. W. Norton, 1987.

———. *Wonderful Life: The Burgess Shale and the Nature of History.* New York: W. W. Norton, 1989.

———. *Dinosaur in a Haystack: Reflections in Natural History.* New York: Harmony Books, 1995.

———. *Full House: The Spread of Excellence from Plato to Darwin.* New York: Harmony Books, 1996.

Grant, Peter R. "Natural Selection and Darwin's Finches." *Scientific American* 265, no. 4 (October 1991): 82–87.

Green, Judith. "Retrieving the Human in Nature." *Environmental Ethics* 17, 4 (Winter 1996): 381–96.

Griffin, Donald R. *The Question of Animal Awareness: Evolutionary Continuity of Mental Experience.* New York: Rockefeller University Press, 1981.

Hahn, Lewis Edwin. *The Philosophy of Charles Hartshorne.* La Salle, Ill.: Open Court, 1991.

Hargrove, Eugene C. *Foundations of Environmental Ethics,* Englewood Cliffs, N.J.: Prentice-Hall, 1989.

———, ed. *Beyond Spaceship Earth: Environmental Ethics and the Solar System.* San Francisco: Sierra Club Books, 1986.

———, ed. *The Animal Rights, Environmental Ethics Debate: The Environmental Perspective.* Albany: State University of New York Press, 1992.

Harrison, Richard J., and William Montagna. *Man.* 2d edition. New York: Appleton-Century-Crofts, 1973.

Hengstenberg, Hans-Eduard. *Philosophische Anthropologie Dritte Auflage.* Stuttgart: W. Kohlhammer Verlag, 1966.

Himmelfarb, Gertrude. *Darwin and the Darwinian Revolution.* London: Chatto & Windus, 1959.

Hjelmslev, Louis. *Language: An Introduction.* Translated by Francis J. Whitfields. Madison: Univ of Wisconsin Press, 1970.

Hollander, Bernard. *In Search of the Soul and the Mechanism of Thought, Emotion, and Conduct.* 2 Vols. London: Kegan Paul, n.d.

Howard, Jonathan. *Darwin.* New York: Hill and Wang, 1982.

Hughes, John P. *The Science of Language: An Introduction to Linguistics.* New York: Random House, 1962.

Hume, David. *An Enquiry Concerning Human Understanding.* LaSalle, Ill.: Open Court, 1958.

———. *A Treatise of Human Nature.* Edited by L. A. Selby-Bigge. Oxford: The Clarendon Press, 1960.

Jespersen, Otto. *Mankind, Nation and Individual.* London: George Allen and Unwin, Ltd., 1946.

———. *Language: Its Nature, Development and Origin.* New York: W. W. Norton, 1964.

Johanson, Donald C., and Maitland Edey. *Lucy: The Beginnings of Humankind.* New York: Simon and Schuster, 1981.

Johnson, Phillip E. "Evolution as Dogma: The Establishment of Naturalism." *First Things,* no. 6 (October 1990): 15–22 and no. 7 (November 1990): 50–52.

———. *Darwin on Trial.* Washington, D.C.: Regnery Gateway, 1991.

———. "God and Evolution: An Exchange." *First Things* (July 1993): 40–41.

Kant, Immanuel. *Critique of Pure Reason.* Translated by Norman Kemp Smith. New York: Modern Library, 1958.

Kaplan, Helmet F. *Warum Vegetarier: Grundlager Einer Universalen Ethik.* Frankfurt am Main: Peter Lang, 1989.

Kaufmann, Stuart A. "Antichaos and Adaptation." *Scientific American* 265, no. 2 (August 1991): 403–25.

Kings, Sallie B. "Buddhism and Hartshorne." In *The Philosophy of Charles Hartshorne.* La Salle, Ill.: Open Court, 1991.

Klubertanz, George, S.J. *The Discursive Power. Sources and Doctrine of the 'Vis Cogitativa' According to St. Thomas Aquinas.* St. Louis: The Modern Schoolman, 1952.

Knoll, Andrew H. "End of the Proterozoic Eon." *Scientific American,* October 1991.

Koerner, Stephan. "Cassirer." In *Encyclopedia of Philosophy,* vol. 2, edited by Paul Edmonds. New York: Collier-Macmillan, 1967.

Landmann, Michael. *Philosophical Anthropology.* Translated by David J. Parent. Philadelphia: The Westminister Press, 1974.

Leakey, Richard E., and Roger Lewin. *Origins.* New York: E. P. Dutton, 1977.

Lehmann, Scott. "Do Wildernesses Have Rights?" *Environmental Ethics* 3 (1981): 129–46.

Leopold, Aldo. *A Sand County Almanac and Sketches Here and There.* London: Oxford University Press, 1949.

Leslie, John. "Modern Cosmology and the Creation of Life." In *Evolution and Creation,* edited by Ernan McMullin. Notre Dame: University of Notre Dame Press, 1985.

Lieberman, Philip. *On the Origins of Language: An Introduction to the Evolution of Human Speech.* New York: MacMillan, 1975.

———. *Uniquely Human: The Evolution of Speech, Thought, and Selfless Behavior.* Cambridge: Harvard University Press, 1991.

Limber, John. "Language in Child and Chimp." In *Speaking of Apes: An Anthology of Two-Way Communication with Man,* edited by Thomas A. Sebeok and Jean Umiker-Sebeok. New York: Plenum Press, 1980.

Linden, Eugene. *Apes, Men, and Language*. New York: Saturday Review Press, 1974.

———. "Talk to the Animals." *Omni*, vol. II, no. 4 (January 1980), pp. 88–90, 107–9.

Linzey, Andrew, and Dorothy Yamamoto, editors. *Animals on the Agenda*. Chicago: University of Illinois Press, 1998.

Locke, John. *An Essay Concerning Human Understanding*. London: E. Holt, 1689.

Lombardi, Louis G. "Inherent Worth, Respect and Rights." *Environmental Ethics* 5, no. 3 (Fall 1983): 257–70.

Lorenz, Konrad. *Evolution and Modification of Behavior*. Chicago: University of Chicago Press, 1974.

Lyons, John. *Noam Chomsky*. New York: Viking Press, 1970.

MacIntyre, Alasdair. *Dependent Rational Animals*. Chicago: Open Court, 1999.

———. *Whose Justice? Which Rationality?* Notre Dame: University of Notre Dame Press, 1988.

Marietta, Don E., Jr. *For People and the Planet: Holism and Humanism in Environmental Ethics*. Philadelphia: Temple University Press, 1995.

Maritain, Jacques. *The Rights of Man and Natural Law*. New York: Scribners and Sons, 1943.

———. *The Person and the Common Good*. New York: Scribners and Sons, 1947.

Martinet, Andre. *Eléments de Linguistique Générale*. Paris: Librairie Armand Colin, 1960.

Marx, Jean. *Research News Science* 207 (March 21, 1980): 1330.

Mayr, Ernst. *The Growth of Biological Thought: Diversity, Evolution, Inheritance*. Cambridge: Harvard University Press, 1982.

———. *Populations, Species, and Evolution*. Cambridge: Harvard University Press, 1970.

McCloskey, H. J. *Meta-Ethics and Normative Ethics* The Hague: Martinus Nijhoff Press, 1969.

———. "The Right to Life." *Mind* 84, no. 335 (1975): 403–25.

———. *Ecological Ethics and Politics*. Totowa, N.J.: Rowman and Littlefield, 1983.

———. "The Moral Case for Experimentation on Animals." *Monist* 70, no. 1 (January 1987): 64–82.

McInerny, Ralph. *A First Glance at St. Thomas Acquinas: A Handbook for Peeping Thomists*. Notre Dame: University of Notre Dame Press, 1990.

———. *Aquinas on Human Action: A Theory of Practice*. Washington D.C.: The Catholic University of America Press, 1992.

McLaughlin, R. J. "Men, Animals, and Personhood." *Proceedings: American Catholic Philosophical Association* 59 (1985): 166–81.

McMullin, Ernan, ed. *Evolution and Creation.* Notre Dame: University of Notre Dame Press, 1985.

Midgley, Mary. *Beast and Man: The Roots of Human Nature.* Cornell University Press, 1978.

———. "Talk to the Animals." *Omni* (January 1980), p. 109.

———. *Animals and Why They Matter.* Harmondsworth: Penguin, 1984.

———. *Evolution as a Religion: Strange Hopes and Stranger Fears.* London: Methuen, 1985.

———. "Persons and Non-Persons." In *In Defense of Animals,* edited by Peter Singer. Oxford: Basil Blackwell, 1985.

Miller, Harlan B., and William H. Williams, eds. *Ethics and Animals.* Clifton, N.J.: Humana Press, 1983.

Miller, Michael. "Descartes' Distinction Between Animals and Humans: Challenging the Language and Action." *American Catholic Philosophical Quarterly* 32, no. 3 (1998): 339–70.

Monod, Jacques. *Chance and Necessity.* New York: Vintage, 1972.

Montague, Phillip. "Two Concepts of Rights." *Philosophy and Public Affairs* 9, 4 (1980): 372–84.

Morris, Simon Conway. *The Crucible of Creation: The Burgess Shale and the Rise of Animals.* Oxford: Oxford University Press, 1997.

Naess, Arne. *Ecology, Community and Lifestyle.* Edited and translated by David Rothenberg. Cambridge: Cambridge University Press, 1991.

Nash, R. F. *The Rights of Nature: A History of Environmental Ethics.* Madison: University of Wisconsin Press, 1989.

Norton, Bryan G. "Environmental Ethics and Nonhuman Rights." *Environmental Ethics* 4 (1987): 17–36.

———. *Why Preserve Natural Variety?* Princeton: Princeton University Press, 1987.

———. "The Constancy of Leopold's Land Ethics." *Conservation Biology* 2 (1988): 93–102.

———. *Toward Unity among Environmentalists.* New York: Oxford University Press, 1991.

———. "Epistemology and Environmental Values." *Monist* 75 (1992): 208–26.

———. "Why I Am Not a Nonanthropocentrist: Callicott and the Failure of Monistic Inherentism." *Environmental Ethics* 17, no. 4 (Winter 1995): 341–58.

Norveson, Jan. "On a Case for Animal Rights." *Monist* 70/1 (1987): 31–49.

Passmore, John. *Philosophical Reasoning.* New York: Basic Books, 1969.

———. *Recent Philosophers.* LaSalle, Ill.: Open Court, 1985.

Poole, Joyce H., "An Exploration of a Commonality between Ourselves and Elephants." In *Etica & Animali,* edited by Paolo Cavalieri, pp. 85–110. Milan: 1998.

Portmann, Adolf. *Animals as Social Beings.* New York: Harper and Row, 1964.

———. *Animal Forms and Patterns: A Study of the Appearance of Animals.* Translated by Hella Czech. New York: Schocken Books, 1967.

———. "Colour Sense and the Memory of Colour." In *Color Symbolism: Excerpts Eranos Conference,* pp. 1–22. Zurich: Spring Publications, 1977.

Rachels, James. "Darwin, Species, and Morality." *Monist* 70, no. 1 (January 1987): 98–111.

Rawls, John. *A Theory of Justice.* Cambridge: Harvard University Press, 1971.

Regan, Tom. *The Case for Animals Rights.* Berkeley: University of California Press, 1983.

———. "The Nature and Possibility of an Environmental Ethic." *Environmental Ethics* 5, no. 1 (Spring 1983): 47–61.

———, ed. *Matters of Life and Death: New Introductory Essays in Moral Philosophy.* New York: Random House, 1980.

———, ed. *Earthbound: New Introductory Essays in Environmental Ethics.* New York: Random House, 1984.

Reichmann, James B. "St. Thomas, Capreolus, Cajetan and the Created Person." *New Scholasticism* 30, nos. 1 and 2 (1959): 1–33 and 202–30.

———. "Logic and the Method of Metaphysics." *The Thomist* 29, no. 4 (1965): 341–95.

———. "The Transcendental Method and the Psychogenesis of Being." *The Thomist* 32, no. 4 (1968): 449–508.

———. "Aquinas, God and Historical Process." *Atti del Congresso Internazionale: Il Cosmo, e la Scienza.* Naples: Edizioni Domendicane, 1974.

———. "Immanently Transcendent and Subsistent Esse: A Comparison." *The Thomist* 38, no. 2 (1974): 332–69.

———. "Hegel's Ethics of the Epochal Situation: Morality and Ethics." *Proceedings of the American Catholic Philosophical Association* 49 (1975): 24–36.

———. *Philosophy of the Human Person.* Chicago: Loyola University Press, 1985.

———. "The 'Cogito' in St. Thomas: Truth in Aquinas and Descartes." *International Philosophical Quarterly* 26, no. 4 (1986): 341–52.

———. "Language and Interpretation of Being in Gadamer and Aquinas." *Proceedings of the American Catholic Philosophical Association* 62 (1988): 225–34.

Richards, R. J. *The Meaning of Evolution.* Chicago: University of Chicago Press, 1992.

Rollin, Bernard E. *Animal Rights and Human Morality*. New York: Promethus Books, 1981.

———. "The Legal and Moral Bases of Animal Rights." In *Ethics and Animals*, edited by Harlan B. Miller and Williams H. Williams, 103–20. Clifton, N.J.: Humana Press, 1983.

Rolston, Holmes, III. *Environmental Ethics: Duties to and Values in the Natural World*. Philadelphia: Temple University Press, 1988.

Roslansky, John D., ed. *The Uniqueness of Man: A Discussion at the Nobel Conference, Gustavus Adolphus College, 1968*. Amsterdam/London: North-Holland, 1969.

Routley, R. and V. "Against the Inevitability of Human Chauvinism." In *Ethics and Problems of the 21st Century*, edited by K. E. Goodpaster and K. M. Sayre, pp. 36–59. Notre Dame: Notre Dame University Press, 1979.

Rumbaugh, Duane M., and Sue Savage-Rumbaugh. "Apes and Language Research." In *Ethics and Animals*, edited by Harlan B. Miller and William H. Williams, pp. 225–42. Clifton, N.J.: Humana Press, 1983.

Runciman, Steven. *The Medieval Manichee: A Study of the Christian Dualist Heresy*. New York: Viking Press, 1961.

Ruse, Michael. *Darwinism Defended: A Guide to the Evolution Controversies*. London: Addison-Wesley, 1982.

Sapontzis, Stephen F. *Morals, Reason, and Animals*. Philadelphia: Temple University Press, 1987.

Savage-Rumbaugh, Sue. *Kanzi, the Ape at the Brink of the Human Mind*. New York: John Wiley and Sons, 1994.

Scherer, Donald, ed. *Upstream/Downstream, Issues in Environmental Ethics*. Philadelphia: Temple University Press, 1990.

Schleifer, Harriet. "Images of Death and Life: Vegetarianism." In *In Defense of Animals*, edited by Peter Singer, pp. 63–76. Oxford: Basil Blackwell, 1985.

Schilpp, Paul Arthur, ed. *The Philosophy of Ernst Cassirer*. LaSalle, Ill.: Open Court, 1949.

Searle, John R. "Animal Minds." In *Etica & Animali*, edited by Paolo Cavalieri, pp. 37–50. Milan: 1998.

Singer, Peter. "Animals and the Value of Life." In *Matters of Life and Death*, edited by Tom Regan. New York: Random House, 1980.

———, ed. *In Defense of Animals*. Oxford: Basil Blackwell, 1985.

Sloan, Philip R. "The Question of Natural Purpose." In *Creation and Evolution*, edited by Ernan McMullin, pp. 121–50. Notre Dame: Notre Dame University Press, 1985.

Sorabji, Richard. *Animal Minds and Human Morals: The Origins of the Western Debate*. Ithaca: Cornell University Press, 1993.

Stanley, Steven M. *Macroevolution: Pattern and Process*. San Francisco: W. H. Freeman and Company, 1979.

――――. *Earth and Life Through Time.* New York: W. H. Freeman and Company, 1986.

Stephens, William O. "Masks, Androids, and Primates.: The Evolution of the Concept of 'Person.'" In *Etica & Animali,* edited by Paolo Cavalieri, pp. 111–27. Milan: 1998.

Sterba, James P., ed. *Earth Ethics: Environmental Ethics, Animal Rights, and Practical Applications.* Englewood Cliffs, N.J.: Prentice Hall, 1995.

Stone, Christopher. "Moral Pluralism and the Course of Environmental Ethics." *Environmental Ethics* 10 (1988): 139–54.

Stringer, Christopher B. "The Emergence of Modern Humans." *Scientific American* (December 1990): 98–104.

Strum, Shirley C. *Almost Human: A Journey into the World of Baboons.* New York: Random House, 1987.

Swinburne, Richard. *The Evolution of the Soul.* Oxford: Clarendon Press, 1986.

Sylvan, Richard, and David Bennett. *The Greening of Ethics: Anthropocentrism to Deep-Green Theory.* Cambridge: White Horse Press, 1994.

Taylor, Paul W. *Respect for Nature: A Theory of Environmental Ethics.* Princeton, N.J.: Princeton University Press, 1986.

Thomas Aquinas, St. *See* Aquinas, St. Thomas.

Thorpe, W. H. "Vitalism and Organicism." In *The Uniqueness of Man,* edited by John D. Roslansky, 73–99.

――――. *Animal Nature and Human Nature.* Garden City, N.Y.: Anchor Press, 1974.

Tooley, Michael. "Speciesism and Basic Moral Principles." In *Etica & Animali,* edited by Paolo Cavalieri, pp. 5–36. Milan: 1998.

Urban, Wilber M. "Cassirer's Philosophy of Language." In *The Philosophy of Ernst Cassirer,* edited by Paul Arthur Schilpp, 403–41. LaSalle, Ill.: Open Court, 1949.

――――. *Language and Reality: The Philosophy of Language.* London: George Allen & Unwin; New York: Macmillan, 1939.

Van de Veer, Donald, and Christine Pierce, eds. *People, Penguins and Plastic Trees.* Belmont, Cal.: Wadsworth, 1986.

Verene, Donald Phillip, ed. *Symbol, Myth and Culture: Essays and Lectures of Ernst Cassirer, 1935–1945.* New Haven: Yale University Press, 1979.

Vitali, Theodore, R. "Sport Hunting: Moral or Immoral?" *Environmental Studies* 12 (Spring 1990): 69–82.

von Humboldt, Wilhelm. *On Language: The Diversity of Human Language-Structure and Its Influence on the Mental Development of Mankind.* Translated by Peter Heath. Cambridge: Cambridge University Press, 1988.

Ward, Peter, and Donald Brownlee. *Rare Earth.* New York: Springer Verlag, 2000.

Warren, Karen J. "The Power and the Promise of Ecological Feminism." In *Earth's Ethics,* edited by James P. Sterba, 231–41. Englewood Cliffs, N.J.: Prentice Hall, 1995.

Warren, Mary Anne. "The Rights of the Nonhuman World." In *Environmental Philosophy: A Collection of Readings,* edited by Robert Elliot and Arran Gare, 109–34. St. Lucia: University of Queensland Press, 1983.

Watson, Richard A. "Self-Consciousness and the Rights of Nonhuman Animals and Nature." *Environmental Ethics* 1 (1979): 99–129.

Weiss, Edith Brown. *In Fairness to Future Generations.* Tokyo and Dobbs Ferry, N.Y.: United Nations and Transnational Publishers, 1989.

Weiss, Paul. *Nature and Man.* Carbondale and Edwardsville: Southern Illinois University Press, 1947.

Weston, Anthony. *Toward Better Problems: New Perspectives on Abortion, Animal Rights, the Environment, and Justice.* Philadelphia: Temple University Press, 1992.

———. *Back to Earth: Tomorrow's Environmentalism.* Philadelphia: Temple University Press, 1994.

Westra, Laura. *An Environmental Proposal for Ethics: The Principle of Integrity.* Lanham, Md.: Rowman and Littlefield, 1994.

Whatmough, Joshua. *Language: A Modern Synthesis.* New York: St. Martins Press, 1956.

White, Thomas I., and Denise L. Herzing. "Dolphins and the Question of Personhood." In *Etica & Animali,* edited by Paolo Cavalieri, pp. 64–84. Milan: 1998.

Wiener, Philip. *Evolution and the Founders of Pragmatism.* New York: Harper Torchbooks, 1965.

Wild, John. *The Challenge of Existentialism.* Bloomington: Indiana University Press, 1959.

Wilson, E. O. *The Diversity of Life.* Cambridge: Harvard University Press, 1992.

Zuckerman, Lord. Review of Jane Goodall, *Through a Window: My Thirty Years with the Chimpanzees of Gombe.* In *New York Review,* May 30, 1991.

Index

abortion, 269
Abraham, 364
act utilitarianism, 266
action: as free, 258; as morally good, 256; as reasonable, 257; various senses of, 258
Alain of Lille, 372
Albigensians, 369
animal intelligence, 81–82, 84–85, 88, 101; compared with human intelligence, 107; instinctive judgments, 107; and past and future, 85; practical judgment of, 156; and reasoning, 103–6. *See also* nonhuman animal
animal liberation movement, 2, 4, 11, 241–42; and Darwinist continuism, 247–49; logical effects of, 302–3; positive and negative aspects of, 247–48; repudiation of human superiority, 246
animals and Christians, 360–69
anthropocentrism, 2, 250, 318, 323; and Darwinism, 290; and ecocentrism, 323, 341; hard and soft, 323; as opposed to biocentrism, 309; and Regan, 290
appetition, 111–12
Aristotle, 14–15, 28, 37; on animal intelligence, 138–40, 142, 145, 157; on crossmodal transfer, 131–32; definition of human, 251–52; on discursive power (practical reason), 146–48, 155; environmental ethic, 252; on free act, 115–16; on freedom in nonhuman animal, 121; on the good, 307; on intellect as tool of tools, 162–63; on language, 168; on nature of intellect, 87, 130; on the phe-

nomenon of life, 64–65; on the practical intellect, 123–24, 157; theory of rights, 256; wonder of nature, 92
Auschwitz, 303

behavior in human and nonhuman animals: animal behavior as qualifiedly free, 127; instinctive judgments of animals, 107ff.; as purposeful, 85, 88–92, 96–97; and sensory awareness, 128; uniformity of animal behavior, 104
behaviorism. *See* determinism
Behe, Michael: critique of Darwinian theory, 33–34
being: as object of intellect, 80, 100; as the unclassified classifier, 79
Belsen, 303
Bennett, David, 340
Bentham, Jeremy, 31
Berlin, Isaiah, 329
Bickerton, Derek: and continuist theory, 220; and Darwinian evolutionary theory, 221–23; human language, uniqueness of, 219–20; import of syntactical aspect of language, 220–21; on intelligence and language in nonhuman animal, 81–82, 107; lack of speech in nonhuman animal, 219; on Locke's theory of knowing, 222; on mental evolution of the human, 225–26; on origin of language, 226–27, 239; on non-oral language, 227, 239; on proto-language, 227–28, 240; on punctuated equilibria, 221; on teleology in nature, 221–22

Evolution, Animal 'Rights', and the Environment was designed and composed in Sabon by Kachergis Book Deisgn, Pittsboro, North Carolina, and printed on 60-pound Writer's Offset Natural and bound by Thomson-Shore, Dexter, Michigan.